Bracing for Disaster

Earthquake-Resistant Architecture and Engineering in San Francisco, 1838–1933

MDCCCXCVII

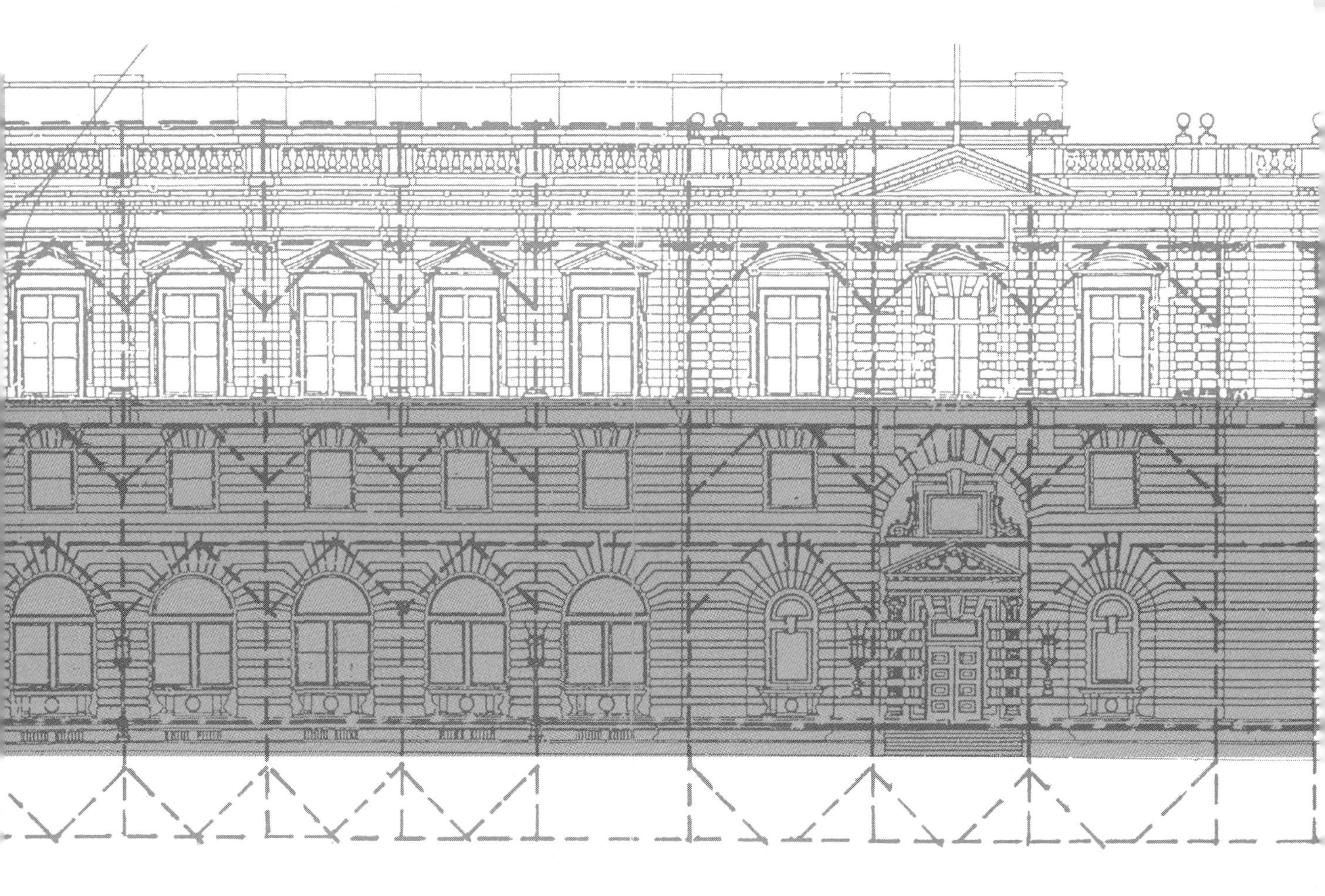

Bracing for Disaster

Earthquake-Resistant Architecture and Engineering in San Francisco, 1838–1933

Stephen Tobriner

The Bancroft Library, University of California, Berkeley
Heyday Books, Berkeley, California

Library of Congress Cataloging-in-Publication Data
Tobriner, Stephen.
Bracing for disaster : earthquake-resistant architecture and engineering in San Francisco, 1838-1933 / Stephen Tobriner.
p. cm.
Includes bibliographical references and index.
ISBN 1-59714-025-2 (pbk. : alk. paper)
1. Buildings—Earthquake effects. 2. Earthquake engineering. I. Title.
TA658.44.T53 2006
624.1'7620979461--dc22
2006003056

Cover photograph courtesy of The Bancroft Library, University of California, Berkeley
Book design by Rebecca LeGates
Printing and Binding: Transcontinental Printing G.P., Louiseville, Quebec

Orders, inquiries, and correspondence should be addressed to:
Heyday Books
P. O. Box 9145, Berkeley, CA 94709
(510) 549-3564, Fax (510) 549-1889
www.heydaybooks.com

Printed in Canada

10 9 8 7 6 5 4 3 2 1

To all the architects and engineers who have made and are making San Francisco safer for us all, including fifth generation Tobriners—my sons, Mike and Armando

The Bancroft Library

The Bancroft Library is the primary special collections library at the University of California, Berkeley. One of the largest and most heavily used libraries of manuscripts, rare books, and unique materials in the United States, Bancroft supports major research and instructional activities. The library's largest resource is the Bancroft Collection of Western Americana, which was begun by Hubert Howe Bancroft in the 1860s and which documents through primary and secondary resources in a variety of formats the social, political, economic, and cultural history of the region from the western plains states to the Pacific coast and from Panama to Alaska, with greatest emphasis on California and Mexico from the late eighteenth century to the present. The Bancroft Library is also home to the Rare Book and Literary Manuscript Collections, the Regional Oral History Office, the History of Science and Technology Collections, the Mark Twain Papers and Project, the University Archives, the Pictorial Collections, and the Center for the Tebtunis Papyri. For more information, see the library's website at http://bancroft.berkeley.edu.

For information on the Friends of The Bancroft Library, to make a gift or donation, or if you have other questions please contact:

Friends of The Bancroft Library
University of California, Berkeley
Berkeley, California 94720-6000
(510) 642-3782

Contents

Acknowledgments

Fred Krimgold, as program director for the National Science Foundation, encouraged me to begin this project in the early 1980s. Although I was an architectural historian who had studied reconstruction after early modern earthquakes, I was not an engineer. I joined the Structural Engineers Association of Northern California as an affiliate and the Earthquake Engineering Research Institute as a member and was privileged to be welcomed into the world of structural engineering. Many engineers took the time to teach me aspects of the field. Just a few can be mentioned here. The late Frank McClure insisted I be a faculty member of the Seismic Review Committee of the University of California, Berkeley, in order to understand earthquake engineering in the context of an institution—as Frank put it, "to know where all the bodies are buried." I have been on the committee for almost twenty years and still haven't found the bodies. During the last year of his life, Henry J. Degenkolb (Degenkolb Engineers) was kind enough to read hundreds of San Francisco blueprints with me, recounting his engineering experiences. The late Constantine (Stan) Chekene (Rutherford & Chekene) educated me, and a small group of students, on steel and concrete in relation to problems in San Francisco. The late Nick Forell (Forell/Elsesser Engineers) was a generous tutor and critic, spending hours teaching me and lecturing to my classes. Eric Elsesser critiqued my work, took me on many tours, and became a friend. Charles Scawthorn recruited me to talk about the Auxiliary Water Supply System in San Francisco and has read my manuscripts over the years, including parts of this book. The late Michael Pregnoff was kind enough to talk to me at length about the old days and wanted me to give engineers a broader perspective in relation to dependence upon computer analysis. Ephraim Hirsch and Filip Filippou read versions of the manuscript. Over the years, William T. Holmes (Rutherford & Chekene) has supported my work, and it was a great honor to have him review this manuscript. This book could never have been written without the enthusiasm and constructive criticism of Peter Yanev (formerly principal of EQE Engineering, and author of *Peace of Mind in Earthquake Country*). In the dark days, his spirit and generosity helped overcome the doubts I had. Scores of other engineers and professors of engineering helped me along the way, among them Daniel L. Schodek; Vitelmo Bertero; Bret Lizundia and Dominic Campi (Rutherford & Chekene); and Ed Zacher and Mike Davies (H. J. Brunnier Associates). Gary Black and Henry Lagorio were tremendously helpful. James Casey, professor of mechanical engineering at UC Berkeley and an unfailing friend, went over the first part of the manuscript with an eye to detail and scholarship.

The late seismologist Bruce A. Bolt heard and read early versions of chapters for this book and was ready to give me his reactions to the manuscript when, sadly, death intervened. Tousson Toppozada, an authority on California earthquakes, read the chapters on early earthquakes, and Carol Prentice of USGS, Stanford professor Greg Beroza, and geologist Doris Sloan tried to save me from making mistakes in seismology and geology.

Dell Upton, my former colleague in architectural history at UC Berkeley and a dear friend, read the early drafts, and the late Spiro Kostof, my senior colleague, encouraged my work. Paul Groth read drafts of the early chapters. Dora Crouch read an early draft of the manuscript. The late Carl Condit, an authority on the history of engineering, critiqued an early draft. I did not heed the advice of the late Gunther Barth to publish the manuscript as it was years ago, but I did incorporate his comments. Michael Corbett, whose *Splendid Survivors* is still the most valuable book on San Francisco architecture, read the final draft. Mary Comerio, my colleague in the architecture department who has made major contributions to seismic safety at the University of California, read the manuscript, and my brother Michael C. Tobriner edited the first chapters with skill and grace. My sons Mike and Armando put together the illustrations for an earlier draft, and Mike (HOK Architects) saved me more than once this time around by drawing diagrams and maps. My friends Bob and Carol Gill took on the task of reading the last draft as a sample lay audience, and if I am fortunate enough to have other readers of such intelligence, I will be blessed.

Gary Goss, who knows San Francisco better than anyone, took me around the city and sifted through blueprints in the Environmental Design Archives. Gray Brechin found material on architecture and technology that would have eluded me. At the beginning of my research in the 1980s, Kevin Powell and E. G. Daves Rossell did exceptional work as research assistants in the archives and libraries of San Francisco. Nina Lewallen and Diane Shaw were outstanding research assistants, as were Eileen Keefe, Christal Elliott, Michael Corbett, and Eric Klocko. Julie Masal did a superb job combing the archives in Washington, D.C. Kimberly Butt's patience, tenacity, and excellent scholarship were in evidence as she painstakingly checked the notes for the last draft. Ki Yeong Kim helped find missing illustrations. Anthony Vizzari not only searched for illustrations but reworked them digitally and produced the annotated photographs for this book.

Special thanks must be given to the scores of building owners, building engineers, and building managers who let me rummage through their structures, in particular Karen I. Epstein, Pat McCarthy, Bryant Farr, Elaine Hui, and George and Alexis Selland. To the many students who took my course on the history of San Francisco in the architecture department, I must say thank you, in particular to Keith S. Tsang, Sumaya Jones, Erica N. Boyd, Nurit London, Lisa Chen, John Coop, Amber Hoffmann, Dawn King, Andrew Ballard, Peter Benoit, and Chloe Clair.

The librarians and archivists of San Francisco collections are the unsung heroes of a book like this. Having been the curator of UC Berkeley's Environmental Design Archives, I know what it is like to serve patrons. Waverly Lowell, the present curator of the Environmental Design Archives, and Carrie McDade, assistant curator, have both been open and accommodating. Elizabeth Byrne and the entire staff of the College of Environmental Design Library have been both efficient and inventive in finding sources for me. Charles James, head of the Earthquake Engineering Research Library, is the most knowledgeable researcher on earthquake subjects I have met. He has helped me on numerous occasions. For my entire career at Berkeley, the staff of the Bancroft Library has helped me find obscure, earthquake-related sources. Most recently Susan Snyder helped to find all of the illustrations from the Bancroft collections used in this book. Chris McDonald helped me sort through the new online visual archives. Former San Francisco Public Library City Archivist Gladys Hansen was helpful in the transfer of rescued blueprints from the building inspector's office, and City Archivist Susan Goldstein and her staff at the History Center have been hospitable and extremely forthcoming. The staffs of the California Historical Society, the California Academy of Sciences, the Society of California Pioneers Library, the Stanford Library, the California State Library, and the Society for the Preservation of San Francisco's Architectural Heritage have all assisted in the research. Stephen L. Quarles, wood durability advisor in agriculture and natural resources at the University of California, helped me understand the fire-resisting qualities of redwood.

Funding for this work came from the National Science Foundation, the Getty Senior Fellowship, and the Al-Falah Foundation, as well as grants and fellowships from the University of California and a generous publication grant from the Earthquake Engineering Research Institute.

This book went through several metamorphoses and title changes before appearing in its present form. It began as a UC Press book. My editor, William J. McClung, and I spent many hours discussing disaster and reconstruction in the study of the house he had rebuilt after the 1991 Berkeley fire. Having studied the field of earthquakes and reconstruction in Sicily, I would call him its godfather. Mariah Bear helped pull the chapters together in their early stages. I owe Charles Faulhaber, director of The Bancroft Library, a great debt of gratitude. He suggested I publish the manuscript as a Bancroft book and introduced me to Malcolm Margolin, founder of Heyday Books and one of the major figures in California studies. Although Heyday does not specialize in architecture and engineering, the entire team has been outstanding, and my designer, Rebecca LeGates, and my insightful editor, Jeannine Gendar, have been incredible.

At this juncture, authors usually thank their partners or spouses. My wife, Frances Tobriner, with whom I discussed every sentence of this book, threatened to delete any dedication which she deemed too lavish. A clinical psychologist/Jungian analyst and a talented editor, she has been helpful in both capacities, and I could not have produced this manuscript without her humor and wisdom.

Introduction

How did San Franciscans respond to the great earthquakes of 1868 and 1906? *Bracing for Disaster* adds something new to the answers usually given from sociological and political perspectives. Here, for the first time, is a history of San Francisco's buildings in relation to their safety in earthquake and fire. But it is more than that. It is the history of San Francisco's built environment from the city's beginnings as a pueblo through the celebration of its rebirth in the Panama-Pacific International Exposition.

The conclusion of this study, which some may find surprising, is that architects and engineers were constructing buildings with earthquake-resistant features in San Francisco after the earthquakes of 1868 and 1906, well before the first mandatory state earthquake code of 1933. In making this assertion, *Bracing for Disaster* contradicts popular authors, some seismologists, some historians, and even a few engineers who have claimed that nothing was done to improve building construction or to promote earthquake safety after the earthquakes of 1868 and 1906; that in both cases there was a conspiracy to deny the reality of earthquake damage in the city, in order to promote investment and to speed recovery.[1]

Prior to *Bracing for Disaster,* no one has studied the seismically resistant aspects of San Francisco's buildings or read technical engineering papers in their historical context. Once the presumption of denial and inaction is set aside, evidence abounds for interest in earthquake-resistant construction practices. Behind the facades of San Francisco's buildings and under its streets, and in hundreds of articles and books, manuscripts and photographs, in libraries and archives across the country, lies evidence of the complexity and diversity of building professionals' responses to earthquake danger. Engineers, architects, builders, and inventors have left irrefutable evidence of their attempts to design and build earthquake-resistant structures. Many of these attempts are still with us today: the first skyscraper in San Francisco, the Chronicle Building (1889), was built to be earthquake-resistant. The present City Hall of San Francisco (1912) was built to be earthquake-resistant. The U.S. Court of Appeals and Post Office (1905) was built to be earthquake-resistant. The Royal Globe Insurance Company (1909) was built to be earthquake-resistant. Most of the tall steel buildings constructed in the decade after the 1906 earthquake were built to be earthquake-resistant. Reinforced concrete, introduced after the earthquake, was touted as being earthquake-resistant. The entire Auxiliary Water Supply System of San Francisco, which was built to protect the city from fire after the 1906 earthquake—every component, from pumping stations and pipes to cisterns and fireplugs—is designed to be earthquake-resistant.

With all the attention to earthquake-resistant construction, why and how were these efforts overlooked? Andrew Lawson, the seismologist who discovered and named the San Andreas Fault, referring to a statement by the geologist George Davidson (a member of the 1868 earthquake committee), wrote:

> Shortly after the earthquake of 1868 a committee of scientific men undertook the collection of data concerning the effects of the shock, but their report was never published nor can any trace of it be found, although some of the members of the committee are still living. It is stated that the report was suppressed by the authorities, through the fear that its publication would damage the reputation of the city.[2]

Statements like this discouraged the search for a history of earthquake-resistant construction. Ironically, the late Bruce Bolt and a team of researchers who in 1986 explored this charge at length found it to be false.[3] There is no evidence to corroborate Davidson's position. Publication was not, as far as the researchers could tell, suppressed by the business community. The report was, in fact, never written, probably because of a dispute among the members of the committee. Nonetheless, the story of the unpublished report supported the impression that because of business pressure, architects and engineers were not interested in earthquake safety.

No doubt greed, denial, corruption, and boosterism played a significant part in San Francisco's recovery. However, *Bracing for Disaster* teases out, from a complicated body of data, the intentions and efforts of architects and engineers before 1933 to make San Francisco safer by using earthquake-resistant construction practices. The phrase "architects and engineers" is used as shorthand here and not meant in any way to imply that all architects and engineers were equally concerned. Among those concerned with earthquake safety, some, like the Chicago architects Daniel Burnham and John Wellborn Root, are famous. But most, like architects David Farquharson, John Wright, John Gaynor, Alfred Mullett, and Frederick Meyer; like engineers Charles Derleth Jr., R. S. Chew, Christopher Snyder, and Marsden Manson; like inventors William Foye, David Emerson, Jules

Touaillon, and Joseph Hofmann, are hardly household names. And there are, of course, many others whose intention was to make San Francisco's buildings safer. This is the element of San Franciscans' response to the earthquakes of 1868 and 1906 that is long lost and radically new.

CHAPTER 1

Location, Location, Location

On October 17, 1989, at 5:04 p.m., as the World Series opened at Candlestick Park, San Francisco was struck by a moderate, magnitude 6.9 earthquake.[1] The epicenter was sixty miles to the southwest, near the coastal city of Santa Cruz, in a sparsely populated area of rolling hills and redwood trees close to a low mountain called Loma Prieta. Although the duration of the Loma Prieta earthquake was a scant fifteen seconds and the epicenter far to the south of the San Francisco Bay Area, damage there was widespread. A section of the roadway of the San Francisco–Oakland Bay Bridge collapsed, severing the eastern approach to the city. A half-mile stretch of the Cypress Viaduct, an elevated freeway in Oakland, collapsed, smashing cars and killing forty-one people. In San Francisco, the Embarcadero Viaduct, built along the waterfront on the same design as the Cypress, shifted, cracked, and nearly collapsed. At Sixth and Townsend Streets, in the South of Market District, a brick parapet fell outward, crashing down on five parked cars and killing some of their waiting drivers. Throughout the South of Market and the Mission District, sidewalks settled and buildings shifted on their foundations. Downtown, the walls of brick, concrete, and steel-frame structures shuddered and moaned as they moved, some pounding against their neighbors and cracking their brick and terra-cotta facades. In the Marina District, on the site of the 1915 Panama-Pacific International Exposition, which had been built on fill, apartment buildings came to rest on partially collapsed ground-floor garages (figure 14.1).

The shaking caused sand to liquefy and compact below the streets, snapping electrical conduit and gas lines and breaking water mains. Gas leaks ignited a fire that quickly burned through the ruins of a collapsed building and began to engulf its neighbor; within minutes, the cloudless afternoon sky was marred by a plume of smoke rising from the Marina District. Fire after an earthquake was no

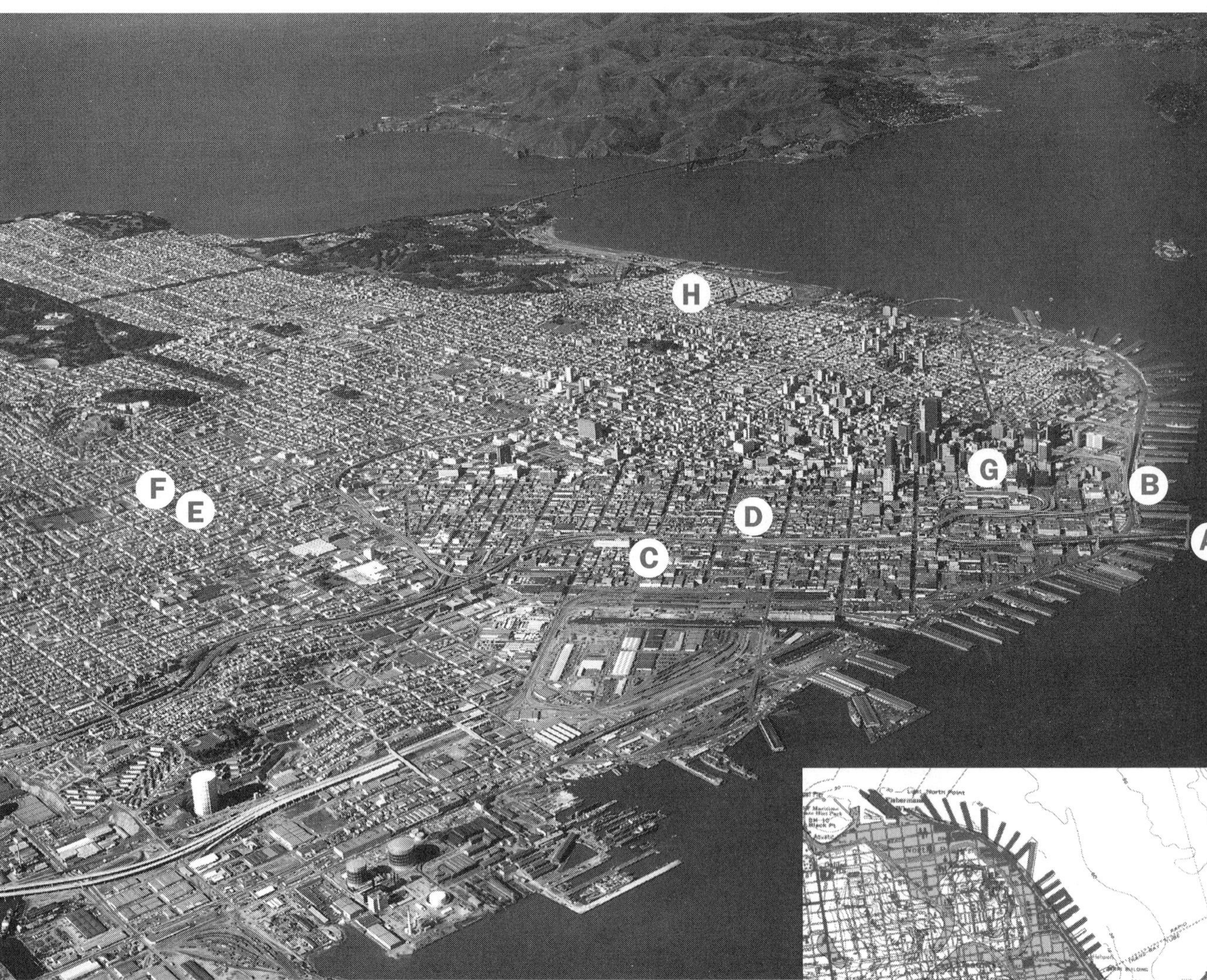

Figure 1.1: Aerial view of San Francisco looking northwest (above)
A. San Francisco–Oakland Bay Bridge
B. Embarcadero Viaduct (built on filled land)
C. Sixth and Townsend Streets (built on filled land on the border of old Mission Bay)
D. South of Market (largely built on filled land over a swamp)
E. The Mission District (indicating the filled Laguna de los Dolores)
F. Mission Dolores
G. Downtown (built on the filled Yerba Buena cove)
H. The Marina District (built on filled land)

Figure 1.2 (right)
Seismic hazard zones in the city of San Francisco, USGS, November 17, 2000, detail. This legal document indicates hazardous ground in San Francisco. Gray areas are where historic occurrences of liquefaction or local geological, geotechnical, and groundwater conditions indicate a potential for permanent ground displacement. Color indicates the original shoreline and the marsh areas.

Map courtesy of th

Historic areas of liquefaction.
Original shoreline and marsh areas.

novelty in San Francisco, and neither was the pattern of damage caused by intensified shaking and catastrophic settling on filled land: the entire waterfront of San Francisco, from the Marina on the north to well past Hunter's Point on the southwest, including the heart of the Financial District, most of the South of Market District, and all of Mission Bay, is built on fill (figure 1.2). During the previous one hundred and fifty years, the people of San Francisco had transformed a bleak, treeless peninsula with sand hills and a jagged waterfront of rock outcroppings and muddy tidelands into one of the most beautiful cities in America. But it is sobering to realize that the Loma Prieta damage would have been much less if the topography of San Francisco had not been so radically altered.

Real estate brokers like to ask their clients this riddle: "What are the three most important qualities of a property?" The answer: "Location, location, location." Environmental health and safety have long been secondary to strategic location. The Roman writer Vitruvius, in his *De Architectura,* counseled caution in choosing the location of cities, advocating that they be placed in very healthy sites, high but "neither misty nor frosty, and in a climate neither hot nor cold."[2] These sensible guidelines have rarely been followed. The center of Rome, the Forum, was constructed in the middle of a swamp, and the district called Campus Martius, where the Pantheon stands, lies in the Tiber's flood plain. St. Petersburg was founded on a swamp so unhealthy that thousands died while building the city. Calcutta, built in yet another unhealthy swamp, still attracts vast crowds of immigrants. Mexico City, with more than twenty-two million inhabitants, shares its site with a former lake. Every year the lake floods portions of the city, causing vast sewage and runoff problems. Buildings settle unevenly as the lake bottom sinks. Although Mexico City is prone to catastrophic earthquakes and enveloped by a mantle of deadly smog, the population increases daily.

Earthquakes in the Old World and the New

By 1776, when San Francisco was chosen as a site for a colonial mission and presidio, the Spanish had a long history of confronting the dangers of earthquakes. In 1693 the Spanish colony of Sicily had been struck by one of the most devastating earthquakes in the history of the Mediterranean; it leveled forty cities and killed more than fifty thousand people. The famous Lisbon earthquake of 1755, which shocked Europe and inspired Voltaire's *Candide,* destroyed scores of buildings as far away as Seville and provoked heated debates about the causes of earthquakes. The church, whether Catholic or Protestant, seized on earthquakes as signs of God's wrath. Architects, engineers, and natural philosophers could not agree on the cause or the mitigation of the destructive shaking. Were earthquakes caused by underground fires, as posited by Aristotle, or by newly discovered electrical conductivity? Could shaking be controlled by digging underground caverns to release or channel trapped air, as was posited by architectural writers from the Roman Pliny the Elder to the famous Italian Francesco Milizia in 1781?[3]

In some reconstructed cities, wide streets and large plazas provided refuge once shaking started.[4] In seventeenth-century Peru, the Spanish had prescribed an earthquake-resistant construction technique that involved reinforcing adobe with bamboo to create a flexible wall called a *quincha*. When Lima was leveled by an earthquake in 1746, authorities limited the height of buildings and required *quincha* construction for all walls above the ground floor.[5] Unlike the authorities in Lima, the Spanish viceroy in Guatemala decided after the earthquake of 1773 that three earthquakes in fifty-six years were too many and evacuated Antigua Guatemala, establishing a new capital. In Portugal, the *gaiola,* a masonry building reinforced by a comprehensive wooden framework, was invented, perfected, and mandated after the1755 Lisbon earthquake.[6] In the seismically active areas of Mexico, the Spanish and Mexican governments instituted height limitations. Churches were built with stout, low towers, broad building profiles, and huge buttresses to mitigate the effects of frequent earthquakes.[7]

A single clue to dangerous sites was first proposed after the Calabrian earthquakes of 1783. Examining the degrees of structural damage, the royal physician Giovanni Vivenzio correctly postulated that shaking is more violent on filled or alluvial soil than on rock. Domenico Campolo, a Sicilian cartographer, had actually grasped the same idea years earlier, after the earthquake of 1726, when he drew a plan of Palermo illustrating the prevalence of structural damage in old filled riverbeds.[8]

The Spanish knew that Alta California was prone to earthquakes. The first expedition northward from San Diego, led by Gaspar de Portolá in 1769, experienced a series of very strong earthquakes in southern California. Chaplain and diarist Francisco Palóu wrote, "We called this place the sweet name of 'Jesus of the Tremors,' because we experienced here a horrifying earthquake, which was repeated four times during the day. The first, which was the most violent, happened at one in the afternoon, and the last one about four." Aftershocks continued for six days.[9]

The San Andreas Fault[10]

The Spanish party had happened on the southern portion of the San Andreas Fault (figure 1.3). The fault, an eight-hundred-mile fracture in the earth's crust, extends south from Punta Gorda and through Point Arena, in northern California's Mendocino County, to Tomales Bay, in Marin County, plunging into the Pacific at Bolinas Lagoon and surfacing at Daly City, south of San Francisco, then running alongside present-day Skyline Drive right through San Andreas Lake and Crystal Springs Reservoir, south through the Santa Clara Valley, and through eastern Los Angeles County and Orange County to the Salton Sea and the Gulf of California.

The San Andreas Fault marks the boundary between the Pacific and North American tectonic plates. Current theory is that earthquakes are caused by movements of the great tectonic plates that form the earth's lithosphere, or outer layer.[11]

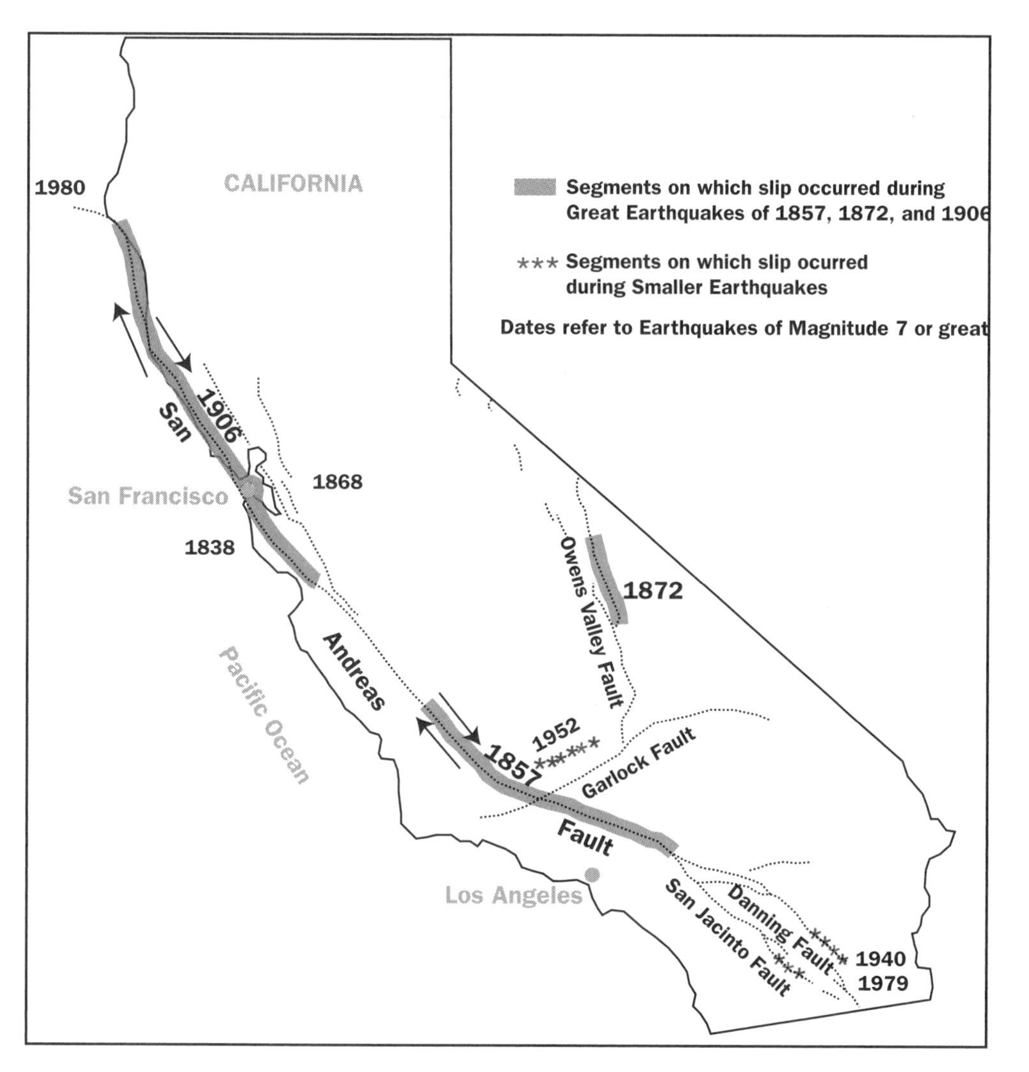

Figure 1.3
A map of the San Andreas Fault system, USGS, 2004. Arrows indicate the directions of the Pacific and North American Plates. Earthquakes with magnitudes greater than 7 are listed. The 1838 and 1906 earthquakes occurred on the San Andreas Fault. The 1868 earthquake occurred on the Hayward Fault, which runs through the East Bay. Map courtesy of the USGS

Each plate is about one hundred kilometers deep and continually moving in relation to neighboring plates. At the edges of these plates, great geological changes occur. The North American Plate stretches across the United States and Canada to the middle of the Atlantic Ocean, while the Pacific Plate underlies the Pacific Ocean, reaching all the way to New Zealand, Japan, and Alaska. In San Francisco's case, the Pacific Plate is moving northwest relative to its eastern neighbor, the North American Plate. The slippage of masses of rocks in the rift between the two plates causes earthquakes, and the surface fracture where the slippage is visible is a fault. Because of the huge forces at work, many subsidiary faults parallel the San Andreas, like the Hayward Fault, which runs north and south on the east side of San Francisco Bay. These faults result from the tearing that occurs as the lithosphere moves, folds, and cracks.

The Founding of San Francisco

By 1775, Spanish officials understood the extent of the San Francisco Bay and saw its importance for control of their northern territories in the face of increasing English and Russian shipping traffic. Having secured Arizona and New Mexico for New Spain, Inspector General José de Gálvez shifted his focus to colonizing California.[12]

In what was to become San Francisco, the three elements of Spanish colonial presence—mission, presidio, and pueblo—were established in three different locations because of the topography of the peninsula (figure 1.4). The mission, San Francisco de Asís, was founded in 1776, inland of San Francisco Bay.[13] This site was protected from the prevailing winds and on good soil near a small stream. The stream fed a small lake, Laguna de los Dolores (the Lake of Sorrows), the outflow of which formed Mission Creek, which meandered out to the bay through a great marsh full of wildlife. The pueblo was expected to form around the mission. To guard the entrance to San Francisco Bay, the Spanish presidio, also founded in 1776, was positioned on the north coast of the San Francisco peninsula.[14] Between the presidio and the mission stretched miles of sand dunes. Sand from Ocean Beach blew east across the peninsula, engulfing the future sites of the Sunset and Richmond Districts and Golden Gate Park, swirling around Lone Mountain and into what is now known as the Western Addition. Sand flowed through Hayes Valley to what would be the Civic Center; it was eighty feet deep at the spot where City Hall now stands.

By 1812 the Spanish presidio and mission were in trouble. Mexico's bitter and protracted revolution against Spain had sapped the strength of the mother country and reduced supplies and reinforcements to the San Francisco garrison. Native American converts at the mission were dying of disease and suffering under a regimen contrary to their former lifestyle. With the

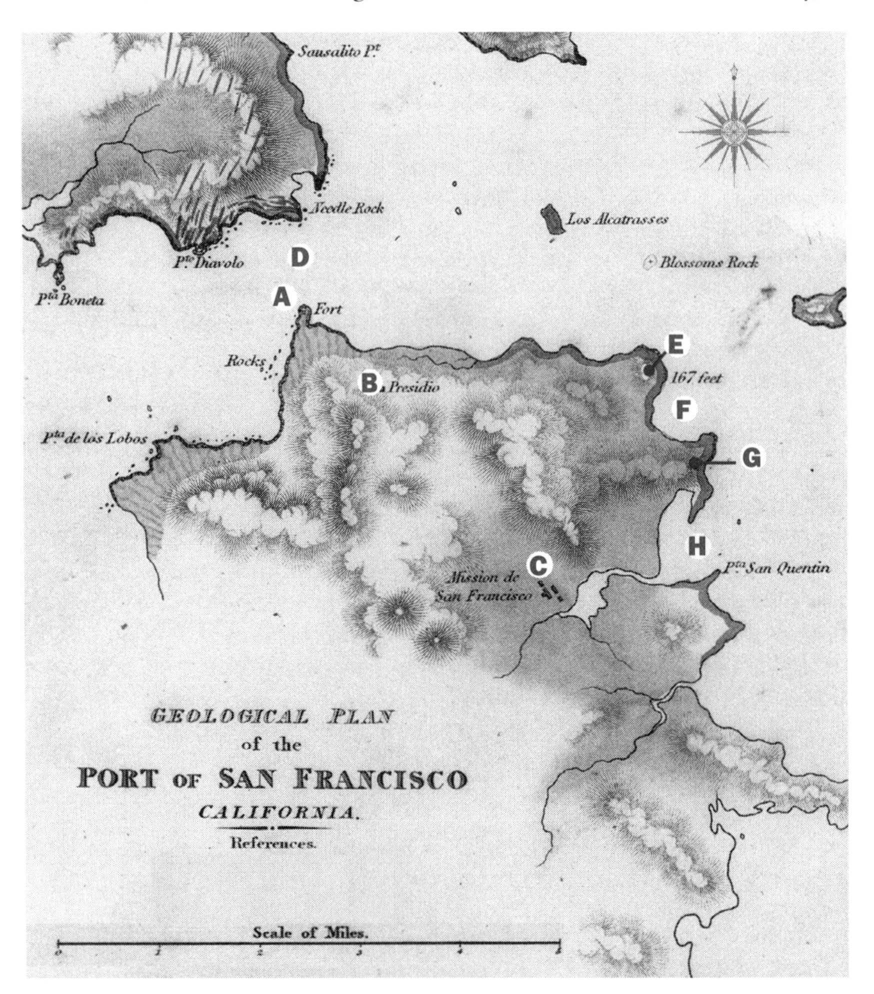

Figure 1.4
Beechey's map of San Francisco, 1826, detail (compass rose added, references deleted)—the San Francisco Peninsula in Spanish times
A. The Fort
B. The Presidio
C. Mission Dolores
D. The Golden Gate
E. Telegraph Hill
F. Yerba Buena cove
G. Rincon Hill
H. Mission Bay

establishment of the Republic of Mexico in 1832, the mission was secularized and its lands dispersed. By 1834, the fort was in ruins and nearly deserted, the bulk of its troops deployed northward to Sonoma.[15]

Under Mexican rule, the locus of development shifted from the mission and presidio to the newly designated pueblo, which coalesced around a cove called Yerba Buena on the eastern side of the peninsula, facing the bay.[16] Settlement began on the shore of the muddy cove below Telegraph Hill, with present-day Portsmouth Square as its center. There the anchorage was more protected than at the presidio, but Yerba Buena had intrinsic problems as well. It was a windy site in a sandy and desolate part of the eastern peninsula. Worse than the nuisance of perpetual grit and chilly discomfort was the constant threat that the wind might whip any small fire into an uncontrollable conflagration.

Water for drinking, and for extinguishing fires, was scarce in the San Francisco area, particularly around Yerba Buena cove.[17] Seasonal creeks within the cove soon became contaminated, and during the late 1840s and early 1850s, the scarcity of good water supported a brisk trade in barreled pure water delivered by steamer from across the bay in Sausalito.[18] Concerned about fires, the city council authorized the digging of artesian wells, but these proved insufficient. A water company was established to pipe in pure water from the best local source, Mountain Lake, on the border of the presidio, but this scheme failed. Only in 1858 did the first dependable water supply arrive by flume, from Lobos Creek near the Golden Gate. Later, water from more distant lakes on the peninsula was piped in by the Spring Valley Water Company, organized in 1860. Water quantity and pressure remained chronic problems for the growing city, and as late as 1905, insurance inspectors reported that San Francisco's waterworks were insufficient.[19]

One of the first Spaniards to explore the bay wrote that "most of the beach of the harbor, according to what I saw when we went around it, is not clean, but is muddy, miry, and full of sloughs, and for this reason is bad."[20] This meant that ships had to lay off the cove, lower boats, transfer goods, and row them in. At low tide, the cove was a miserable marsh. Still, there was no more protected bay in the three hundred miles of northern California coast, and the settlement along Yerba Buena cove soon took shape; it was identified on cartographers' maps as early as 1826.

In 1846 Captain John B. Montgomery landed American forces in Yerba Buena from his sloop, the USS *Portsmouth,* and raised the American flag (figure 1.5).[21] Montgomery appointed one of his lieutenants, Washington A. Bartlett, to be the first American alcalde, or mayor, of the city. Because of a bid by the present-day city of Benicia to take the name "Francesca" and thus connect itself to the San Francisco Bay, Bartlett changed the name of Yerba Buena to San Francisco.[22] As he had hoped, the city's name became linked with the famous bay; San Francisco's meteoric rise in 1848 and 1849 made real the dreams of the Yankee settlers and traders who inhabited Yerba Buena cove.

Figure 1.5
"View of San Francisco, Formerly Yerba Buena, in 1846-7." This souvenir print was meant to give an impression of the city before the gold rush. It combines the ships of Montgomery's occupation in 1846 with the resurveyed O'Farrell plan begun in 1847. The stream running down present-day Sacramento Street on the left is soon to be buried, as is the small saltwater lake to the right. This is the nucleus of O'Farrell's Bartlett plan (see page 9).

The Flattening and Filling of San Francisco

Before the American presence, the site of the city had barely been altered. Native Americans had changed it in subtle ways, heaping discarded seashells in huge mounds and occasionally burning grasses in what was to become the South of Market District. The Spanish and Mexican impacts had also been light. They had not cut, leveled or filled, changed watercourses, or built on marginal land. But by 1839 it had become clear that the street plan of the original village of Yerba Buena, with its lanes converging on Calle de Fundación, which ran north from the present intersection of California and Kearny Streets approximately to Pacific and Stockton, was outmoded and inadequate. The Mexican mayor Alfonso de Haro commissioned Jean-Jacques Vioget, a Swiss immigrant, to resurvey the occupied area of Yerba Buena cove.[23]

The Spanish had established the grid as the preferred form for organizing new cities in an official directive, the Laws of the Indies, in 1573, and although it was not always followed to the letter, this document guided the foundation of Spanish cities throughout the Americas.[24] It stipulated that each town should be founded around a rectangular plaza which generated a grid with streets intersecting at 90-degree angles. The resultant plaza and grid plan can be seen as the nucleus of Spanish and Mexican towns in California from Los Angeles to Sonoma. Vioget, who may have known Spanish planning practices from his earlier stay in Chile, or who may have been directed to survey a grid by de Haro, platted a twelve-block grid in the cove from the water's edge at present-day Montgomery Street uphill to the west to Grant Avenue, south to California Street, and north to Pacific Avenue. The blocks were laid out in Spanish varas, one vara equaling 33 inches. Each block was 100 by 150 varas (about 92 by 137 yards) and contained six 50-by-50-vara lots. Later surveyors condemned this first survey because the streets were not precisely laid out, did not intersect at 90 degrees, and were not oriented toward the cardinal points. The plan also expressed a cavalier disregard for water: the bed of the small freshwater stream that once flowed down Sacramento Street was simply deleted from the landscape, never to be mapped again. This was the first of countless such alterations which served to distance cartographers and settlers from the reality and particularity of place.

In 1847, before the treaty of Guadalupe Hidalgo formally ratified the property rights of Mexican citizens, the United States military governor abolished the Mexican communal landholding system and opened the inside lands—the lands owned by the pueblo—for sale, ushering in a period of land speculation. Because the outside lands, roughly to the west of present-day Van Ness Avenue, were still under Mexican land grants, development concentrated in the eastern border of the bay. Also in 1847, Bartlett commissioned a trained civil engineer, Jasper O'Farrell, to correct Vioget's original surveying errors and to extend the established grid.[25] O'Farrell first produced a larger grid superimposed over Vioget's (figure 1.6). But this was too limited. According to one source, O'Farrell at first balked when asked to extend the grid without reference to topographical features, but in the end he did. In accordance with standard practice in nineteenth-century America, O'Farrell platted across tidelands that were regularly under water, over marshes, and up the slopes of steep hills, without a single concession to topography, in order to facilitate land sales (figure 1.7). In the area of old Yerba Buena, he used the 100-by-150-vara grid established by Vioget. East of the original grid, in the tidelands of Yerba Buena cove, he used the same street widths but reduced the size of the blocks, and for the

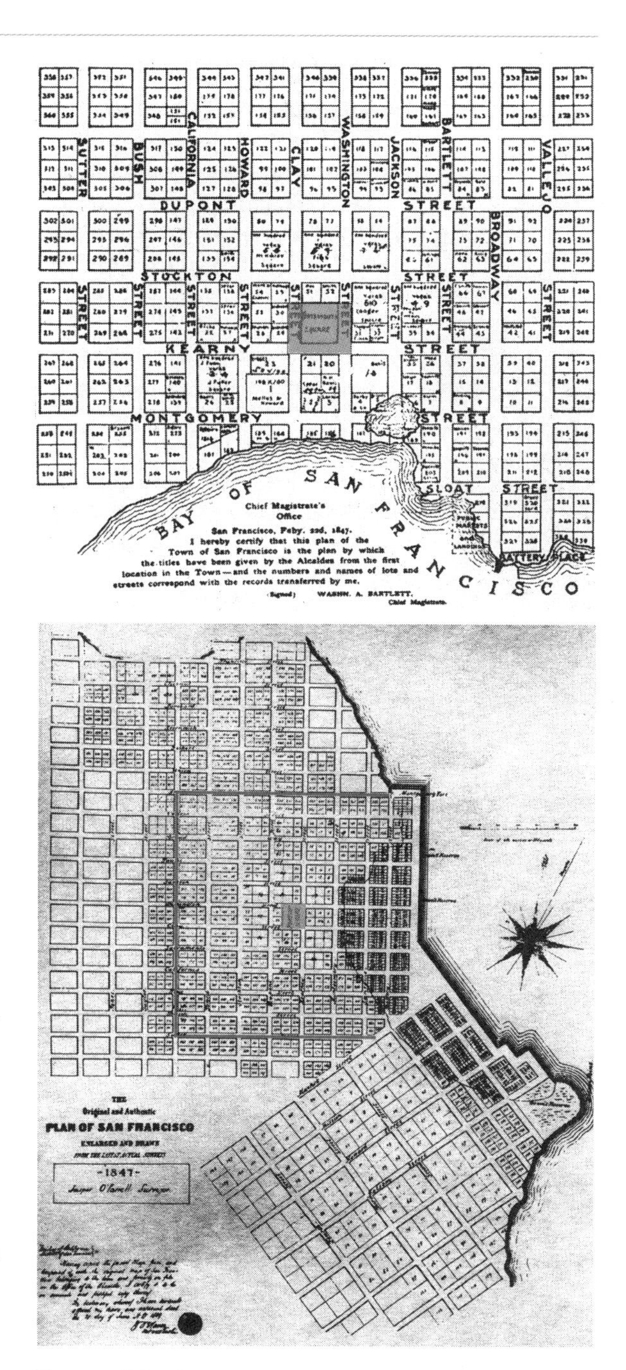

Figure 1.6 (top)
The so-called Bartlett plan, Jasper O'Farrell, 1847. Vioget's plan has been regularized but the grid is still small. Portsmouth Square is marked in color.

Figure 1.7 (bottom)
The O'Farrell plan, Jasper O'Farrell, 1847. The water lots are in place and the new South of Market grid created. Although it runs just two blocks southeast from First Street, the new Market Street, where the two grids join, is the backbone of the new city. Color indicates Portsmouth Square and the old Bartlett plan.

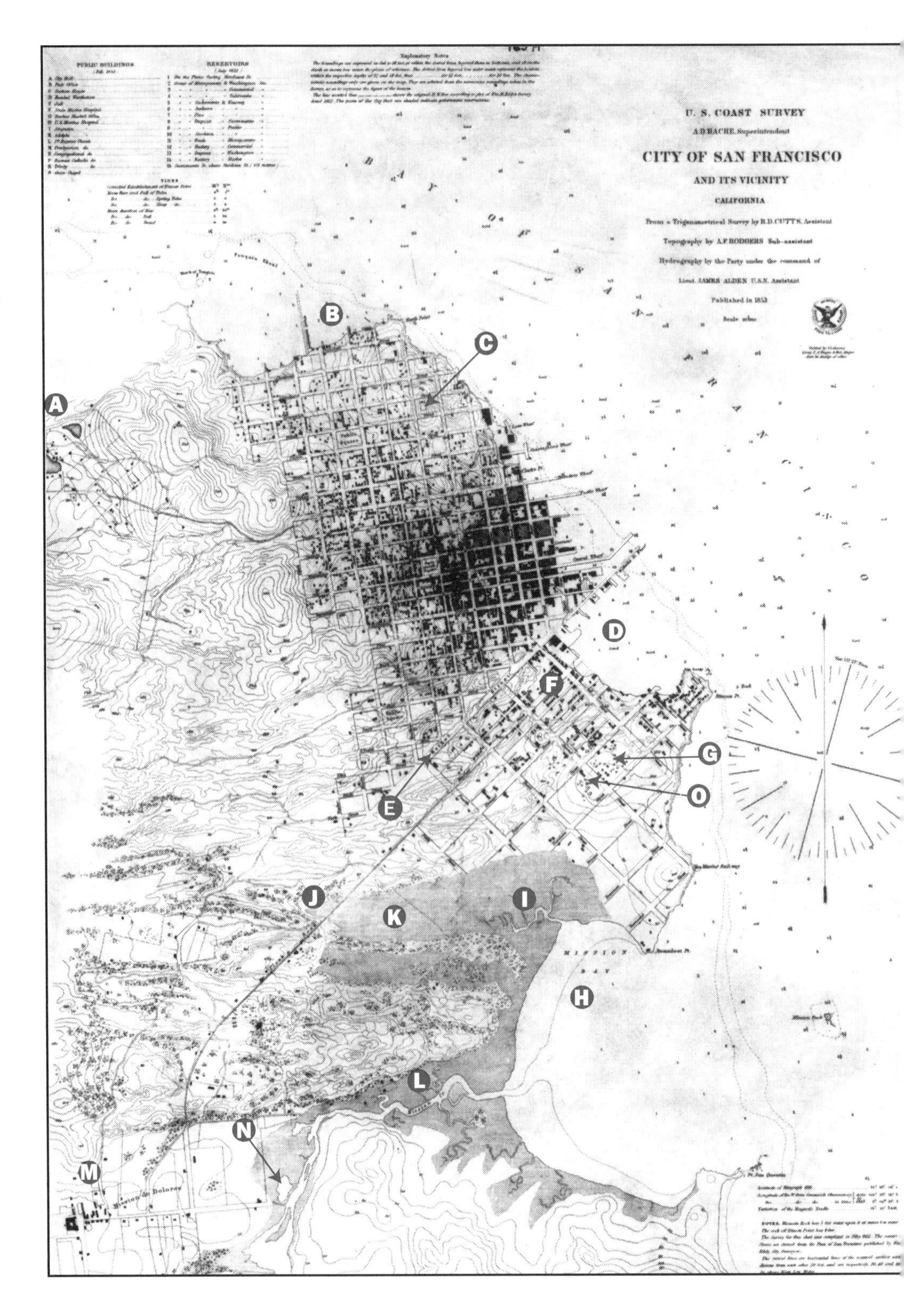

Figure 1.8
U.S. Coast Survey of the City and County of San Francisco, 1852. Dark squares mark the extent of the city. Yerba Buena cove is nearly half filled. As the grid moves westward, surveys confront ever more difficult topographic conditions.

A. Washerwoman's Lagoon
B. North Beach
C. Telegraph Hill
D. Yerba Buena cove
E. Market Street
F. "Happy Valley" and "Pleasant Valley"
G. Rincon Hill
H. Mission Bay
I. Hayes Creek
J. Mission Plank Road
K. Swamps south of Market
L. Mission Creek
M. Mission Dolores
N. Banks of Laguna de los Dolores

first time employed American measurements. This grid's southern boundary was formed by a 120-foot-wide diagonal street paralleling the old trail and plank road to the mission. To the southeast of this new street, called Market Street, he platted a larger grid containing lots that were 100 by 100 varas, four times bigger than the lots north of Market. These larger blocks joined the earlier grid at Market, creating a street pattern like an odd fish skeleton. The lack of through streets and the different look of the grids defined two different areas: the business district to the north, and the more industrial district to the south, which came to be called South of Market.

The topographical features O'Farrell ignored were significant (figure 1.8). Market Street could be traversed for only a few blocks before it ended in an eighty-foot sand hill at present-day Geary Street. In the South of Market, Second Street ran into an even steeper slope at the edge of Rincon Hill. The marshes throughout the South of Market, southwest of Rincon Hill, were impassable and so unstable that Mission Street began as a plank road. The marshes were actually underground lakes forty to eighty feet deep, overlaid with ten feet of peat. It was not unusual for animals grazing in the marsh to disappear when the peat gave way.[26] Between the rock outcropping of Rincon Hill, which rose above angular Rincon Point, and Telegraph Hill, which lies to the north, lay the muddy cove of Yerba Buena.

O'Farrell's plan had already been drawn and approved when gold was discovered in the foothills of the Sierra Nevada in 1848. San Francisco was the closest established deepwater port and thus the natural place to transfer men and provisions from deepwater ships to the shallow-draft vessels that plied the Sacramento River. The promise of gold attracted an army of largely male adventurers bent on the campaign for riches, and between 1848 and 1850, San Francisco metamorphosed from a trading village with a population of one thousand to a major harbor and supply center with a population of around twenty-five thousand.

Modern scholars have called western American boomtowns like San Francisco "instant cities," cities built with great speed by enterprising businessmen to serve a temporary population with no stake in the city's future.[27] As a San Francisco resident observed in 1849, "It's an odd place, unlike any other place in creation, and so should it be, for it is not created in the ordinary way, but hatched like chickens by artificial heat."[28]

By 1851, buildings covered all the level land near Yerba Buena cove. Immediately to the west, steep hills surrounded the cove and stymied development. Huge dunes dotted the downtown area. The only remaining level sites were tidelands, attractive for their proximity to shipping. These tidelands had already been platted for development four years earlier, in O'Farrell's plan.

The tidelands and wharves became a magnet for development. Like Boston and Charleston, San Francisco was a peninsula town that could not expand without filling in its commercial shoreline area. The city capitalized on this situation by selling water lots to finance civic expenses.[29] New landowners filled in the water lots, moving the land out into the water, toward the ships at anchor. Raised streets followed the lines of the wharves out into the tidelands, and other streets were built at right angles to them. Buildings on stilts adjoined the streets built of boards, which crisscrossed blocks of stagnant water. The filled-in area of the city began to settle almost immediately, causing severe problems for recently constructed buildings. Heavy rains returned the uncompacted streets to their former marshy consistency. Lieutenant (later General) William T. Sherman saw "mules stumble in the street and drown in the liquid mud."[30] At the corner of Clay and Kearny Streets, near the old bay shore, a sign warned, "This street is impassable—not even jackassable."[31]

As property in Yerba Buena cove became scarcer, developers and speculators turned to the hills and marshes south of Market and to Mission Bay, a muddy, crescent-shaped inlet south of Yerba Buena cove. During the gold rush, tent cities had sprung up in the protected valleys south of Market.[32] But not for long: Mission Bay was ripe for filling, and the terrain had to be leveled for envisioned residential and industrial development. By 1852 the sand from the hills had been removed by marvelous new machines, steam shovels, and put into hopper cars to be dumped into Yerba Buena cove (figure 1.9). One by one, the valleys disappeared, and today it is hard to imagine that they ever existed. By 1853, San Francisco had moved almost four blocks into Yerba Buena cove, and present-day Davis Street was partially completed.

Mission Bay, bordered on the west by a vast marsh, temporarily obstructed the southwestern progress of the South of Market grid.[33] A mile-wide finger of this marsh extended northwest, tapering down to a half-mile between present-day Seventh and Eighth Streets. Aquifers near present-day City Hall, just across Market Street, fed underground creeks and surface streams like Hayes Creek. As noted earlier, solid ground lay more than forty to eighty feet below the surface. A smaller marsh to the southwest, paralleling the first, reached as far west as Folsom Street. On the southern border of the marshland was Mission Creek, an estuary of Mission Bay fed by the Arroyo de los Dolores, which descended from Twin Peaks to Laguna de los Dolores, in the vicinity of Eighteenth and Mission Streets, before meandering through the present-day intersection of Treat and Eighteenth Streets, through Alabama Street to Division Street and the San Francisco Bay, where the ballpark is today. It is ironic that by the 1880s, not a trace remained of the beautiful laguna and healthy freshwater creek around which the Spanish had focused their settlement.

Figure 1.9
A portrait of the city surveyors and the "Steam Paddy" owned by David Hewes, who may be pictured here, in front of George Gordon's Pacific Sugar Refinery at Eight and Harrison Streets, after the 1868 earthquake. (Gordon himself owned a one-horse carriage and may also appear here.) The chimney was shortened because of partial collapse in the earthquake and the iron reinforcements added to the facade.

A privately owned plank road was laid along the old trail from Yerba Buena to Mission Dolores in 1849.[34] This road would become present-day Mission Street. Because peat could not bear heavy loads, the road was subject to shifting and sinking. In 1860 the city required the toll-road owner either to repair the existing bridge across the worst part of the marsh between Seventh and Eighth Streets or fill it in. Another toll-road owner successfully built a road across a half-mile of marshland between Fourth and Eighth Streets. These plank roads were an impetus for development, which of course meant filling the marsh. Mission Bay itself began to disappear with the construction, in 1862 to 1863, of a four-mile bridge across its mouth. Known at the time as Long Bridge, it marked the outer boundary of Mission Bay, which was gradually filled in its entirety. In 1849 the Willows, a resort on Eighteenth Street bordering the old Laguna de los Dolores, had been established, and in 1870 the Willows Land Association filed a lot map with the country recorder.[35] The laguna had been renamed Lake McCoppin after former mayor Frank McCoppin, whose homestead was on the southwest corner of Seventeenth and Valencia, and who had made his fortune as president of a local earthmoving company. By February 1873 the *San Francisco Real Estate Circular* reported:

> Of all the stagnant lakes and swamps which once extended irregularly from Dolores Street to Mission Creek, between Seventeenth and Nineteenth Streets, only a small patch is left—south of Eighteenth and east of Dolores: the remainder has been filled in, much to the relief of the nostrils of those who live Missionward. The last of Lake McCoppin is now disappearing from view. A wooden sewer large enough for a six-footer to promenade through was lately laid on Eighteenth…[along] Dolores Street toward Mission Creek; it was to carry off the surplus water, which flows down the adjacent Mission hills.[36]

Despite these apparent changes, Mission Creek still flowed underground, and the marsh and the old lake still existed under the fill. The gap between surface appearance and underlying reality continued to widen as entrepreneurs tried to use O'Farrell's grid to master San Francisco's topography.

As the city grew in population, assaults intensified on nonconforming topographical features that blocked development. In 1849 O'Farrell's survey was extended west into the sand hills by William W. Eddy. But this survey had so little correspondence with topography that Eddy's assistant, Mil Hoadley—soon to become city engineer—could not lay down the hypothetical streets on the actual site. Directed to construct a rectilinear grid of streets, regardless of elevation or other natural obstacles, he proposed a grading plan that included the planing down of Rincon Hill, Russian Hill, Nob Hill, and Telegraph Hill.

To planners and speculators of the time, this seemed to be a rational way to prepare the site for settlement, but to citizens who were beginning to embrace a picturesque aesthetic that included an appreciation of varied topography and natural features, wholesale leveling seemed needlessly destructive. When wealthy citizens who had begun to build view houses on these hills opposed Hoadley's proposal, the city council appointed a committee of two military engineers to study the impact of filling and leveling.[37] The report, submitted in May 1855, lamented the destruction of San Francisco's hills and the continuation of O'Farrell's grid up Telegraph Hill. The committee members felt the grid's insensitivity to topography was a threat to the city's natural beauty. They wanted to stop the quarrying of stone from Telegraph Hill, the summit of which they described as a beautiful promontory, a perfect spot from which to gaze down at San Francisco. This was an unusual conclusion for the time. There was money to be made, and constraints on property speculation were seldom considered. However, the city council voted to modify Hoadley's plan by instituting a program of grading streets rather than leveling hills.

As grading projects accelerated in the late 1850s, an editorial in the *Alta California* noted:

> The city was laid out by those who believe that there is no beauty in anything topographical but dead level, and streets running at right angles...a wiser community would have tried to make their streets suit the topography of the site...even Telegraph Hill is beginning to disappear.[38]

This lament stopped neither the quarrying of Telegraph Hill nor the near obliteration of Rincon Hill. In 1867 tons of rock to be used in the construction of the city's seawall were quarried from the side of Telegraph Hill at Battery Street near Vallejo.[39] As blasting increased, frequent landslides occurred. In the 1880s the Gray brothers established a quarry for extracting aggregate to be used in sidewalks and curbs. Local residents fighting the blasting were not pleased to hear co-owner George Gray declare that one of his goals was to annihilate Telegraph Hill. Despite lawsuits and ordinances curtailing their activities, the Gray brothers continued excavating, causing landslides and building failures, until 1915, when George Gray was shot dead by a disgruntled employee.

Across the old Yerba Buena cove, Rincon Hill met a worse fate. Rising a hundred feet above the South of Market District on the shore of the bay, the broad summit and steep slopes of this rock outcropping stretched from Spear Street to Third Street between Second and Bryant Streets.[40] Flat-topped, protected from westerly winds and situated well above the marshy and congested Yerba Buena cove, Rincon Hill had a beautiful view of the bay and had quickly become the focus of much upscale residential development. In 1852 the English entrepreneur George Gordon gambled on a real estate development, modeled on London's residential squares, which he called South Park, a private park in the shape of an elongated oval surrounded by fashionable row houses. Although its appeal was in keeping with the rest of the development of Rincon Hill, half the lots were never sold. The final blow to this project was the savage Second Street cut.

The cut was the brainchild of John Middleton, who had come to San Francisco in 1849 and owned property on Second Street, on Rincon Hill.

Middleton embraced the idea of connecting the South of Market to South Beach on the bay by lowering Second Street so that heavily laden wagons could travel directly back and forth from the Pacific Mail wharves without having to go around Rincon Hill. Middleton fought for this plan after he was elected to the state legislature in 1868. Spearheaded by Middleton, Bill 444 passed the same year, authorizing the grading of Second Street between Howard and Bryant Streets. By 1869 a deep cut tore through the flank of Rincon Hill, cleaving it in two. A photograph vividly illustrates the massive site alteration in progress (figure 1.10). Homes on either side of the cut began to shift on their foundations. Property values plummeted and owners sued the city, but the end result was the destruction of the once fashionable neighborhood. More than fifty years later, what remained of the hill's desirability would be further diminished by its use as the approach and anchorage for the Bay Bridge. But by 1869 the deed was done. That same year, the *Alta California* editorialized:

Figure 1.10
Rincon Hill, the Second Street cut, 1869

> We have done more in a score of years in changing the topography of the city than Venice did in five centuries or Amsterdam in two, and those cities, like ours, were built up partly in defiance of Nature.[41]

Tons of sand, dirt, and rock from the former hills of San Francisco had been dumped into the South of Market marsh, Yerba Buena cove, the Mission District's streams and lake, and the South Beach mud flats, creating what San Franciscans called "made land." But even as filling was underway, San Franciscans began to understand the price they would have to pay for their new real estate. Filled land was extremely dangerous in earthquakes.

Early Earthquakes

Before the 1850s, Spanish, Mexican, and Yankee settlers had already experienced earthquakes in San Francisco.[42] Some of the first adobe buildings were damaged in a series of earthquakes from June 21 to July 17, 1808. The first major earthquake to strike the settlement of Yerba Buena was in June 1838. The earthquake occurred along the San Andreas Fault near Santa Clara, causing serious damage to the walls of San Francisco's Mission Dolores and destroying adobe walls in Yerba Buena.

The shaking in Yerba Buena harbor was described as very severe. At the time, Yerba Buena boasted fewer than seven or eight buildings, and reports of this earthquake are sketchy. After the great shock of 1838, a twenty-two-year lull ensued, with no major earthquakes until 1851. Whatever fear or wisdom the earthquake of 1838 might have inspired was lost. Some of the fifty thousand people who arrived in San Francisco in the late 1840s and early 1850s probably had heard of the earthquakes, but only a handful had experienced their force.

With the earthquakes of the 1850s, the story of seismicity and the city begins. After the earthquake of May 15, 1851, the *Alta California* reported, "In many parts of the city the shock was so marked as to cause for a short time great alarm. Yet with all the tremor we have heard of no walls falling and of no serious accident."[43] Nevertheless, the anonymous author grasped the possibility that San Francisco's earthquakes might be linked to others on the Pacific rim, and he had the imagination to understand the threat of regional danger. Just days before the 1851 temblor, an earthquake had destroyed Valparaiso, Chile. The author concludes his article with a warning:

> It was rumored that Valparaiso and a portion of Chile surrounding it have been recently subjected to a terrible earthquake, which in a few moments had crumbled walls and caused death to a frightful extent. The report may not be well founded yet the shock of yesterday morning would seem to confirm it. Indeed, it would not be unreasonable to anticipate on the whole extent of the eastern Pacific occasional tremblings of the earth during this coming season, slight in this latitude, but at many points causing frightful and extensive ruins.[44]

The next earthquake that San Franciscans believed to have occurred was in 1852. It was purported to have been a violent but localized shock on the coast near present-day Lake Merced which broke open a passage between the lake and the ocean, quickly draining the lake.[45] Strangely, no shaking was felt in the city proper. Modern researchers doubt an actual earthquake took place. Four years later, on February 15, 1856, a strong earthquake with its epicenter near Fort Point (in the Presidio) struck the city and caused extensive damage, particularly in filled areas.[46] Goodwin and Co., on filled land at Front Street, lost one hundred feet of fire wall. Two buildings on filled land at Battery and Washington Streets settled, opening a gap between them. Plaster fell in Wilson's Exchange, the International Hotel, Tremont House, the St. Nicholas Hotel, the Merchant's Exchange, and City Hall. All were positioned on filled ground or alluvium at the edge of the old cove of Yerba Buena. The same area would be shaken again in January 1857.[47]

San Francisco's First Earthquake-Resistant Building?

The multiple fires and two earthquakes of the early 1850s and the widespread settling problems caused by filled ground seem to have led one of the capitalists investing in downtown San Francisco to build a seismically safe, "fireproof" building. In 1853 Captain Henry W. Halleck commissioned Gordon P. Cummings, an English architect, to build a massive brick structure at Montgomery and Washington Streets.[48] The building would become known as the Montgomery Block (figure 1.11). Halleck, trained at West Point as a civil engineer specializing in military fortifications, undoubtedly took a personal interest in the building's design and was responsible for its innovative structural features. He specified that it be constructed on a deep raft of redwood logs that would float on the bay mud below it. The raft foundation ensured that if the building settled, it would settle evenly. Such a raft foundation could move as a unit when shaken by an earthquake. The building was symmetrical, with a circumference of tied brick walls which later authors thought contributed to its earthquake resistance. The building contract states that "transverse and longitudinal rods of iron will run through each wall and also through partitions," and "every 8th joist will have an iron anchor." The bills for iron confirm the building was tied by a quantity of iron rods, bars, and anchors.

Perhaps Halleck learned about reinforcing with iron during his military engineering training. Iron reinforcing rods and tie-rods had been used for centuries in Europe, so their appearance is not wholly unexpected. In Europe, as early as the

Figure 1.11
The Montgomery Block (formerly the Washington Block), on Montgomery Street, erected on a wooden raft foundation with iron reinforcement, probably designed to resist earthquakes as well as settlement. To the left of the Montgomery Block, with a segmental roof, is an iron building. G. F. Fardon, 1855

Middle Ages, exposed iron rods in tension (tie-bars and tie-rods, *catene* or *tiranti* in Italian) had been placed at the rafter level of masonry buildings from one outside wall to the opposite wall. They were secured on the outside of the building by crosses of iron called keys (Italian *chiave*). There were also horizontal iron rods or bars laid in the walls themselves, or sometimes along the window mullions, to hold the in-plane wall in tension. For example, iron rods and belts tie the outside walls of the beautiful late gothic Sainte-Chapelle in Paris. (I am calling both the rods in the walls and the exposed tie-rods "reinforcing rods" here.) Scholars examining the Montgomery Block before its demolition have also described "diagonal iron bars" in the building. These diagonal bars would have been tied to the corners of the building to ensure the exterior wall would not fail.

The use of all these reinforcing devices in the Montgomery Block illustrates an understanding, rare at the time, of the threat of out-of-plane lateral bending (discussed in chapter 4) and possible collapse of exterior brick walls, which occurs in extreme settlement and earthquakes. It is impossible to prove that Halleck was concerned about the threat of earthquakes, but the Montgomery Block incorporated earthquake-resistant concepts and survived all subsequent earthquakes, including 1906, with little damage.

Disaster and the Specter of Economic Decline

Although San Franciscans were more worried about fire, the earthquakes of the 1850s raised serious questions. An *Alta California* editorial of February 18, 1856, just a decade after the American occupation of San Francisco, castigates the fearful and pessimistic for their concern with the dangerous nature of San Francisco's location. The writer correctly correlates state of mind and property value:

> Since the terrible shock of Friday last, speculation has been rife as to the probable effect it would have upon the minds of the inhabitants of San Francisco, and upon the property interests of the city. Any amount of *croakers* were found who could at once see something awful in this visitation... They predicted that scores of families would remove from San Francisco for fear of more serious convulsions hereafter, and take up their abode in places unfrequented by earthquakes. They anticipated a wonderful decline in real estate. It is well that these people can only talk. Their influence is exhausted with their breath, and sensible people pay little heed to such croakers and panic makers.[49]

The writer continues by minimizing the damage to buildings:

> From all that we have yet heard, we can learn of no material damage done to any of the most frail brick buildings in the city, further than the cracking of the plaster walls, and if this shock, which is admitted to have been the most powerful of any for nearly a half century, has had so little effect

upon the structures of the city, there is but little reason to fear more serious shocks in the future. This has been a sort of test, and has proved the capability of our structures to withstand such vibrations of the earth as are liable to occur in the future.[50]

Few San Franciscans had experienced the large earthquake of 1838. There were no memories to deny. Public opinion remained optimistic, and real estate values were undiminished by the earthquakes of the 1850s. The *Alta California* forecast that "the city will long continue to have a name and place among those of the earth, and retain a very fair population, notwithstanding she may be occasionally jarred a little by the internal convulsions of the earth."[51] The city would prosper on its site, but ten years later another series of earthquakes would force the issue of seismic danger, and sixty years later the name "San Francisco" would become synonymous with seismic catastrophe. San Francisco's earthquakes and San Franciscans' reaction to them illustrate that the threat of natural disaster rarely if ever overrides commercial interests—and San Francisco had location, location, location.

CHAPTER 2

Fire, a Compelling Danger, 1849–1851

Figure 2.1
The first seal of the City of San Francisco, adopted November 4, 1852. The phoenix, wings outstretched, rises from a pyre blazing on the surface of San Francisco Bay. In 1852 the memory of San Francisco's reconstruction after six fires was still vivid, evoking parallels to the mythological phoenix. After a long life, the phoenix is said to build its own funeral pyre, burn itself, and regenerate from its ashes.

On the first official seal of San Francisco, adopted in 1852, a phoenix rises with wings outstretched from a blazing fire on the surface of the bay, with the city's hills and the Golden Gate in the background (figure 2.1). San Franciscans chose the phoenix as the centerpiece of their seal in commemoration of the city's recovery after six major fires in three years.

In the nineteenth century, fires caused more damage to cities in the United States than any other kind of urban catastrophe. Unlike the more widely spaced, uncanny, and unpredictable earthquakes, fires occurred in San Francisco at such close intervals that the need for change and regulation was all too obvious. The citizens of San Francisco—the boomtown, the instant city, a place first created with little thought for the future—were forced to either confront the threat of fires or endure their repetition.

Responses to Disaster

Earthquakes and fires are linked in San Francisco's history, and their relationship is at the heart of a dilemma that architects and engineers have had to face: wooden structures, generally safer in earthquakes, are extremely flammable; brick buildings, safer in fire, are more dangerous in earthquakes. Which danger is greater, earthquake or fire? The fact that earthquakes cause fires complicates choices still further. According to Charles Scawthorn, coeditor with Wai-Fah Chen of the *Earthquake Engineering Handbook,* "In both Japan and the United States, fire has been the single most destructive seismic agent of damage in the twentieth century. The fires following the San Francisco 1906 and the Tokyo 1923 earthquakes, which were both terribly destructive, rank as the two largest peacetime urban fires in man's history."[1] Both cities were built mainly of wood.

More than fifty years passed between San Francisco's sixth major fire, in 1851, and the disastrous fire of 1906. What happened to make this possible, when six fires had occurred in three years, 1849 to 1851? A community must consider how likely a disaster is to be repeated in the future. How much would it be worth to mitigate the impact? San Francisco's response to the fires provides both a model and a foil for assessing the city's response to earthquake danger and the difficulty of considering multiple dangers simultaneously.

J. W. Powell, a sociologist, was the first to enumerate the stages of response to catastrophes like the San Francisco fires:[2]

1. The warning period, when the conditions leading to the danger become apparent
2. The time when people become aware of the imminence of the impending catastrophe
3. The impact, when disaster strikes
4. The inventory, when victims comprehend the scope of damage
5. Rescue
6. The remedy, when organized relief efforts occur
7. The recovery, when citizens try to reconstitute or reform the community to protect again the repetition of the catastrophe

Other analysts view the sequence differently. For example, J. Eugene Haas, Robert W. Kates, and Martyn J. Bowden add stages to the end of Powell's list that include restoration of services, reconstruction, commemoration of the disaster, a period of betterment, and finally, a developmental reconstruction period.[3]

These stages are simply conceptual frameworks that help to classify very complex societal behavior. It doesn't always turn out this neatly. In Taiwan, during the earthquake of 1999, rescuers, after realizing in the midst of the disaster that a collapsed building had been constructed of rice tins, angrily pursued the contractor. This "should" have happened in the recovery period. Likewise, in Turkey, only a month after the Kocaeli earthquake, also of 1999—in what might be called the remedy period—architects and contractors accused of substandard design were arrested and the government was accused of negligence in its code enforcement. In San Francisco in 1851, vigilantes singled out foreigners, accusing them of arson, and hung one of them immediately after the fifth fire.[4]

San Francisco at Midcentury

In the winter of 1849, the urban problems facing San Francisco were staggering. City streets platted in O'Farrell's survey were disappearing in the mud. Writing home, William S. Jewett described his experience:

> Yet we must pick our way! Pick, jump, stride and totter and we got somewhat into something that no doubt looks very like a street on a map but it was not recognizable in its natural form although they called it "Broadway"; it proved so to us for...all succeeded in getting stuck.[5]

The streets had no surface, no grade for drainage, and no storm drains or sewers. Except for rain, there was no civic water supply. There was neither fire department nor police department. The city building stock was completely unregulated. A vast majority of structures were built of canvas and wood. Taking office in 1849, Alcalde John W. Geary lamented the lack of any civic funds or services: "Public improvements are unknown in San Francisco...you are without a single requisite necessary for the promotion of prosperity, for the protection of property, and for the maintenance of order."[6] All the more surprising then, that San Franciscans effectively confronted the danger of fire by the mid-1850s.

The First Fire: December 24, 1849[7]

The population was wholly unprepared when the first fire broke out, on Christmas Eve day in 1849 in a hotel called Dennison's Exchange on the east side of Portsmouth Square.[8] Because the hotel's walls and ceilings were painted canvas, the whole structure burned quickly. The fire spread to the west side of the square, then burned down Washington Street to Montgomery Street, where a group of citizens managed to defeat it by blowing up and pulling down adjoining buildings. There was no fire department.

The morning after the fire, on Christmas Day, citizens who had been members of volunteer fire companies on the East Coast gathered to organize San Francisco's first volunteer fire company. The recovery period had begun. On January 28, 1850, the town council formally elected one of these founding volunteers, Frederick D. Kohler, the first chief engineer of the San Francisco Volunteer Fire Department.[9] Alcalde Geary ordered two side-stroking fire engines from New York, and a local businessman, William D. Howard, bought a third. The department under Kohler consisted of over a hundred volunteers: a definite improvement, but limited by water shortages.

Although ancient Rome had been protected by a disciplined, paid fire-fighting force, municipal fire departments were rare in nineteenth-century Europe and America. Instead, cities relied on either private firefighters employed by insurance companies or volunteer organizations. Volunteerism was part of the community ethos in the early United States; indeed, volunteer fire companies and ambulance crews still exist in rural communities today. In the nineteenth century, volunteer engine companies were often powerful political and social clubs to which entire communities looked for protection. In Philadelphia and New York, it was not uncommon for these clubs to be allied with particular parties or ethnic groups, and to fight one another as well as fires. With axes and hooks and uniforms at their disposal, volunteers could become an instant army, or at least a dangerous armed gang.[10] The companies fostered a fierce rivalry, each priding itself on its speed and endurance. Members often paid for their own uniforms, engines, and engine houses. As in other western cities, San Francisco's volunteer companies associated themselves with specific East Coast cities or regions. Men who came from New York, the "Empire State," organized Empire Number 1 after the first San Francisco fire. Later, the Social Number 3 attracted Bostonians. Unionists joined the Knickerbocker Number 5, while Southern sympathizers joined the Monumental Number 6.[12]

Figure 2.2 illustrates a typical engine house in San Francisco, Columbian Engine Company Number 11, founded in 1852. Standing in front is a hand-drawn and hand-pumped fire-fighting engine, much prized by company members. Until

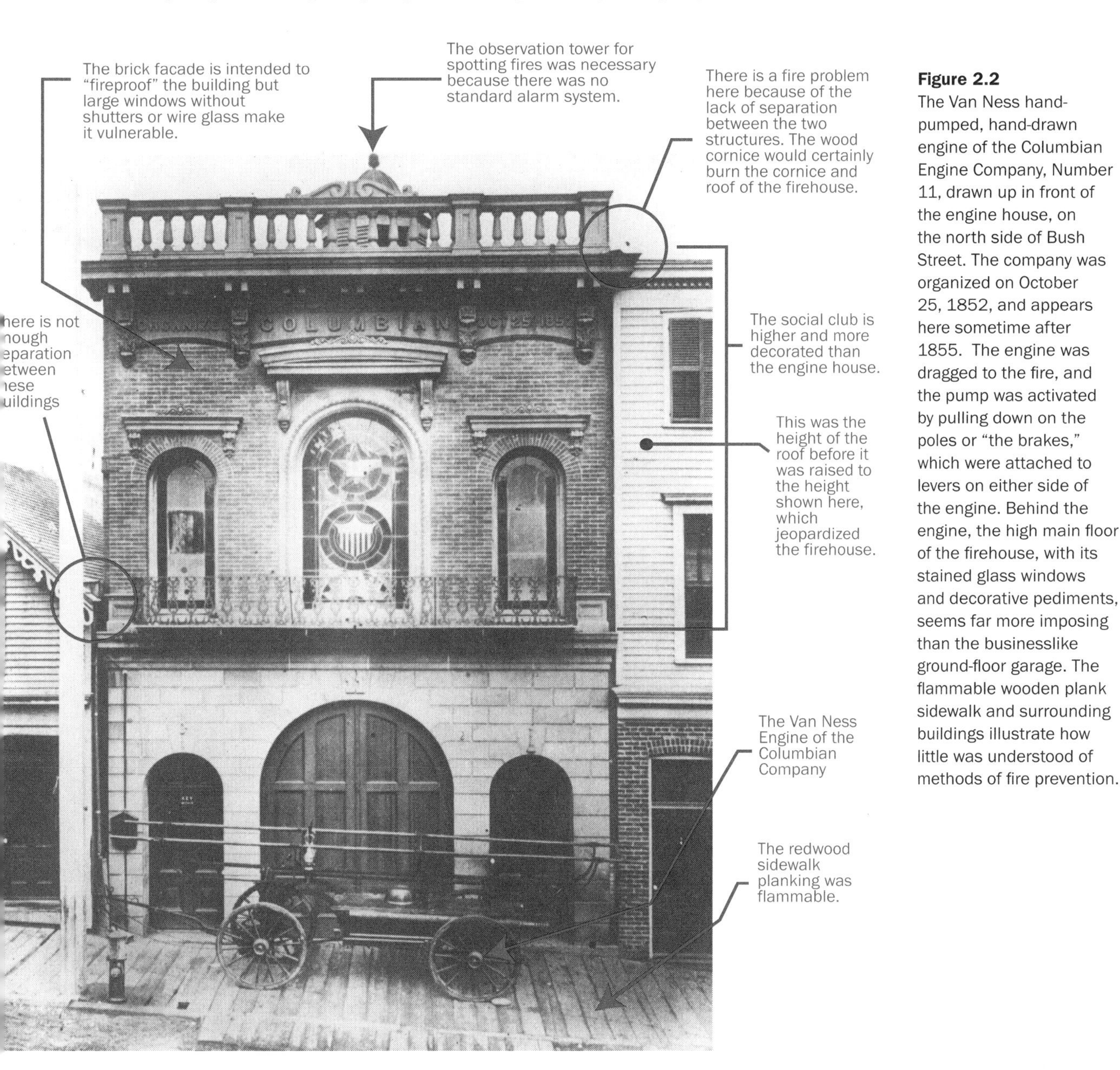

Figure 2.2
The Van Ness hand-pumped, hand-drawn engine of the Columbian Engine Company, Number 11, drawn up in front of the engine house, on the north side of Bush Street. The company was organized on October 25, 1852, and appears here sometime after 1855. The engine was dragged to the fire, and the pump was activated by pulling down on the poles or "the brakes," which were attached to levers on either side of the engine. Behind the engine, the high main floor of the firehouse, with its stained glass windows and decorative pediments, seems far more imposing than the businesslike ground-floor garage. The flammable wooden plank sidewalk and surrounding buildings illustrate how little was understood of methods of fire prevention.

the mid-nineteenth century, engines had to be pulled by the volunteers to the site of a blaze and pumped by hand to build up enough pressure to force water through the hoses. Even men in top condition were soon exhausted "manning the brakes," as this was called. Fire departments also had hook-and-ladder companies whose job it was to rescue people from endangered buildings and then, if no water was available, pull off cladding to expose the fire within. Setting their hooks on the roofs of a frame house, the entire company would man the ropes and pull out the external supporting walls, dropping the floors and roof to the ground to remove fuel from the path of the conflagration.

Within weeks of the December 1849 fire, the entire area had been rebuilt. But the new buildings, according to reporters of the time, were "like those that had just been destroyed, and like nearly all around...chiefly composed of wood and canvas."[13] Changes to construction practices and building materials as a means of mitigating fires had yet to be implemented. However, the volunteer fire department was a definite improvement, though still limited by water shortage.

Three Fires in 1850 and the Importance of Water

The second major fire, on May 4, 1850 (figure 2.3), broke out at about four a.m. in the United States Exchange, a saloon and gambling house that had been built, by coincidence, on the burned-out site of Dennison's Exchange.[14] The fire grew quickly into a conflagration: a very large fire generating high heat and towering flames. It burned fiercely for seven hours, destroying three blocks and nearly three hundred houses, four times the damage caused by the first fire.[15]

Figure 2.3
"Great Fire in San Francisco, May 4, 1850"—the second fire

This time the city, or at least its aldermen, understood the urgency of a better water supply. The aldermen ruled that six buckets of water had to be at the ready in every building in the city, and property owners who were able to do so were required to dig artesian wells and reservoirs.[16] Further, any person who refused to fight fires when asked could be fined. The council also pledged city funds to construct cisterns around the city, one of which was to be on Portsmouth Square.[17]

The Third Fire: June 14, 1850

The third fire (figure 2.4) broke out at eight a.m. in a bakery behind the Merchant's Hotel, on Montgomery Street between Sacramento and Clay.[18] Spectacularly destructive, it was similar in intensity to the fire of 1906, causing temperatures as high as 2700° Fahrenheit, melting iron, and vaporizing structures in its path.[19] The city's characteristic high winds were blowing from the west, and the fire soon spread through the whole district bounded by Clay and California Streets, Kearny Street, and the waterfront. The hand-drawn fire engines were helpless against the blaze. Merchants raced frantically to save their goods. The wharfs were deliberately cut off from the land in order to save them from the fire. About five hundred men were left on the wharfs to wait out the fire. One of them recalled a blaze so fierce that "at every blast the hurricane came surging down the wharf in clouds of smoke and cinders, obliging us to lie flat on our faces."[20] Metal and glass melted in the heat, gun barrels twisted and knotted, tons of nails were welded into the shapes of the kegs that contained them, and molten glass pooled in small lakes.

In spite of their good intentions after the previous fire, the property owners of San Francisco had provided fuel for the next conflagration by again rebuilding quickly and poorly. The *Alta California* had smugly reported, "In ten days, more than half the burned district was rebuilt."[21] This time, two and a half blocks had been destroyed. And again, "In forty-eight hours after the fire the whole district resounded to the busy din of workmen."[22]

The Fourth Fire: September 17, 1850

A little over three months later, the fourth fire broke out at the Philadelphia House, on the north side of Jackson Street between Kearny and Dupont (later Grant Avenue), burning one hundred and fifty houses on four city blocks.[23] As had happened after the first three fires, the district's real estate values increased because fire-cleared lots could be rebuilt on even more densely.[24] This windfall did not entirely blind San Franciscans to the amply demonstrated danger of fire. The city council decided to ensure that water would be on hand to combat future blazes by building four underground

Figure 2.4
"Great Fire in San Francisco, June 14, 1850"—the third fire

cisterns. These reservoirs of caulked redwood were a significant step forward in fire protection. Unfortunately, because of lack of rain, they were empty when they were called into service in June 1851 and remained so for two more months.[25]

The Fifth Fire and a Turning Point: May 4, 1851

The fifth fire (figure 2.5) was also fanned by a high wind from the west, and it became so intense that people in Monterey, a hundred miles south, reportedly saw flames in the sky.[26] It "destroyed more buildings than all the previous fires combined. Eighteen blocks were in ashes, three-quarters of the city gone."[27] Like the third fire, it burned along San Francisco's plank streets and out into the wharves on the bay. It must have been eerie and upsetting to see the buildings in the water lots of Yerba Buena cove burning on their pilings (figure 2.6). This time the tenor of the reaction changed. The magnitude of the damage surpassed the worst that could be imagined. Realizing that another fire might doom the city, the citizens wanted answers, and they wanted blood.

San Franciscans, already suspicious of arson, were convinced the fifth fire had been intentionally set: after all, it started exactly one year after the city's second major fire.[28] A rash of recent false alarms and robberies confirmed their conviction

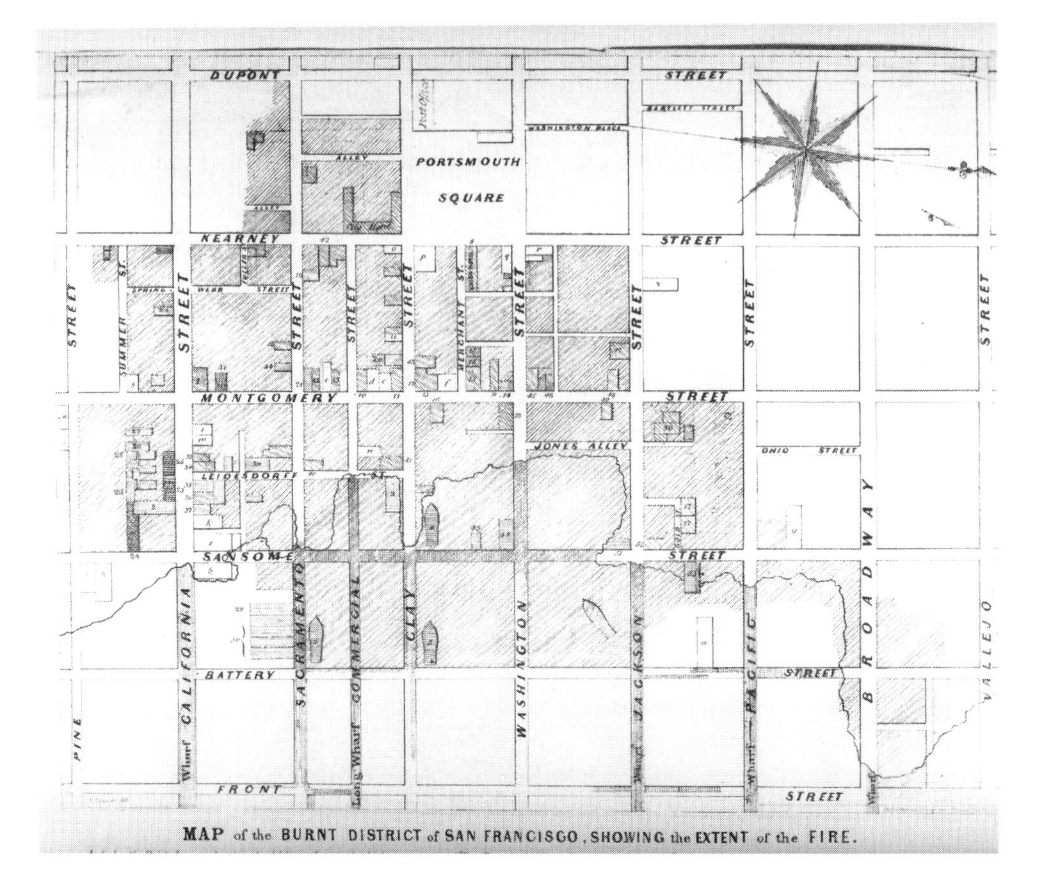

Figure 2.5
Burned area of the fifth fire, May 4, 1851

that their city was in the terrifying grips of "incendiaries."[29] Five days after the fire, the editors of the *Alta California* concluded that it had been set as a diversion for robberies, although they never unearthed definite proof, and one month later the paper published its summation of the situation under the heading "Another Attempt to Fire the City":[30]

> This could not possibly have been the result of accident, and it is now rendered positive and beyond a doubt, that there is in this city an organized band of villains who are determined to destroy the city. We are standing as it were upon a mine that any moment may explode, scattering death and destruction.[31]

Outraged by what they saw as collusion between these gangsters and the police, worried about future arsonists, and spurred on by newspaper editorials, a group of men formed a secret vigilante society.[32] The vigilantes, along with many others, believed that the Sydney Ducks—former Australian convicts living at the base of Telegraph Hill—were responsible for starting the fires and looting while they burned. On the night the vigilante society was formed, June 10, 1851, a hapless Sydney Duck named Jenkins stole a safe from a merchant's storeroom and was caught by citizens who heard the alarm. Instead of handing the robber over to the police, the vigilante committee seized Jenkins, tried him, and hanged him in Portsmouth Square.

Scholars debate whether Jenkins' lynching was a deterrent to future arsonists. The sixth fire occurred eleven days after his death, and according to the *Alta California,* arson accounted for 31.6 percent of the city's small fires between 1851 and 1856.[33] The citizens had been pushed to the limit, and perhaps this episode of blind and brutal fury helped to focus subsequent attention on the need to do something to mitigate the threat of fire.

In the midst of the tension, city officials accepted a proposal to bring a dependable supply of water to the city from Mountain Lake (now at Park Presidio Avenue and Lake Street), about four miles from the heart of the city. Filling cisterns and keeping them filled had been a constant problem. The city granted a private entrepreneur, Arzo D. Merrifield, the license to build an aqueduct and lay pipes under streets. This difficult, costly project did not, in the end, succeed. Only one-third of the aqueduct was finished by 1855, and it was still incomplete in 1857 when the newly constituted San Francisco Water Works took over the

Figure 2.6
The fifth fire, May 4, 1851, as seen from the California Street wharf

project. By tapping into Lobos Creek, which drained into the Golden Gate just south of Baker Beach, they were able to build a flume around the cliffs, through the present-day Presidio, through the Marina District and around Fort Mason to a pumping station at the foot of Van Ness Avenue, where the water was pumped uphill to reservoirs on Francisco and Lombard Streets, on Russian Hill. They finally delivered an adequate supply of running water to the city in 1858.[34]

Ultimately, the poor quality of its buildings—not arsonists—was at the core of San Francisco's problem. Canvas was still used in construction in 1850; wood was at a premium and bricks still a luxury. The vast majority of structures were shanties, poorly built wooden buildings and wooden frames covered with canvas, packed together in the center of the city.

A few San Franciscans had seen the danger and embraced the promising technologies of the time, brick and iron. For hundreds of years, masonry had been recognized as a dependable fire-resistant building material. At least as early as the twelfth century, observant officials in London, realizing that urban fires could be eliminated only by controlling and ultimately forbidding the use of flammable building materials, enacted laws prohibiting straw roofs and requiring the use of nonflammable masonry. These laws were poorly enforced until the great fire of 1666 destroyed much of London. The use of masonry was strictly enforced after that, and it was still considered the most fire-retardant material available in the 1850s.[35] Less expensive to buy and assemble were the largely untried but promising building components of iron that would be popularized by the famous Crystal Palace, built in London in 1851.

Prefabricated iron structures had been imported from Britain to provide quick and dependable shelter during the gold rush, and these easy-to-assemble buildings were sometimes described as fireproof.[36] After the first fire, an article in the *Alta California* reported the superiority of iron to wood:

> Messrs. Moore & Co., who were heavy losers by the fire, have very quietly framed a new store on Washington Street, and above them a few cinders stands an iron house, with a seeming pertinacity, equivalent to a sturdy interrogatory of "who's afraid?"[37]

Merchants selling these iron buildings ran advertisements in the *Alta California* after the fire to draw attention to their wares:

> FIRE! FIRE! In consequence of the late alarming fire it will be evident to all that IRON Houses have superior advantages to wood... The sad disasters which so lately befell our town and many of its most enterprising citizens might have been avoided if the roof, rear, and front sides of their buildings have been covered with metal.[38]

Far from being fireproof, cast iron is brittle at room temperature and melts at 2300°F, a temperature commonly reached in major fires.[39] The iron buildings, which might have seemed safe in small blazes, failed gruesomely in the fifth fire. Heinrich Schliemann, the German merchant who later gained fame as an archaeologist and discoverer of Troy, visiting San Francisco at the time of the fire recounted:

> Neither the iron houses nor the brick houses (which were previously considered as quite fireproof) could resist the fury of the element: the latter crumbled together with incredible rapidity, while the former got red-hot, then white-hot, and fell together like card houses. Particularly in the iron houses, people considered themselves perfectly safe and they remained in them to the last extremity. As soon as the walls of the iron houses got red-hot, the goods inside began to smoke, the inhabitants wanted to get out, but usually it was already too late, for the locks and hinges of the doors were no more to be opened.[40]

Five brick buildings did survive in the heart of the blaze. These included the Dewitt and Harrison warehouse, which was saved by employees wetting it down with eighty thousand gallons of vinegar that had been stored inside.[41] Other brick buildings that survived the blaze, like the El Dorado Hotel and the Naglee Building, became exemplars of safe construction, and many new buildings were modeled on them. (Henry Naglee, a local merchant, was burned out four times before he constructed the Naglee Building of brick in 1851.)[42] The low flammability of neighboring structures and the effectiveness of fighting fire from inside the building were probably significant in their survival as well.

Although iron houses fared poorly in this conflagration, iron was not entirely discredited as a fire-resistant material; iron could resist flame and heat for a longer time than any contemporary material other than masonry or cement. In the 1860s, both cast and wrought iron became standard fire-resistant elements in factory interiors and building fronts in fire districts. In San Francisco, iron roofs, shutters, moldings, cornices, and entire facades were all used to increase fire resistance.

The Sixth Fire: June 22, 1851

Eleven days after the fifth fire, two hundred and fifty buildings were already rising from the ashes. Unfortunately, the new brick buildings did not prevent or stop the sixth fire (figure 2.7), which broke out a month and a half later, on June 22, 1851. This fire nearly equaled the destruction of the fifth fire; thirteen blocks were destroyed. Fanned by westerly winds, the blaze spread from a frame house on the north side of Pacific near Powell and destroyed four hundred and fifty buildings, including City Hall and the city hospital.[43] The fifth fire had burned from Portsmouth Square east to the bay, while in the sixth fire most of the damage occurred to the north. Several blocks were burned by both the fifth and sixth fires (figure 2.8).

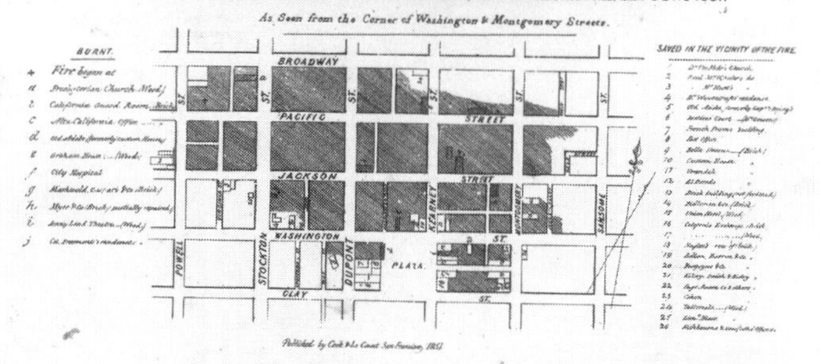

Figure 2.7 (top)
An illustration of the sixth fire, June 22, 1851, made from a reused plate of the fifth fire, which had taken place six weeks before. (The date "June 22," in a different typeface than the rest of the caption, does not completely obscure the "4" of the erased "May 4" date.) The printer can be excused for economizing: the fires were very similar.

Figure 2.8 (bottom)
A view of the aftermath of the sixth fire, June 22, 1851, with a map and score list indicating which buildings were burned and which saved

Buildings that survived both the fifth and sixth fires included the previously mentioned Naglee Building and the rebuilt El Dorado House, as well as the Verandah House and the California Exchange, on Portsmouth Square. Subsequent builders took their cues from these successes. In 1852, a successful merchant, John Parrott, commissioned an all-granite building on the northwest corner of California and Montgomery (figure 2.9).[44] Stones for the foundation were quarried from Goat Island (now Yerba Buena Island); granite for the rest of this three-story structure came by boat from China, along with Chinese workmen who were capable of assembling the precut granite blocks on the site. Care was taken to isolate more flammable roofs by extending parapets above the cornices on buildings and along sides and rear facades. These walls around the perimeters of the roofs were called "fire walls." The Montgomery Block (see chapter 1) was not only built of stone and brick with fire walls, but also protected by iron shutters. The Douglas Building at Montgomery and Jackson Streets, designed by Thomas Boyd and completed in 1857, is a good example of fire-resistant (and inadvertently earthquake-resistant) construction:

> Built in the side walls from front to the rear, and from side to side in the front and rear walls, are heavy iron rods anchored in the solid masonry work, which binds the walls together in the most secure manner. The openings are all guarded from fire by heavy iron shutters. The sidewalk in front is of iron set in with dead-lights. The iron columns, sills, doors and window shutters were all made here... Notwithstanding the buildings adjoining are all fire-proof, this has

been protected from all possible accidents, and fire-walls have been carried six feet above the roof, to give, if possible, additional security.[45]

The First Building Code and Fire District: 1852

San Franciscans had begun to see that the material well-being and safety of even rich merchants was a communal problem and depended upon the surrounding built environment (see, for example, figure 2.2). Not even a four-sided brick building could withstand a fire if it was surrounded by flammable, wooden buildings. The roof would be vulnerable, as would any decorative wooden details. Little thought had been given to how easily windows and doors could be breached, letting fire into the flammable wooden interior, which was often laden with equally ignitable furnishings and goods.

Each fire had been a lesson in the cause and prevention of civic conflagrations, and after each, San Franciscans had tried to introduce some form of mitigation. After the fifth fire, there was a shift in the city's attitude toward its own safety. Citizens understood the necessity of restrictions; the boomtown attitude, anything goes, was no longer appropriate. Building codes stipulating fire-resistant construction were adopted, and a fire district was established, expanding as the city grew. In the first fire code, a May 22, 1852, ordinance, the fire district was to be bounded by Union, Powell, Post, Second, and Folsom Streets (figure 2.10).[46] No open fires were allowed there, lanterns had to be used instead of torches and candles, and tents and canvas had to be removed. Eight months later, on December 6, a stronger ordinance forbade new construction of frame buildings within the new fire district bounded by Dupont (present-day Grant Ave.), Pacific, Pine, and Front Streets.[47]

Figure 2.9
The northwestern corner of California and Montgomery Streets as it appeared in 1855. The building on the corner with the Wells Fargo sign is the Parrott Building, erected with granite from China and a Chinese crew to assemble it in 1852. It was designed to be fireproof, with a stone exterior, iron window surrounds, and iron shutters. Each of the buildings constructed on this side of Montgomery Street was a fire-resistant unit. The side walls of each, called "fire walls," extended to the top, and sometimes above, the decorative facade cornice. Parapets encircled the top of the cornice and sometimes the fire wall as well, protecting the more flammable roof. G. F. Fardon, 1855-56

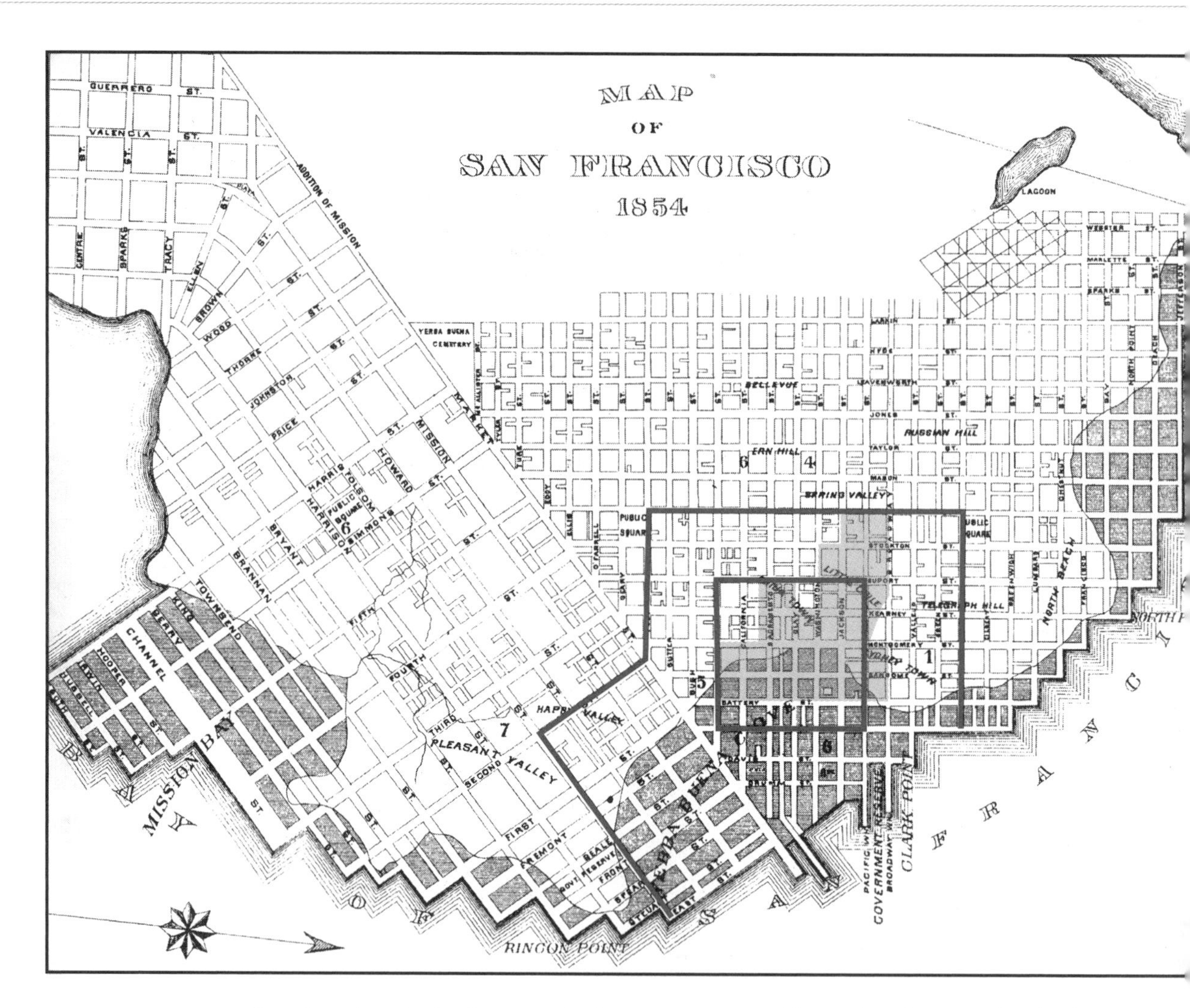

Figure 2.10
The first fire district, May 22, 1852 (in which canvas was forbidden), and the second, smaller, 1853 fire district (in which new wooden buildings were forbidden). The tinted section shows the accumulated area burned in all six fires.

San Francisco had taken the basic measures to mitigate fire damage that had been pioneered elsewhere and were well known in the United States: the establishment of a volunteer fire department with fire-fighting equipment; the imposition of fire laws governing fire-resistant construction; the establishment of fire districts; the installation of water systems; and the construction of more advanced fire-resistant buildings.

In 1856 the photographer George R. Fardon compiled an album of photographs of San Francisco.[48] Possibly meant to be propaganda for the vigilante movement, Fardon's photographs, which include the temporary vigilante stronghold of "Fort Gunnysacks" against the backdrop of numerous new brick buildings, show a stable, prosperous city in the charmed interlude between the fires of 1849 to 1851 and the earthquakes of the 1860s (see figure 2.11).

In 1853, San Francisco boasted 626 brick buildings, half of which had been built that year. By 1860, 1,461 brick buildings had been built, but the city's 18,603 wooden buildings made up the majority of the building stock.[49] Nonetheless, as a result of fire laws, the establishment of a fire district, and an increasingly modern fire department, conflagrations were prevented for fifty-four years, until the fire of 1906. San Francisco's safety was built on an understanding of the history of urban fire and fire reduction efforts nationwide. But what distinguished San Francisco from other cities would be the risk of earthquakes—a risk that was easy to ignore, because most of the city's population in 1852 had yet to experience one.

Figure 2.11
San Francisco, 1855, detail. The brick survivors of the fifth and sixth fires, and new brick and iron buildings. This is the heart of the fire district, with many brick buildings but also flammable wooden structures appearing here as well. G. R. Fardon, 1855

A. The City Hall, combining the El Dorado Hotel (just outside the photo to the left), the Jenny Lind Theater (with the tower), and the Union Hotel, on the right. The pedimented face faces Portsmouth Square.
B. Former Post Office, also facing Portsmouth Square
C. The Montgomery Block, on Montgomery Street
D. Iron buildings
E. The new Custom House and Post Office in construction on Battery Street
F. The Merchant's Exchange Building

CHAPTER 3

Earthquakes, Weighing the Danger, 1863–1869

Earthquakes are becoming a permanent institution in California. Scarcely a week passes without one or more shocks being felt; this season they seem to be more numerous and severe than usual.[1]—Alta California, *1864*

Just as multiple fires forced San Franciscans to confront fire risk in the 1850s, so multiple earthquakes forced them to face earthquake danger in the 1860s. This is the stage in the development of disaster response that Powell would characterize as awareness of the impending threat of catastrophe. Unlike fire, for which mitigation practices were being developed throughout the United States, earthquakes were mysterious, their prediction impossible, protection problematic, and danger uniquely San Francisco's.

Many scholars and popular historians have accepted and repeated the myth that San Franciscans of the 1860s denied the existence of seismic danger. Philip Fradkin writes in *Magnitude 8,* "The 1868 shock had all the principal elements of what was to come in 1906, but unfortunately *nothing was learned from the earlier experience*" [emphasis added].[2] However, historical records show that architects, engineers, and even everyday citizens understood the consequences of the earthquakes of the 1860s and tried to inventory the damage, to understand what had happened, to retrofit buildings to resist future earthquakes, and to build earthquake-resistant structures.

San Francisco in the Money

San Francisco in the 1860s and 1870s had far more to lose than it had in the 1850s. In 1859, just as the influx of gold that had driven the economy of the city for a decade was declining, silver was discovered in Nevada. Once again fabulous wealth flowed directly into San Francisco. San Franciscans owned not only the mines but the mills extracting and processing the silver strikes around Virginia City and the other towns of the Comstock Lode. Silver transformed investors like restaurant owner James L. Flood, one of the "silver kings," into instant multimillionaires. Bankers like William C. Ralston amassed huge fortunes. In addition to mining, increased maritime trade and local manufacturing also brought wealth to the city. It was natural that the wealthy would pour their profits into real estate. Ralston financed the largest luxury hotel west of the Mississippi, the Palace Hotel. Large multistory brick buildings replaced wooden structures that had survived the fires of the 1850s. New banks, retail stores, theaters, bars, restaurants, churches, synagogues, and fine homes went up throughout the city. Machine shops and foundries supplying the silver mines flourished in the South of Market. The population grew from less than 35,000 in 1850 to 56,800 in 1860. By 1865, San Francisco was the largest and most populous city on the West Coast, fifteenth in the United States. During the Civil War, the city funneled much-needed cash into the coffers of the federal government. At the war's end, in 1865, the city's population had nearly doubled, to more than 100,000. Buildings were capital and their survival was paramount. By 1868 the city had grown to 18,009 buildings: 4,097 in brick and 13,912 in wood.[3]

Citizens were alerted to the imminent danger of earthquakes by a series of escalating tremors that began on December 19, 1863.[4] This first earthquake was greeted in the press with curiosity and humor. Three months later, on March 5, 1864, another earthquake struck the city.[5] The walls of many buildings cracked, some glass was broken, and people ran out into the streets in alarm. A little over three months later, on May 21, 1864, there was another earthquake. Journalists described each one of these quakes as "the heaviest ever felt in San Francisco," but they would be dwarfed by the earthquakes of 1865 and 1868.[6]

The Earthquake of 1865

On Sunday, October 8, 1865, an earthquake struck the Bay Area that would jolt architects and engineers into focusing on earthquake damage and the possibility of mitigation.[7] San Francisco and San Jose, thirty miles to the south, suffered most, and small East Bay towns, including Oakland, Hayward (then called Haywards), and Berkeley, were also shaken. San Jose residents reported that this was the worst earthquake since Anglo-Americans had settled in the valley. One reporter wrote, "A very slight increase in violence would have thrown every brick building in the city into a mass of ruins." Another wondered "how any building could withstand such fearful throes...To one standing, or attempting to stand, upon the street, the buildings appeared to be swaying and pitching like ships in a rolling sea."[8]

Buildings in San Francisco were severely shaken, even though the epicenter was approximately sixty miles southwest, in the Santa Cruz mountains.[9] Witnesses reported a slight, initial, five-second shock just before 12:45 p.m. on Sunday. San Franciscans took great pride in experiencing minor earthquakes—a common occurrence—without being frightened, unlike "greenhorn" easterners. Many people were in church, and apparently no one went outside after the first small shock. Then, as the reporter and author B. E. Lloyd wrote:

> A second shock came—not a simple jar or tremulous motion, but a rapid shake, powerful and convulsive, like the crash of a terrific concussion; increasing in intensity...accompanied by a frightful roaring sound...This shock continued for the space of twenty seconds...
>
> The walls of buildings swayed to and fro, or parting in places, closed and opened again. Cornices and fire walls came tumbling to the sidewalk; chimney-tops went crashing through the roofs and ceilings; window glass snapped and cracked...Whole walls crumbled and fell...and the dropping of plaster and the clatter of crockery added to the fearful racket.
>
> Horses pricked up their ears, snorted, and dashed away at full speed, as if to flee the danger, or stood trembling in their tracks...Dogs skulked in their kennels growling, or crouched to the ground whining and howling from fright...and even the birds were bewildered...
>
> The people were panic-stricken. Those who were indoors rushed to the street...In the churches, there was utmost consternation...for a moment all were paralyzed with awe—then all with sudden impulse rushed for the door...all clambering for egress, wild with fright, not knowing what they did. All was excitement and alarm—yet, strange to say, amid all the threatened dangers...there was not a single life lost, and no one seriously injured.[10]

Another journalist, Mark Twain, just fired from his job at the *San Francisco Morning Call*, also experienced the quake. Although he was more interested in burlesquing people's reactions than describing building damage, his account of the failure of Popper's Building at Third and Mission Streets is notable:

> As I turned the corner, around the frame house, there was a great rattle and jar...there came a really terrific shock; the ground seemed to roll under me in waves, interrupted by a violent joggling up and down, and there was a heavy grinding noise as of brick houses rubbing together...at that moment a third and still severer shock came, and as I reeled about on the pavement trying to keep my footing, I saw a sight! The entire front of a tall four-story brick building in Third Street sprung outward like a door

> and fell sprawling across the street, raising a dust like a great volume of smoke.[11]

It takes more than literary descriptions to comprehend an earthquake of this magnitude. Engineers, architects, and citizens need quantitative information if they are to build structures that can survive future temblors. How long will an earthquake last? How strong will it be? What kind of damage can it do? What is the probability of future earthquakes? The risk of future earthquakes must be weighed against the cost of earthquake-resistant structural improvements. None of these questions could be answered in 1865. Seismology was in its infancy, with few viable theories explaining earthquakes and no instruments in the Bay Area to record them.

A Brief Introduction to Seismology[12]

Using what we know about earthquakes today, we can judge how well San Franciscans understood the earthquake of 1865 and what they did to prepare for the effects of earthquake shaking. Well before 1865, scientists and lay people had tried to explain why earthquakes occurred. They had struggled with theories of "earthquake weather," electricity, subterranean fires, and wind from underground caves; it was commonly acknowledged that earthquakes were impossible to predict and difficult to understand. Today, with the hypothesis of plate tectonics, we have a conceptual framework for understanding why earthquakes happen in the San Francisco Bay Area and around the world. The earthquake of 1865 occurred because the adjacent Pacific and Continental tectonic plates, made of masses of rock pulling in opposite directions along the San Andreas Fault, suddenly broke apart and moved, releasing a huge burst of energy.

The place under the earth's surface where a break first occurs is called the hypocenter, or focus, and the corresponding spot on the surface closest to the hypocenter is called the epicenter (see figure 3.1). The sudden release of energy at the hypocenter creates seismic vibrations or waves. To paraphrase Berkeley seismologist Bruce Bolt, a rock shattering underground spreads waves through the rocks around it like the ripples created by a stone thrown into a pond. And like pond ripples, seismic waves can be measured: (T) is the time between the beginning and end of one complete oscillation, (A) is the amplitude, or size, of the wave's oscillation, and (F) is the frequency of oscillation. Seismologists have distinguished between underground, or body waves, and surface waves (figure 3.2). The primary, or P wave, travels through the earth, pulsing out from the initial shock, alternately pushing and pulling the rock along its path of travel. These horizontal waves resemble sound waves and, in fact, can be heard when they reach the surface.

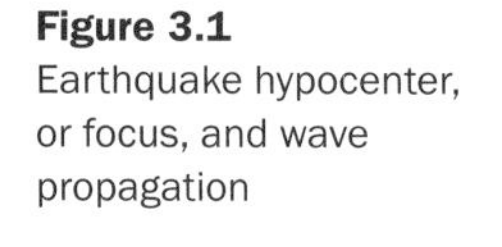

Figure 3.1
Earthquake hypocenter, or focus, and wave propagation

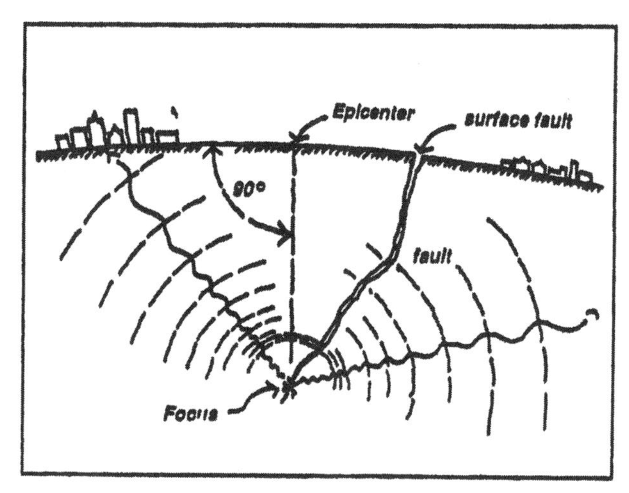

Animals react strangely to these sounds, as B. E. Lloyd observed in 1865. The S, or secondary wave, slower than the P wave, shears the earth while vibrating perpendicular to the path of travel, but it cannot travel through water. The P waves are felt first, sometimes with an effect like a sonic boom (note the noise mentioned in Lloyd's account); then the side-to-side and up-and-down S-wave motion is felt. As the P and S waves travel through the earth, encountering materials of different consistencies, like rocks, hard soils, mud, and water, they behave differently. Sometimes they are reflected, bouncing backwards or to the side; sometimes they are refracted, changing direction as they transfer from one substance or material to the next. As they move through loose soil, the waves slow down and are amplified; as they move through rock, the amplification diminishes and the velocity increases.

When the P and S waves reach the surface of the earth, they give rise to surface waves, which travel along the earth's surface, causing its distinctive ground motion. Surface waves are divided into two subtypes, the Love wave and the Rayleigh wave. The Love wave, like the S wave, causes the side-to-side shaking so destructive to buildings. The Rayleigh waves rise up and roll forward like ocean rollers, causing both vertical and horizontal movement along the land. The "waving structures" and ground motion similar to ocean swells in many earthquake accounts, including those of the reporter in San Jose and Mark Twain quoted earlier in this chapter, were probably due to Rayleigh waves. Within seconds after the waves strike a particular location, the result is a random wave motion from reflection and refraction that occurs in horizontal and sometimes vertical directions.

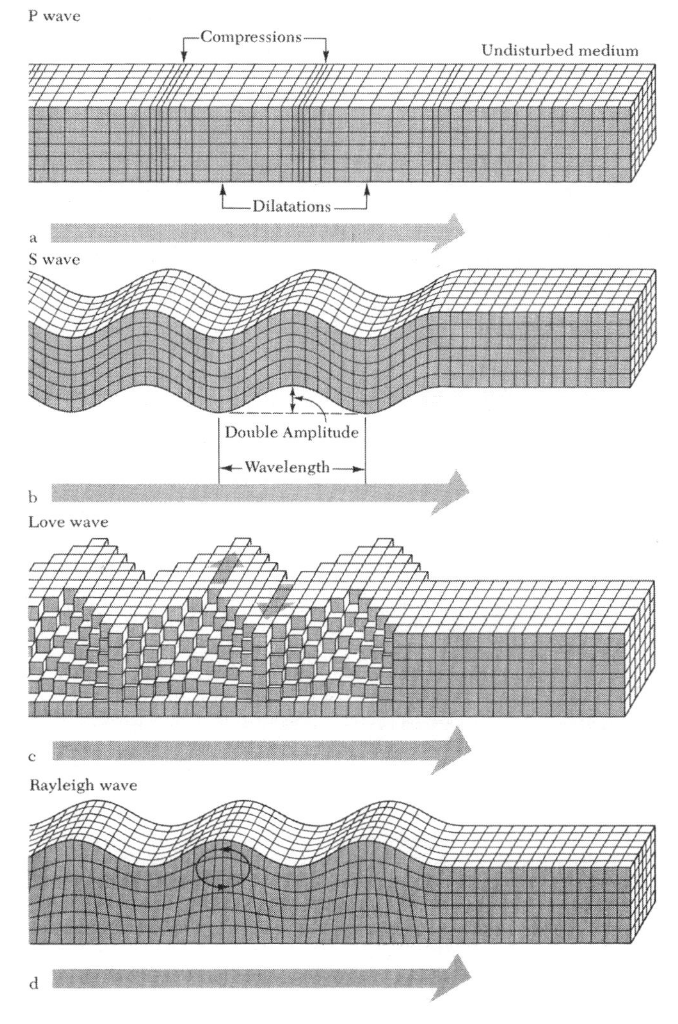

Figure 3.2
Earthquake waves. P waves, or compression waves, and S waves travel within the earth and are called body waves. The P waves are the fastest, followed by the S waves. Slower still are the surface waves, the Love waves, which shear side to side, and the Rayleigh waves, which are like ocean rollers. Wavelength is measured from crest to crest, and amplitude from crest or trough to center point. The measurement from trough to crest is called double amplitude.

Measuring Earthquakes

The first modern scale for measuring the relative strengths of earthquake shaking intensity and destruction was developed in the 1880s by an Italian, Michele Stefano de Rossi, and a Swiss, François-Alphonse Forel. This Rossi-Forel scale evaluates intensity—degree of damage and severity of shaking—on a scale of I to X in a particular location. Refined in 1902 by Italian seismologist Giuseppe Mercalli to show twelve gradations or values, the Mercalli scale was further modified in 1931 by Frank Neumann and Harry O. Wood. They were studying California earthquakes, including the shock of 1865, and they wanted a scale that could be used to rate buildings of wood and stucco, not just the Italian masonry upon which Mercalli based his efforts. Their twelve-value Modified Mercalli Intensity Scale (hereafter MMI) begins with I (not felt) and goes to XII (damage nearly total).

No scale of either magnitude or intensity was available in 1865. In order to gauge the strength of the 1865 earthquake today, seismologists try to estimate its intensity over the felt area by using accounts such as Lloyd's. Geologists have reconstructed both the intensity of the 1865 earthquake and its magnitude. It had an estimated intensity of MMI VIII on filled ground in San Francisco, which indicates considerable damage to ordinary substantial buildings, with partial collapse; great damage to poorly built structures; and collapse of chimneys and factory stacks typically built of brick. The intensity of an earthquake differs from one location to the next, even in a single seismic event. While the 1865 quake's intensity shaking may have been MMI VIII on filled land in downtown San Francisco, it was less in the hills.

Our modern measurement of *magnitude*, or the amount of energy released by an earthquake, relies on records produced by seismographs, machines that measure and record ground motion. The forerunner of the modern seismograph was invented in 1892 by an Englishman, John Milne, a professor of geology at the Imperial College of Engineering in Tokyo. Like Milne's machine, the modern seismograph measures movement of a freely supported pendulum within a frame attached to the ground (figure 3.3). The ground movement activates the pendulum, which is attached to a stylus or an electronic mechanism that draws sharply pointed waves on paper, magnetic tape, or computer readouts. Motion, vertical and lateral, caused by the earthquake is thus translated into a record called a seismogram (figure 3.4).

Professor Charles Richter, working at the California Institute of Technology in Pasadena in the 1930s, wanted to establish a uniform scale to measure earthquake size and magnitude anywhere in California, even if the seismographs were far removed from the shock. He devised his first scale, the so-called local magnitude scale, by deriving magnitude from amplitude (figure 3.4). He based his calculations on a hypothetical seismograph 62 miles, or 100 kilometers, from the epicenter of the earthquake and devised mathematical tables for adjusting the results in all other cases. Richter calibrated all earth movements on a logarithmic scale, with each

Figure 3.3
Principles of the seismograph. This simple model illustrates the workings of a pendulum seismograph. The pendulum, here represented as a sphere, must be damped to separate seismic impulses. The pendulums represented here record vertical or horizontal directions of ground motion on a seismogram.

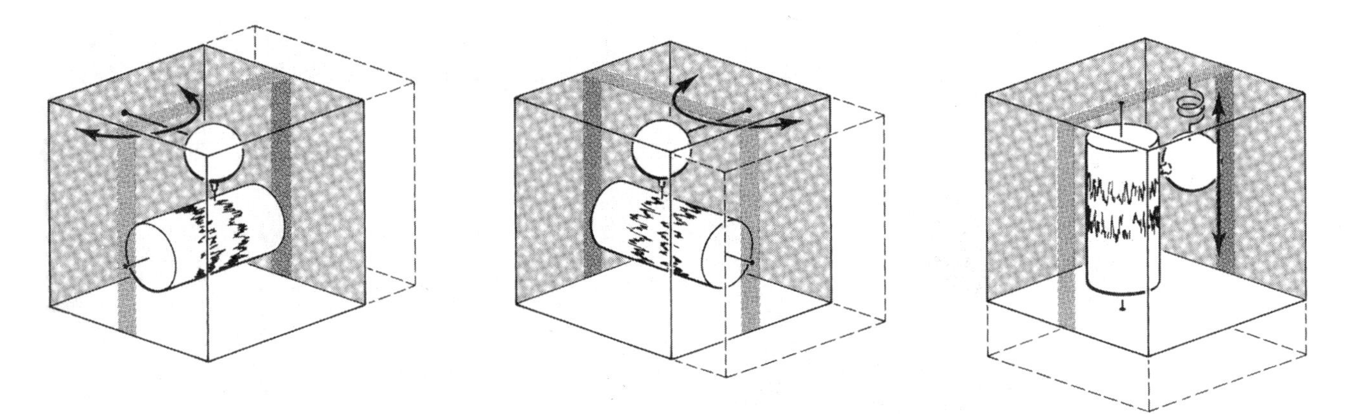

whole number representing an amplitude of earthquake waves 10 times greater and energy release approximately 31.5 times greater than the preceding number. The size of the waves on a seismogram, from crest or trough to their midline, is called their *amplitude*. By measuring the amplitude of the waves written on the seismogram, Richter derived magnitude. The estimated magnitude of the 1865 earthquake is 6.5.

While Richter's magnitude (M_L) is popular with the general public, seismologists have developed several newer scales to help them more accurately measure earthquakes. For example, one problem with the scale is that it does not specify which wave (P, S, Love, or Rayleigh) is to be measured for its maximum amplitude. On the new scale applied to deep-focus earthquakes, amplitude is denoted by M_b; distant and larger earthquakes by M_S; and the so-called moment magnitude, M_W, is being used to measure the world's largest earthquakes. These new scales often alter the magnitudes established for historic earthquakes. In this book, I will be using primarily the Richter magnitude.

In addition to magnitude and intensity, there are two other parameters of earthquake strength, namely maximum *acceleration* and *duration*. Accelerometers measure an earthquake's acceleration, from which velocity and displacement are computed. Acceleration is commonly used to indicate the destructive power of an earthquake in relation to a building. Acceleration is measured in "g," the acceleration of a free-falling body due to earth's gravity (approximately 32 feet/second/second). The level of acceleration generally taken as sufficient to damage buildings is 0.10g, or 10 percent of gravity, roughly equivalent to a MMI VII earthquake. The 1865 earthquake, at MMI VIII, is thought to have had accelerations between 0.25g and 0.30g.

If high accelerations and intense shaking continue for a long period of time, the potential for damage increases. The strong shaking in the 1989 Loma Prieta earthquake lasted for just ten to fifteen seconds, whereas the duration of shaking in the San Francisco earthquake of 1906 was about forty seconds or more.

The duration of earthquakes in the past is hard to establish for sure. According to Lloyd, the second and strongest shock of the 1865 earthquake had a duration of twenty seconds. Observers in San Francisco and in

Figure 3.4
The derivation of Richter magnitude (M_L) from a seismographic record

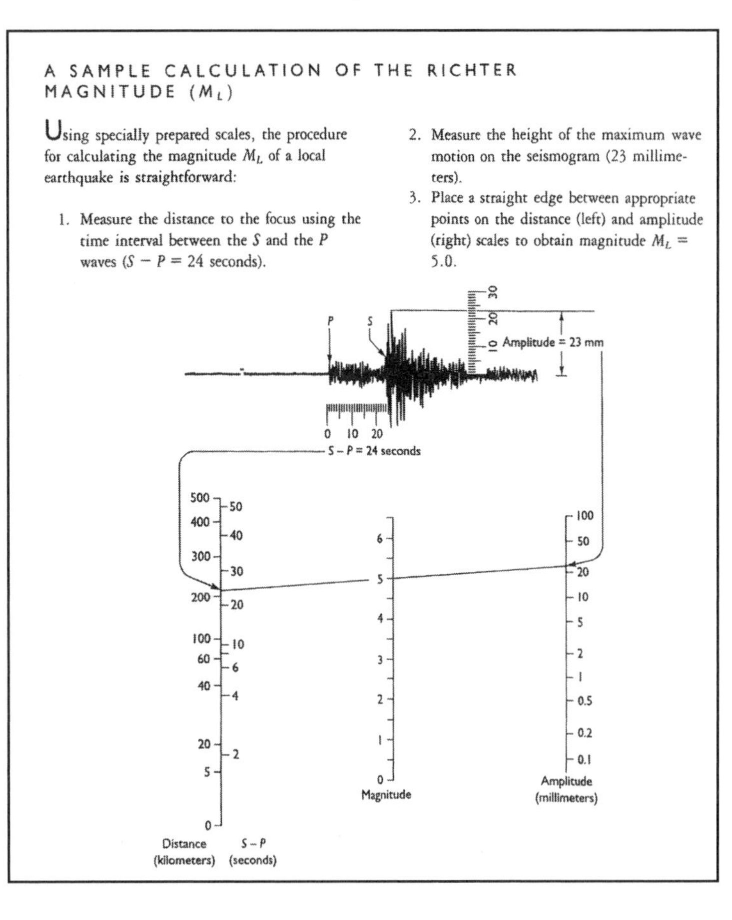

other cities reported both longer and shorter durations, which illustrates how difficult it is to use anecdotal evidence.

The Hazards of Filled Ground in 1865

What, if anything, did Lloyd and other San Franciscans learn about future earthquake danger in San Francisco? The day after the earthquake of 1865, during what might be called the recovery period, the *Alta California* editorialized that brick buildings in San Francisco should no longer be considered at risk in earthquakes.

> The earthquake has come and gone, and San Francisco still stands...tall brick buildings need no longer be considered hazardous during the prevalence of earthquakes, for the five- and six-story hotels, covering the frontage of an entire block, large banking houses, factories, theaters, churches, public halls, and elegant private dwellings have received the shock without injury while buildings of poor construction, even if no higher than two stories, have alone been affected.[13]

The newspaper also noted differing structural damage around the city, citing the poor performance of buildings constructed on filled land:

> No house built on hard ground has suffered, or the damage, if any, is too slight to deserve notice...In those parts of the city which were formerly part of the Bay, and have been filled with earth, few of the foundations are firm, and there the most damage has been done.[14]

B. E. Lloyd explains, with telling detail, that it is not the poor quality of the building stock which caused the damage, but the poor quality of the land itself:

> In the low grounds of the city, and at those parts where the ground had been reclaimed from the water of the bay by filling in, the devastation was great. Water-pipes were broken, the ground elevated or depressed in places and cracked open for a considerable distance. In one place the sewer was warped and twisted and lifted entirely above the surface. Lampposts were bent from a perpendicular to an inclined position, and everything had more or less suffered some displacement.[15]

The problems of O'Farrell's first survey, which platted the city over tidelands and marshes, were surfacing—literally. The earthquake of 1865 caused severe shaking on unconsolidated fill, which amplifies earthquake waves far more than rock. When fill, composed of sand, soil, and debris, is dumped into the waterlogged coastal marsh, it too becomes permeated with water. During intense shaking, the friction breaks down between particles, releasing the water trapped between them. The ground loses its solidity, and it drops. This condition, called liquefaction, damaged numerous buildings. It probably accounts for the cave-in that occurred at Howard between Fifth and Eighth Streets, the openings of ground through which water surfaced at Sixth and Howard, Mission and Beale, and the fissure west of Fillmore Street. If the reports can be believed, the corner of Seventh and Howard, directly over the old South of Market marsh, dropped fourteen feet. This could have been caused by compaction or more likely lateral flow, which happens near the banks of creeks that have been filled. At Tehama, Howard, and Mission Streets, the formerly level ground was shaped into waves. Buildings constructed on soil that liquefies, subsides, or shakes violently are obviously at risk. Post-earthquake accounts and assessments made no mention of the fact that gradual settlement had been causing similar, if not identical, damage on this "made land" for years. The American Theater, which had sunk on its inaugural night, suffered similar damage in the 1865 earthquake.[16]

Figure 3.5
Popper's Building, San Francisco, corner of Third and Mission streets, 1865. This was the most spectacular failure in the 1865 earthquake and the only photographic image of building damage in San Francisco. The reasons for its failure are multiple. It was constructed on filled land in the South of Market. It was in construction at the time of the earthquake, as the broken scaffolding illustrates—its brickwork may not have been cured. Its internal wooden skeleton, where the initial failure seems to have occurred, looks exceptionally light.

Severe shaking along Third between Market and Mission indicated loose soil and drainage problems. The most spectacular building failure in the 1865 earthquake was the four-story Popper's Building (figure 3.5) at the corner of Third and Mission Streets, which shed half its brick facade and a portion of its rear wall. According to one eyewitness, the building fell straight down, as if an internal portion had given way, perhaps suggesting that settlement was to blame rather than quake-related swaying.[17] But Mark Twain's account seems to describe a wall dropping outward from the building. In either case, Popper's Building was vulnerable because it was still under construction.

The list of major damage confirms that buildings on filled land suffered most.[18] A wall of a grocery store at Fourth and Harrison

*This map represents an approximation of building damage as compiled from contemporary eyewitness accounts
In addition the map does not illustrate exact building footprints, rather they are representations of a building's size and location.

Figure 3.6
The first comprehensive map of damage in the 1865 earthquake, based on newspaper descriptions of the time. I have used the descriptive terms developed by Lawson et al. for damage in their map of the 1868 earthquake: A—Buildings wrecked; B—Buildings shattered with walls down; C—Buildings with walls cracked.

fell, probably because it was built on mud near a stream (figure 3.6). Sachs's Building on Sacramento Street (where my great-grandfather Mathias Tobriner worked), on the boundary of the old shoreline, was dangerous enough to be cordoned off. The facade of the Morrill Block, at the southeastern corner of Washington and Battery, which had rested on bay mud, completely detached itself from its side walls. City Hall, on Portsmouth Square very close to the old shoreline, was badly shaken, as was the Custom House two blocks to the east, built on bay mud. Typical damage included cracked and detached brick walls, downed cornices, parapets, and fire walls, and broken or destroyed chimneys.

The First Seismic Retrofits in San Francisco

Few architects or engineers had experienced a series of earthquakes before 1865. After the earthquake, they quickly realized they had to evaluate and strengthen the city's damaged buildings or face the specter of collapse in future quakes. Several important retrofits illustrate how professionals now began to confront damage, repair, and prevention in San Francisco.

Gridley Bryant, an architect who presided over the largest architectural office in late-nineteenth-century Boston, had designed the Custom House (figure 3.7), which was built between 1855 and 1856. Bryant is credited with designing nineteen state capitols and city halls, ninety-five courthouses, asylums, and schools, and sixteen custom houses nationwide.[19] His dependence on standardized design and his failure to understand the specific problems of San Francisco explain why the handsome Greek Revival Custom House suffered repeatedly in earthquakes, finally facing demolition in the 1870s.

Because the U.S. Custom House was an important federal building, Charles James, the tax collector for the port of San Francisco, commissioned William Craine, an English architect who had come to San Francisco in 1849, and Thomas England, the contractor who had built St. Mary's Cathedral—one of San Francisco's few large brick buildings, and one which still survives today—to write a report describing the condition of the Custom House after the 1865 earthquake.[20] In the final report George Cofran, another builder, replaced Thomas England. Craine and Cofran described the Custom House as:

> A massive and substantial brick building, with a stone basement and solid foundation resting upon piles, of various length from twenty-five to sixty feet. It has always been regarded by competent judges as one of the most superior and substantially built structures on the Pacific coast.

The reason for the wooden piles—twenty-six hundred of them—was that the building rested on bay mud. They had been pounded into the mud and were held in place by friction. The stone foundation was then laid on top of the pile caps, providing a base for the brick walls. Because stone was scarce, the upper stories of the facade were brick coated with plaster. Although the many pilings were obviously a response to the possibility of settlement problems, the building performed poorly because it was too rigid, too heavy, and lacked any iron reinforcement. According to Craine and Cofran, the "considerable damage" for the most part consisted of "cracks, and fractures running through the brickwork of the exterior, as well as through the walls of the safes; and other portions of the interior walls adjoining the stairways and landings thereof." They wrote that "a considerable portions of the plaster ornaments and moldings in the Custom House room and other places have fallen," and the remainder were "liable to fall at any time upon the slightest jar or concussion." They concluded that "the late earthquake has demonstrated the fact that the style of plaster finish of the Custom House room...is entirely unsuited for the future of this City."[21]

Craine and Cofran wanted to retrofit the Custom House by using iron reinforcement and eliminating plaster ornament, replacing it with copies in illusionistic frescoes. This clever solution to falling hazards is important to note. To reinforce

Figure 3.7
Photograph of the Custom House and Post Office, San Francisco, Battery and Washington Streets, before the earthquake of 1865

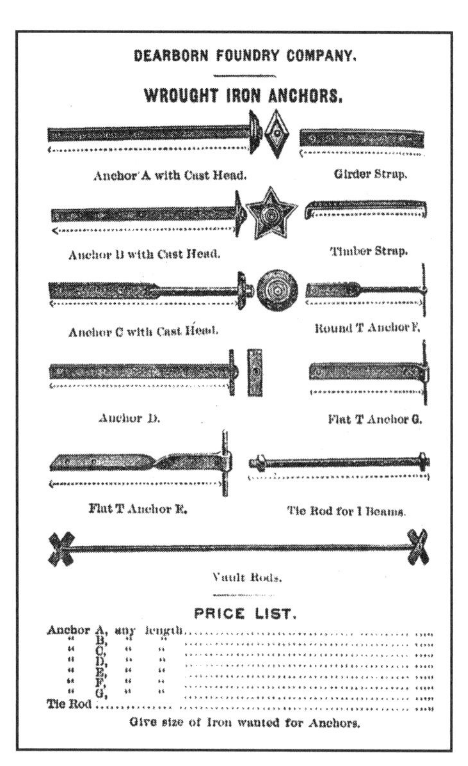

Figure 3.8
Advertisement for iron anchors and rods from Dearborn Foundry Company brochure, Chicago, 1887. The anchors shown here are for securing brick walls to interior wooden joists, or vice versa. These are the anchors specified by Craine and Cofran. The rods they recommend are the "Vault Rods" in the advertisement.

the whole building, they turned to the method already used to tie buildings together in San Francisco, iron rods and anchors (figure 3.8).[22] A bill for the iron reinforcement dated November 29, 1865, mentions "drilling walls" and "iron tie-rods for tying walls of Custom House Building through each of second, third, and attic stories—with plates, washers, nuts, screws."[23]

Using iron to tie masonry buildings together was standard practice in Medieval and Renaissance Europe.[24] Tie-rods (briefly mentioned in chapter 1) are ubiquitous in Early Renaissance architecture, from the arcades of Brunelleschi's Ospedale degli Innocenti to Michelozzo's San Marco Library in Florence. Two centuries later, Claude Perrault used iron tie-rods to counteract the outward thrusts of interior vaults in the east facade of the Louvre. Iron rods were routinely used to strap domes in order to eliminate costly, heavy, and unsightly exterior buttresses. Three rods cinch the dome of St. Peter's in Rome. In New York and Chicago loft buildings, rods were used to control uneven settling, and they became standard practice in all construction in the nineteenth century. But they came to be seen as especially effective in reinforcing masonry buildings shaken by earthquakes. They were extensively used in Sicily after earthquakes in 1726, in Umbria in 1751 and 1832, and later recommended by the United States government after the Charleston earthquake of 1886, years after the earthquake of 1865 in San Francisco.

Craine and Cofran not only understood in 1865 how iron ties worked to stabilize a building but appear to have maximized their use for earthquake-resistant purposes, demonstrating a sophisticated understanding of how buildings respond to earthquakes. Perhaps following their example, other builders in San Francisco retrofitted with iron rods. The old Merchant's Exchange Building, on the northeast corner of Battery and Washington, was "completely ruined and in a dangerous condition and unfit for occupancy."[25] Damage included a four-inch gap separating the rear wall from the rest of the building. An 1868 newspaper article recalls that after the earthquake of 1865, "A considerable sum of money was spent in bolting and anchoring the building."[26] In fact, three years later three experts charged with examining the Merchant's Exchange Building wrote, "We find [it] well ironed throughout, the walls being tied together with 1 and ½ inch rods running through the building in each story."[27] The Portsmouth Square City Hall was similarly reinforced with iron. According to the *San Francisco Bulletin*, it had "received injuries which may necessitate the rebuilding of the front portion...On the upper floor the wall has been sprung from the building several inches...throughout the entire front portion of the building the walls have cracked and opened at the joints."[28] One day later, the board of supervisors authorized the building committee to make appropriate repairs to safely secure the walls. Mr. Torrey, a member of the committee, recommended that the structure be "strengthened throughout and anchored in front."[29] A year later, eight 1½-inch iron rods had been installed in the rear of the building: "four [iron rods] on the Police Court floor running from the outside of the east wall to the quadrangle, and four [iron rods] under the floors of the Fourth District Court and Health office. They are anchored on the outside with broad plates and will give much greater stability to the building."[30]

In addition to the use of iron ties, other simple earthquake-resistant interventions helped damaged buildings. Popper's Building was extensively repaired with better-quality masonry and would resist the 1868 earthquake. The chimney of the Spreckels sugar refinery outside the city was shortened and repaired after it swayed precariously during the earthquake. Architects seemed to take to heart the concepts of improved construction quality and commonsense expediency counseled by articles like this one that appeared in the *Alta California:*

> What does it matter that a few battlements and old walls, built on shaky foundations with poor mortar, have cracked? More careful masonry, better material, less pretentious cornices, and more substantial architecture will become the fashion and will do no harm.[31]

In 1866 the city enacted new fire codes that included guidelines for building construction. Though similar to building codes on the East Coast, the new code had particular relevance for the West Coast, stipulating for example that "all walls shall be securely anchored with iron anchors to each tier of beams."[32] The code further required that anchors of iron secure front and rear walls; that stone veneers be anchored with iron bars; that party walls be anchored to beams with wrought iron anchors; and that beams and girders be strapped together with wrought iron. Foundations in poor soil had to be of brick or stone, laid on driven piles to resist settlement. These modest requirements are embryonic antecedents of the first explicit state earthquake code in the United States, which was not enacted until 1933.

The Earthquake of 1868

If the earthquake of 1865 was a wake-up call, the earthquake of 1868 was a double espresso. On October 21, 1868, at 7:45 a.m., the Bay Area suffered an earthquake estimated at MMI X and Richter magnitude 7.2; the earthquake of 1868 ranks today as one of the six "great earthquakes" of San Francisco Bay Area.[33] It was on the Hayward Fault running north-south through the East Bay. It destroyed more property than the quake of 1865 and killed at least six people in San Francisco and thirty statewide.[34] Falling masonry and a runaway team caused the deaths in San Francisco. Although local newspapers proudly claimed that no one had died inside a falling structure, it was only luck that saved the people who were in the streets from falling cornices and exterior walls. Reporters compared it to the 1865 temblor:

> At precisely six minutes to eight o'clock yesterday morning, the most severe earthquake which has occurred since the occupation of California by the Americans shook our city. The general excitement prevailing through the city renders it difficult to give anything like an accurate account of the amount of damage done or the number of casualties. This is the first earthquake that had ever caused loss of life in San Francisco, and the amount of damage caused is unquestionably greater than that caused by the great shock of October 8, 1865...
>
> The motion was purely horizontal, not perpendicular, as in the great earthquake of 1865...In October 1865 glass was broken and shivered to atoms in all the lower part of the city by the perpendicular oscillation, while comparatively few walls were shaken down or badly injured. The earthquake yesterday broke very little glass in any part of the city, but the damage by the falling of cornices, awnings, and walls was immense.[35]

The account of Mr. F. C. Dodge, who experienced the earthquake of 1868 in the upper floor of the newly reinforced Custom House warehouse, gives an impression of its force:

> His first impression was that a loaded team was passing under the window and he looked out. In a second or two the corner of the building in which he stood appeared to lift up, and plaster began to fall around him. He moved farther into the room, and stood between two desks with the counter steadying himself. In this position he could see obliquely across the long room, and observe the phenomena. The whole building appeared to be hurled south, but not all at once, by what seemed an undulatory motion. As the wave passed along plaster fell, and shortly the dust darkened the room.[36]

Catastrophic settlement, liquefaction, and strong surface motion had caused the most extensive seismic damage San Franciscans had ever seen. Citizens took this quake more seriously than that of 1865, giving more attention to building failures and seismic weaknesses. Again, the worst damage had occurred in the part of the city constructed on fill:

> The shock was principally felt on "made ground" and the flats where the foundation is known to be unreliable at all times...Along the old water line of the Bay, running just back of Macondray's old place on the corner of Pine and Sansome Streets, and thence diagonally north-eastward toward the corner of Front and Jackson Street, something like a slide occurred and buildings suffered severely from the slipping of the made-ground foundation on the old mud bottom of the Bay...
>
> Filling, owing to the deep substratum of mud, was essentially unsubstantial and unsafe, and even the piles driven through it have proved, as in

the case of the Custom House, not to be a sufficient foundation for larger brick or stone houses in a place liable to earthquakes.[37]

The few professional photographers in the city were drawn to the most severely damaged structures. These few photographs of even fewer buildings are all that have come down to us, along with extensive newspaper reports. The first map of damage in the 1868 earthquake was compiled by the State Earthquake Investigation Commission in 1908 to illustrate the continuing problem of "made land."[38] The map divides damage into three categories: buildings wrecked, buildings badly shattered with walls down, and buildings with cracked walls (figure 3.9).

The most dramatic failure to be photographed was the collapse of the Goffery and Risdon Building (figure 3.10) at Bush and Market Streets, where half the facade and interior collapsed. The collapse probably occurred because walls were shaken down, and these released floor and ceiling joists. The building was under construction at the time of the quake, so the honeycomb of interior wood walls, which would have provided additional bracing, was not yet installed. The roof, which would have tied the building together, might not have been completed.

Two buildings, at Front and Clay Streets (figure 3.11) and California Street (figure 3.12), illustrate brick failures. At Front and Clay the brick parapet has fallen away into the street. At 319 California, the Pacific Pump Building has moved laterally, perhaps because of an open, poorly braced ground floor, shattering the facade cornice and collapsing the roof.

The most iconic photographs were of ground

Figure 3.9
The first comprehensive map of the earthquake damage in 1868. An earlier, more cursory map was prepared for the 1908 Carnegie report on the 1906 earthquake. Lawson et al. wanted to illustrate the correspondence between filled land and earthquake damage in 1868 and 1906. I have used the descriptive terms they developed: A.—Buildings wrecked; B—Buildings shattered with walls down; C—Buildings with walls cracked

Figure 3.10 (top left)
Goffery and Risdon Building, at Bush and Market Streets, as it appeared shortly after the 1868 earthquake. The building was in construction at the time of the earthquake. The collapse was probably caused by the fire walls collapsing and dropping the floor joists and roof trusses.

Figure 3.11 (bottom left)
Front Street corner of Clay Street, San Francisco, shortly after the earthquake of 1868: the failure of the parapet of Crane and Brigham druggists, 322, 324, 326 Front Street. The brick parapet fell forward, destroying the awning of the building.

Figure 3.12 (top right)
319 California Street, shortly after the 1868 earthquake: the collapse of the Pacific Pump Manufacturing Building. The reason for this building failure is still unclear, but it was probably due to shaking rather than settlement. Around the collapsed building stand taller brick buildings tha have survived. The arrows illustrate the positions of wall anchors. Far mo are visible in the photograph. These anchors may have helped nearby br buildings survive the earthquake.

Figure 3.13 (bottom right)
Commercial Street or Clay Street between Front and Battery Streets, shortly after the earthquake of 1868: the failure of the Railroad House, 316–320 Commercial Street, due to settlement. Extreme settlement, no shaking, is the cause of the distorted facades of these two buildings. Th was the most popular subject to photograph after the earthquake.

Figure 3.14
Battery Street corner of Sacramento, days after the earthquake of 1868: Esberg and Co., 300, 302 Battery, has lost its cornice and parapet and is shored up with bracing and awaiting repair. Although the buildings on Sacramento Street appear to have their cornices intact, the fact that some of their facades are braced indicates that they are probably out of plumb and threatening collapse.

failures like the one at Commercial Street between Front and Battery (figure 3.13). The shot of Railroad House, leaning precariously, was a favorite. The photograph illustrates what happens to a well-built structure when liquefaction occurs directly under it. The brick facade held together well, considering that the foundation dropped several feet. But cornices and fire walls were major problems (figure 3.14).

Approximately seven buildings were wrecked in the quake and about thirty-five were seriously damaged. Three large commercial chimneys fell, as did many residential chimneys. Although no pictures were taken of them, buildings that had been damaged in previous earthquakes, like the Custom House, the Portsmouth Square City Hall, and the old Merchant's Exchange Building, were badly damaged again, even though they had been reinforced. Even the new Bank of California, built to code and tied together with iron, lost a parapet.

Postquake Reconstruction and Investigation

As the city began to repair itself, during what Powell calls the assessment and reconstruction periods, controversies erupted over the safety of several buildings. The board of supervisors argued over the fate of the Portsmouth Square City Hall, which had been damaged in both 1865 and 1868. They eventually decided to abandon the building in favor of a new, "earthquake-proof" city hall (see chapter 4). The U.S. Custom House had survived both quakes. Some experts found it on the verge of collapse, while others thought it could be repaired.[39] While the strapped rectangular base of the building survived more or less intact, the portico, oscillating at a different frequency, detached itself from the rest of the building:

> Columns and pilasters were cracked at the base or capital...The upper tier of pillars, brick covered with stucco, over three feet square, are all cracked through in horizontal lines between six inches and two feet from the base.

It would appear that the motion of the southern portico and the staircase had not been the same as that of the main building.[40]

Figure 3.15 (left)
Custom House and Post Office, San Francisco. Although this photograph is supposed to have been taken after the 1868 earthquake, it was probably taken after 1865 and before 1868. The anchors and rods of the 1865 seismic retrofit are in place and can be seen in the frieze of the entablature, the second floor level, and the main floor level below the front facade windows.

Because of the building's repeated poor performance, two federal engineers, Colonel Mendell and Major George Elliot, toured it and questioned whether any earthquake-resistant building could be built on that site.[41] After a heated debate, experts convinced the federal government that the building could be retrofitted once again. A photograph of the Custom House taken before the earthquake shows the pediment with its columns intact and iron anchors inserted after 1865 at each floor (figure 3.15). This time aesthetics were set aside; the building was unfortunately disfigured. In a second photograph, after the 1868 earthquake retrofit, the pediment and the columns of the portico have been removed, the chimneys rebuilt, and the whole building extensively braced with rods and anchors (figure 3.16).

Figure 3.16 (right)
Custom House and Post Office, San Francisco, sometime after the earthquake of 1868. Earthquakes have destroyed the dignity of the once proud building. The post-1865 iron rods and anchors have rusted and the projecting colonnaded pediment has been removed because it was damaged in the 1868 earthquake.

The Joint Committee on Earthquakes

Citizens realized that the city had suffered a major blow, and they demanded a full-scale scientific investigation as well as practical advice on earthquake danger. George Gordon, the English entrepreneur who had established a successful sugar refining business, owned a large foundry, and had pioneered the high-end South Park real estate development, wrote a letter asking the chamber of commerce to investigate the earthquake. San Francisco newspapers did a good job cataloguing the damage, and many articles appeared on earthquake-related problems. On November 25, 1868, the chamber of commerce established the Joint Committee on Earthquakes, divided into five formal subcommittees:[42]

On bricks, stones, and timber: Lieutenant Colonel Alexander, U.S. Corps of Engineers; Colonel Mendell, U.S. Corps of Engineers; Colonel Morse, U.S. Architect; A. W. Schmidt, civil engineer

On limes, cements, and other bonds and braces: Colonel Mendell; Major Elliott; Calvin Brown, civil engineer; John P. Gaynor, architect; George Gordon; and later, David Farquharson, architect

On structural designs: Lt. Col. Alexander; Major Elliott; William Patton, architect; Colonel Morse; Calvin Brown; George Gordon; Joseph Moore

On scientific inquiry and collection of facts: James Blake, M.D; Professor John Boardman Trask; William H. Rhodes; Robert B. Swain; "Professor" Thomas Rowlandson; John P. Gaynor; W. Frank Stewart. In December 1869, J. D. Stillmand and George Davidson were added to the committeee.

On the laws governing building: John T. Doyle and Alfred Rix

Two additional committees were formed, one on testing apparatus, chaired by Joseph Moore, and one on finances, made up of lay citizens who vowed to raise $20,000 for investigation into every aspect of construction and earthquake danger. George Gordon was elected president.[43] Louis Sachs, a cloth merchant, was elected treasurer of the joint committee. It might be argued by some that because he owned the Vulcan Iron Works, Gordon's interest in earthquakes was self-serving: he had everything to gain from a committee report that might suggest using iron to reinforce buildings. Yet from his brilliant observations it seems clear that Gordon was personally fascinated by the problem of earthquakes and tried passionately to find answers to the problems of earthquake-resistant construction. He died unexpectedly in 1869, only months after taking the chairmanship of the joint committee, but not before he had provided invaluable insights.

The joint committee declared that it was dedicating itself to the safety of San Franciscans and not to the pursuit of abstract knowledge. The plain language and clear intention ring true even today:

> Resolved, that we, the joint Committee of Professional and Mercantile Citizens...declared the object we view as the ultimate result of our labors to be the protection of our fellow citizens in their dwellings, stores and workshops from the new element of danger which presents itself in the earthquake, and to advise them how they can, in an effective and economical way, secure themselves from harm upon its recurrence, as they have already learned to provide against rain, wind, cold, heat, conflagration, or other elemental disturbances.[44]

By December 1868, all of the subcommittees were hard at work.[45] A history of earthquakes in California was being written and the subcommittee on buildings was calling for experiments on fireproofing wood, conducting their own experiments on a primitive shake table, and asking for an earthquake-proof building competition. Inexplicably, these promising beginnings produced no formal report. Some members published their own partial conclusions. One member, Thomas Rowlandson, an English mining and agricultural engineer who had immigrated to California in 1855, claimed that the committee botched its charge because of its disorganization, unprofessionalism, unprepared members, and lack of funds. He then published his own book, *Treatise on Earthquake Dangers, Causes and Palliatives*, in May 1869.[46]

Although the committee's formal report did not appear, fragments of what would have been published can be reconstructed. Newspapers described the work of the subcommittees, and individual members did publish portions of what would have been in the final reports.

The subcommittee on bricks, stones, and timber left the fewest records, but it most likely endorsed wood as the most earthquake-resistant material available, and encouraged efforts to make wood fireproof through electrical and chemical processes. Since several of the members of this subcommittee were military engineers, they probably knew about recent military tests of a new fireproof wood on the East Coast. This subcommittee would probably have emphasized that brick buildings could survive earthquakes if built on piles capped by concrete, surmounted by properly bonded walls, and constructed with well-wetted bricks cemented with richly mixed mortar.

The emphasis of the second subcommittee, which overlapped the first, was on limes, cements, and other bonds and braces. Thomas Rowlandson and architects David Farquharson and John Gaynor all pointed to the poor mortar used to bond brick buildings in San Francisco as a major cause of damage. Rowlandson noted that sand on the San Francisco peninsula tends not to be good for mortar because it is loamy and insufficiently sharp. Farquharson added that the mix of mortar was often too lean, lacking enough lime or cement, and applied unevenly. For the first time in San Francisco history, the use of iron ties to bond buildings together in earthquakes was discussed heatedly and at length, though iron ties had already been widely employed after the earthquake of 1865. There was disagreement about the best use of iron, but architect John Gaynor's Murphy and Grant Co. Building (figure 3.17) was suggested as a model.

The Murphy and Grant Co. Building was the most recent in a series of successful Gaynor buildings.[47] An Irishman who emigrated to the United States in 1849, committee member Gaynor opened an office in Brooklyn in 1851 and in 1856 built the landmark Haughwout building in Manhattan, which with its iron facade, interior iron columns, and steam elevator was a precursor of the modern skyscraper. Gaynor told those assembled that his Murphy and Grant building had two hundred and fifty tons of iron bars in it. We know from Gaynor's later Grand Hotel and Palace Hotel that he used iron bars and flat iron pieces, called bond

iron, running horizontally through the brick to bind the entire building together. The courses of iron ran as close as every four vertical feet. Committee member David Farquharson, a Scottish architect who had immigrated to San Francisco in 1850, used a similar method, running bars horizontally above and below apertures in a wall to tie a whole structure together. This kind of iron bonding, first suggested in 1865, had been partially adopted in the first San Francisco building codes in 1866. Joint Committee President George Gordon suggested that tie rods should be used to tie a building together, as Farquharson had in his 1864 Bank of California (figure 3.18).[48] Tie rods ran from front to back and side to side in this structure, making it exceptionally strong. While the committee met, Gordon retrofitted his sugar refinery with iron rods and anchors, as a post-1868 photograph

Figure 3.17 (above)
Murphy and Grant Co. Building. This iron-fronted building by John Gaynor, secured with iron anchors and rods, was considered an exemplar of earthquake-resistant building by the 1868 Joint Committee on Earthquakes.

Figure 3.18 (left)
Bank of California. This building by David Farquharson was said to be tied together by rods, making it another of the 1868 Joint Committee's exemplary earthquake-resistant buildings.

clearly shows (figure 1.9). Gordon was also a member of the subcommittee studying structural designs. He owned two wooden houses in Menlo Park, south of San Francisco, seemingly of equal strength. He observed that one house had experienced far more motion than the other. Investigation showed that the latter house was bolted to its foundations. It took the city of San Francisco almost one hundred years to put into law the bolting Gordon suggested in 1868.[49] Gordon also recommended offering a prize to the person who devised the best bracing method for earthquake country. His suggestion sparked inventors to produce a series of patents, although the competition was never held.[50] Gordon joined the fray himself by describing an earthquake-proof wooden building that was radical for his time but absolutely sound by today's standards:

> **First:** That extreme care be taken with foundations, no matter whether on solid or made ground; let the entire bed-frame on which the building rests be a unit, like a ship's keel, and strong enough to bear twice the weight of the building itself up on posts ten or twelve feet apart, and so tied together that you could lift it bodily by a derrick from any point and swing it about.
> **Second:** Dispense with the use of brick, stone or cast iron, except as an exterior protection against fire. Give these materials nothing else to do.
> **Third:** Rely on timber and wrought iron entirely to carry the load and resist motion. Mortise all timbers and rivet all iron. Use boiler plate with angle iron riveted to it, above and below all openings, as sills and caps. No form of iron compatible in this market is so simple, cheap and strong as these combined.
> **Fourth:** Dispense with lath and plaster, and face the inside walls, and make the ceilings with tongued and grooved lumber. Put every board, and lay the floors so as to form diagonals—bracing in every direction. In nailing, put every nail square through the face of the boards and discard the silly carpenter's foible of blind nailing.
> **Fifth:** Lower the ceilings, and the more numerous the rooms in the building, the stronger is the structure.[51]

These ideals, which he published himself after the subcommittee had met, may have been originally set down by the committee, which included an army architect, two architects in private practice, and a civil engineer.

The fourth subcommittee, on scientific inquiry and data collection, was instructed to read the Spanish archives and visit other localities to compare damage.[52] In 1906, George Davidson remembered some of the main points this group had made in their 1868 report. Most important was the discovery that major structural damage occurred not only on "made ground" in general, but particularly in the areas where filled ground met the old shoreline. It is possible, too, that this report discussed the causes of earthquakes—still unknown at that time—mapped the direction of the shock, and speculated on the likelihood of another earthquake.

The most frightening thing about this subcommittee report may have been its complete and honest assessment of the enormous damage suffered throughout the Bay Area.

Little is known about the subcommittee studying laws governing building, but given the comments from other committees, one can imagine that codes could have been drafted, but were not: codes including the correct mixture for mortar and bricklaying, or perhaps the kinds of provisions for wood houses that Gordon had proposed.[53]

We are unsure what kind of equipment the subcommittee concerned with testing apparatus had in mind, but we do know that George Gordon conducted his own tests of brick by erecting brickwork without mortar in different bonding configurations and shaking them to compare their stability.[54] This simple test highlighted significant problems with California bond, which was commonly used in Gordon's time. Brick walls are made of several vertical rows (wythes) of bricks that have been laid longitudinally, end-to-end, next to one another. The wall is bonded or held together by mortar or by the pattern in which the bricks are laid. Bricks laid longitudinally along the row are called stretchers. In order to join two wythes of brick and form a wall, a brick may be laid between them at a right angle to the stretchers. A common pattern is to alternate headers with stretchers. If this is done in every course of bricks, "Flemish bond" results (figure 3.19). But a cheaper way to build walls, called variously California bond, American bond, or common bond, is to run stretchers for five courses without headers, relying on the strength of the mortar rather than the bond of the bricks. Not surprisingly, Gordon found that California bond walls were far less resistant to shaking than Flemish bond. Unfortunately, California bond became as ubiquitous as its name implies and can still be seen in San Francisco buildings to this day.

Figure 3.19
Brick bonds: California bond, American, or common bond, as opposed to Flemish bond. The superiority of Flemish bond is that every row of bricks is bonded to the next with "headers" (in white); these are bricks which bridge two rows of wythes. "Stretchers" (in black) are the bricks laid along the plane of each row. In the weaker California bond, the headers attach the two rows together only every seventh course.

FLEMISH BOND

CALIFORNIA, AMERICAN, COMMON BOND

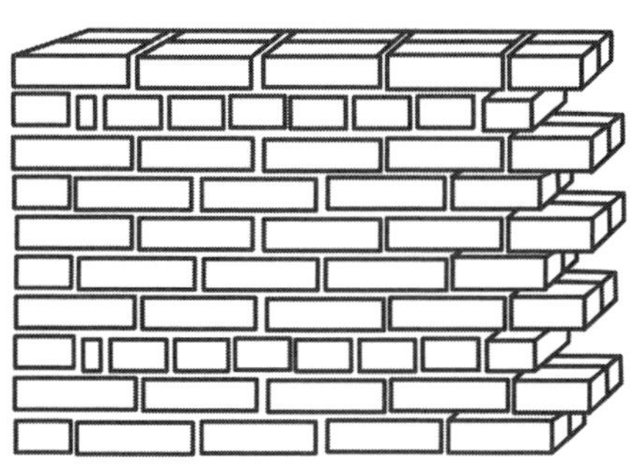

Why the Report was Never Published

Allegations of foul play regarding the joint committee's failure to publish a report have echoed through the history of earthquake studies in California. The renowned seismologist Andrew Lawson, writing a comprehensive report on the 1906 earthquake, was frustrated by the lack of state or city support and recalled hearing from Professor George Davidson that a similar report on the 1868 earthquake had been actively suppressed. Lawson wrote:

> Shortly after the earthquake of 1868 a committee of scientific men undertook the collection of data concerning the effects of the shock, but their report was never published nor can any trace of it be found, although some of the members of the committee are still living. It is stated that the report was suppressed by the authorities, through the fear that its publication would damage the reputation of the city.[55]

Years later, in 1947, the seismologist Harry O. Wood, remembering back to his discussions with Davidson, said, "I well remember Davidson's indignation about the suppression of the report of 1868."[56]

After a lengthy investigation into the allegation that the report was suppressed, Michele Aldrich, Bruce Bolt, Alan Leviton, and Peter Rodda concluded in 1986 that no suppression occurred, and that the report was never completed.[57] The reasons for this are the same as for countless other reports. The first is a lack of agreement between all parties as to the conclusions of the report. Rowlandson belittled the credentials and discussions of his fellow committee members. Gordon had wonderful ideas, but they seem to have remained his alone. One has to wonder whether members of the commission might have doubted their own expertise in relation to their charge. With Rowlandson criticizing them, amateurs such as Gordon might have worried about publishing their observations in a scholarly format. Second, financial support for experiments and the compilation of data never materialized. If the community was not going to support the joint committee, then how could they go forward? Rowlandson's publication of his own report in 1869 may have dampened enthusiasm to continue with the committee's own publication. Finally, the death of Gordon, the driving force behind the formation of the committee, only five months after the earthquake, probably doomed the enterprise. Whereas fire suppression was championed by insurance companies, they provided no financial support for earthquake investigation or mitigation. No scientific or commercial institution supported seismic investigation. Given the complexity of the joint committee's project, it would have been unlikely, even miraculous, for a report sponsored by the city of San Francisco to appear.

Regardless of the reasons why the joint committee's report was never published, the message had penetrated public consciousness. Scholars and lay people alike had a new awareness of seismic danger and of earthquake-resistant retrofit and design. The press and city officials recognized seismic danger but did not emphasize it, perhaps to calm themselves and their public, perhaps to keep easterners interested in investing in California's future. But even though the earthquake of 1868 produced no new seismic legislation, the citizens of San Francisco had seen what worked and what failed. They wanted "earthquake-proof" buildings. It is significant that as a result, many of the most prominent buildings in San Francisco—the new U.S. Mint, the new City Hall, the new Palace Hotel—were built to be earthquake-resistant. Earthquakes had found a place, with fire, at the top of the hierarchy of urban dangers.

CHAPTER 4

Innovation: "Earthquake-Proof" Systems, 1868–1880

San Franciscans who had lived through the earthquakes of 1865 and 1868 knew that there could be another destructive earthquake at any time. If they were over the age of fifteen, they would have experienced catastrophic fire as well. Attention now turned to making purportedly fireproof buildings constructed with bricks, iron shutters, and other materials "earthquake-proof" as well. Decisions about construction materials became more complicated as their performance in earthquakes was added to the equation. Builders, inventors, and architects developed techniques, devices, and entire structures that they claimed would be earthquake-proof, a guarantee no modern engineer would make. As we shall see, by the early twentieth century, engineers saw that "fireproof" structures burned under the right conditions. Today structural engineers use terms like "earthquake-resistant" because they cannot know for sure whether their buildings will survive future earthquakes unharmed. But in the 1860s, the goals of "fireproof" and "earthquake-proof" still seemed attainable. Everything from the simple to the complex, from ties for wooden members to base isolation, was suggested.

In 1868 the regents of the new University of California at Berkeley were so concerned about earthquake danger, considering the risk of earthquake greater than that of fire, that they made an unusual choice for South Hall, their first public building. It would be wood:

> We publish pamphlets to demonstrate that earthquakes in California are not so destructive of human life as lightning and tempests are in the Atlantic States. But still, the historical fact is well established that earthquakes have occurred in California which have caused a fearful destruction of life and property...We are of opinion that we have no right to

disregard these warnings, that one of our first cares should be to make our buildings as safe as possible for the youths who may be confided to our charge...The building which we propose to erect for the purpose of instruction is to be filled three times a day, for eight months a year, with California youth and as we trust, with the flower of that youth, and to be occupied most of the time by the professors and their assistants. Any great calamity which should happen to a large portion of that youth on the site of the University would not only be a great calamity to the State and to the nation, but also create a great prejudice against the University itself.[1]

At this time, all major public buildings were of brick or stone construction, which was considered the most advanced building technology: permanent, prestigious, and fireproof. The regents begged to differ, citing the failure of stone buildings in California and South America, and they were not alone.[2] According to John S. Hittell, a historian of San Francisco writing in the 1860s, "The fear of earthquakes prevents the erection of high buildings for show, and...the same motive induces many wealthy families to reside in wooden houses, which are considered better fitted to resist the shocks of earthquakes." [3]

Wood-Frame Construction

The majority of San Francisco's wooden buildings were built using the Chicago, or balloon-frame system, which appeared in San Francisco in the 1850s and 1860s (figure 4.1). Houses were assembled by placing uniformly produced lumber of

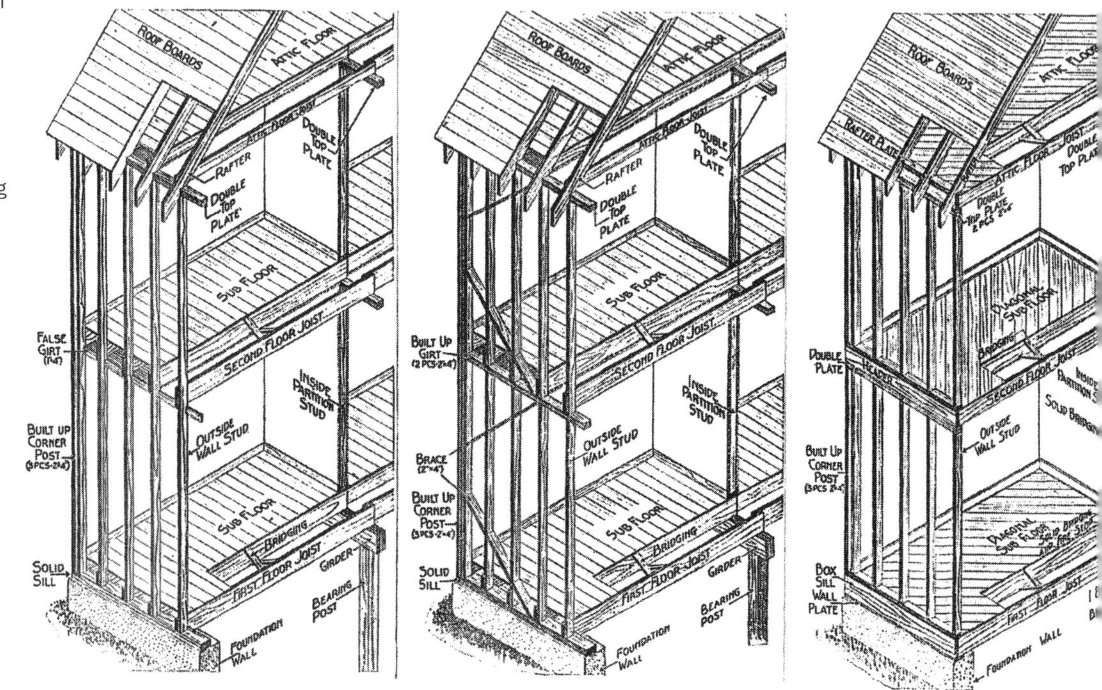

Figure 4.1
Three versions of wood-frame construction prevalent in San Francisco in the nineteenth and early twentieth centuries: A—Balloon-frame construction; B—Braced-frame construction; and C—Platform-frame construction.

Although wood is an ideal earthquake-resistant material, it must be appropriately designed to resist earthquakes and transfer forces. The structure must be designed with distributing elements—diaphragms (floors) and joists—which are tied to resisting elements—walls, columns, and braces—which are in turn connected to a foundation. The earthquake-resistant capacity of the entire system as it existed in the nineteenth century depended on connections between the members provided by nailing, blocking, and metal strapping.

regular sizes—2 x 4 inches, 2 x 6 inches, 2 x 8 inches—at regular intervals and fastening them together with mass-produced nails. Light, cheap construction replaced the heavy, hand-hewn timber framing held together with mortise-and-tenon joints that was prevalent on the East Coast and in Europe. Hittell wrote of the Chicago frame that "though introduced solely because of cheapness and simplicity," it was considered "by far the most secure against earthquakes."[4]

Surviving plans for balloon framing show that the studs or vertical supports nailed to horizontal wood plates ran up the entire length of their exterior walls at sixteen inches on center, tying them together. Studs held up the sills and floor plates. The strength of the system was that it could bend and not break. But it had weaknesses: if it was pushed sideways, it depended on the nails driven through the exterior cladding to hold the studs vertical. If floorboards were not properly placed on the diagonal or nailed well, the diaphragms—horizontal elements like floors and roofs—might distort. If the building was not attached to its foundation, it could easily slip off in an earthquake and collapse.

More effective than the balloon frame was the braced frame, which became common toward the end of the nineteenth century. Because it had diagonal braces at the corners, its strength was somewhat enhanced. A third kind of wood frame, the platform frame, became popular in the late nineteenth century because the supply of timber, and hence of long studs, was diminishing. In this type of construction, each floor is framed separately. Whether a builder used balloon-frame, braced-frame, or platform-frame construction, he had to build well and understand, like George Gordon (see chapter 3), the vulnerabilities of wooden structures in earthquakes. Gordon's ideas illustrate that an attentive nonprofessional could empirically understand seismic problems and propose sound construction principles for an earthquake-resistant structure (figure 4.2).

Figure 4.2
Earthquake vulnerabilities of balloon- and platform-frame construction. George Gordon would have disapproved of this building as illustrated (see his recommendations in chapter 3). He would have approved of the diagonal boards of the subfloors, which, with sufficient nailing, would have helped produce a stiff diaphragm. But he would have recommended bolting the structure to its foundation. He would have insisted on using angle irons and iron ties around openings to strengthen connections, and he surely would have strapped the beam and bearing post in the basement. In this building, wall boards are to be nailed to the studs for channeling shear forces. On this point Gordon is silent, but today plywood with a conservative nailing schedule is considered the best solution for strengthening walls. In Gordon's time, diagonal bracing (similar to the braced frame in figure Figure 4.1) would have been the best solution.

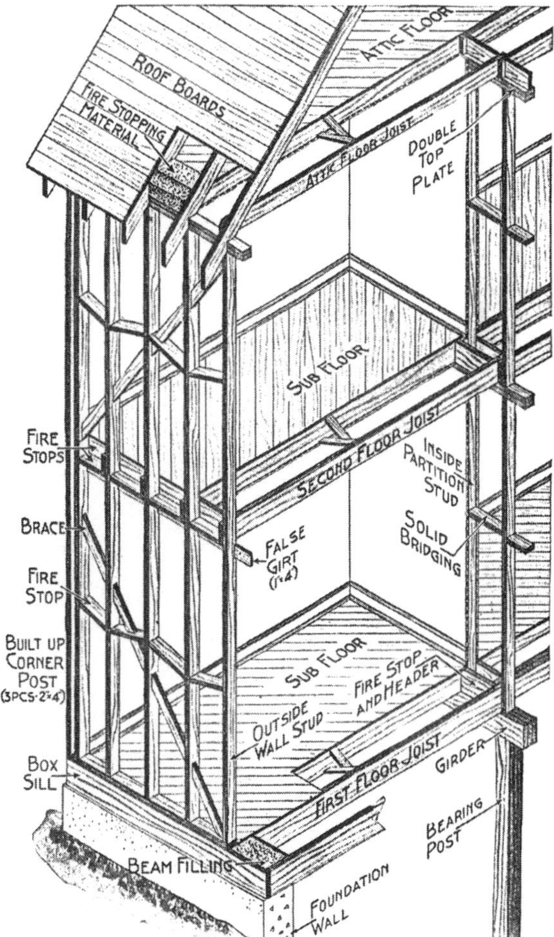

Buildings in Earthquakes

There is a saying: "Earthquakes don't kill people, buildings do." San Franciscans knew that wood resisted seismic forces more effectively than masonry. Buildings are designed to support static or vertical loads. These include the weight of the materials in the walls, floors, and roof (dead loads), and whatever rests on the floors

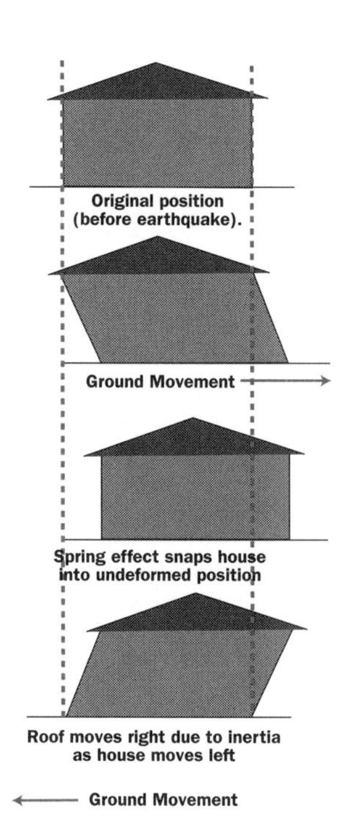

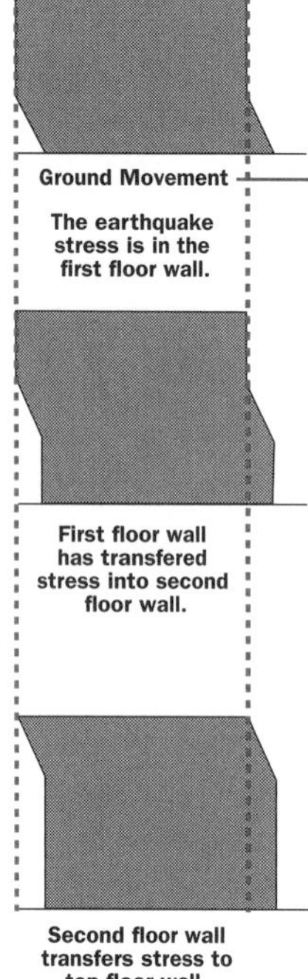

Figure 4.3
This diagram represents the response of a wooden building to earthquake ground waves. The waves cause vertical and lateral movements, or vibrations, which are transferred through the foundations to the building. While buildings are designed to withstand vertical loads, lateral loads can be very destructive, causing walls to bend past their breaking point, destroying themselves or shattering their connections. The diagram illustrates how the inertial movement is transmitted to the roof resulting in an out-of-phase motion of the roof relative to the ground. (Diagram and text adapted from Peter Yanev's *Peace of Mind in Earthquake Country)*

and can be moved, like people and furniture, as well as whatever falls on the roof, like snow (live loads). These loads are usually applied to the structure slowly and evenly, pressing down vertically. But the waves generated by an earthquake create dynamic forces that vibrate the structure and change rapidly (figure 4.3). As the building vibrates in response to seismic ground motion, inertial forces are created within it. When it is pushed to one side, it rebounds, but because of inertial forces it continues past its vertical position to bend in the opposite direction. Because buildings are primarily designed to resist vertical forces, lateral forces are the most dangerous in earthquakes.

Shear forces, which tend to distort the shapes of walls, occur when lateral forces push a wall along its length. The lateral forces are transferred from the ground through shear walls (typically wood-framed walls covered with a structural sheathing material) to diaphragms, and then back to the ground again through the walls. If a brick wall is pushed sideways by lateral forces, it will resist until the bond breaks between the bricks, or the bricks themselves break. A diagonal crack, called a shear crack, will appear, or sometimes an X-shaped crack (figure 4.4).

When a wall cracks along its length, it is said to fail "in plane." It loses its strength and is in danger of complete collapse. If it is pushed sideways and falls, it is said to fail "out of plane." The wider the brick wall in relation to its height, the fewer apertures, and the better the bonding, the more it can resist shear. The regents intuitively understood the problem of shear walls in masonry structures, observing:

> No stone structure standing alone by itself can possibly survive the shock, although a block of buildings in a city may be only shuddered by it...Besides, all our buildings will be weakened by many and large openings on all sides.[5]

Even on an extensive uninterrupted brick wall, bricks are problematic in earthquakes. The greater the mass—the heavier the wall—the greater the inertial forces an earthquake will create within it. In accordance with Newton's Second Law of Motion, F = M x A, inertial force is equal to the mass of the building (equivalent to its weight at ground level) times the acceleration. When shaken side to side, a properly braced, square, wooden, three-story structure on a sound foundation with well-tied diaphragms will bend because of wood's elasticity and ductility. A similar masonry building is heavier, stiffer, and more brittle, and instead of bending to dissipate energy, the brittle masonry will crack, or the walls may rupture and collapse. Because of their poor performance in many

Figure 4.4
Masonry buildings in earthquakes: the Estudillo residence, San Leandro, immediately after the earthquake of 1868. The wooden interior walls (B) moved back and forth on both the ground floor and the second floor. The exterior walls shook too. The diaphragms (A) transferred the load to the exterior walls, but the whole interior assembly of floors and partitions was more flexible than the brittle exterior walls. The walls (D) resisted the shear forces and then broke along the diagonal. The shear cracks are highlighted. Torsion, or the direction of the shaking, might explain why the end wall (C) collapsed. It may have been kicked out of plane by the moving diaphragm. Adobe is the weakest kind of masonry, being both weak and brittle.

earthquakes, unreinforced masonry buildings (called URMs by engineers) were banned by law in California in 1933. Figure 4.5 illustrates the ways in which unreinforced masonry can fail. San Franciscans had seen failures like these in 1865 and 1868.

In spite of the regents' sound reasoning that masonry was dangerous in earthquakes, the *San Francisco Bulletin* launched an attack against the decision to build with wood. The proposed building was called an "educational tinder box":

> Are we to begin the work of liberal education by such architectural antics in wood, and end it by a libel on the country, in assuming that the danger from earthquakes is so great that nothing more enduring than this is warranted?...We protest against the erection of monuments to earthquakes, and to the whole significance of the standing falsehood repeated every day to the youth of the State, and to mature men at home and abroad.[6]

The editors of the *Bulletin* suggested the regents of the university were out of touch with the latest advances in masonry technology, which were widely publicized in the San Francisco Bay Area. San Francisco architects and engineers were working to address the two major problems in masonry construction: how to create a shear-resistant masonry wall which would neither bend sideways nor crack along its length; and how to avoid a global collapse by tying the masonry building together and bracing it so that neither the external walls nor the internal frame or floors would fail.

Figure 4.5
The vulnerability of brick buildings in earthquakes:

A. Parapet and fire-wall failure. The most prevalent failure in masonry construction is the loss of parapets, which fall on neighboring buildings or in the street.
B. In-plane wall failure resulting in diagonal shear cracks. Notice shear cracks in the form of X's along the second story facade as well.
C. Nonstructural falling hazards
D. Wall failure in bending between diaphragms, or floors and roof
E. Facade failure, in which the wall-to-diaphragm tie fails, or does not exist
F. Roof and/or floor collapse
G. Wall failure
H. Soft-story or other configuration-induced failure

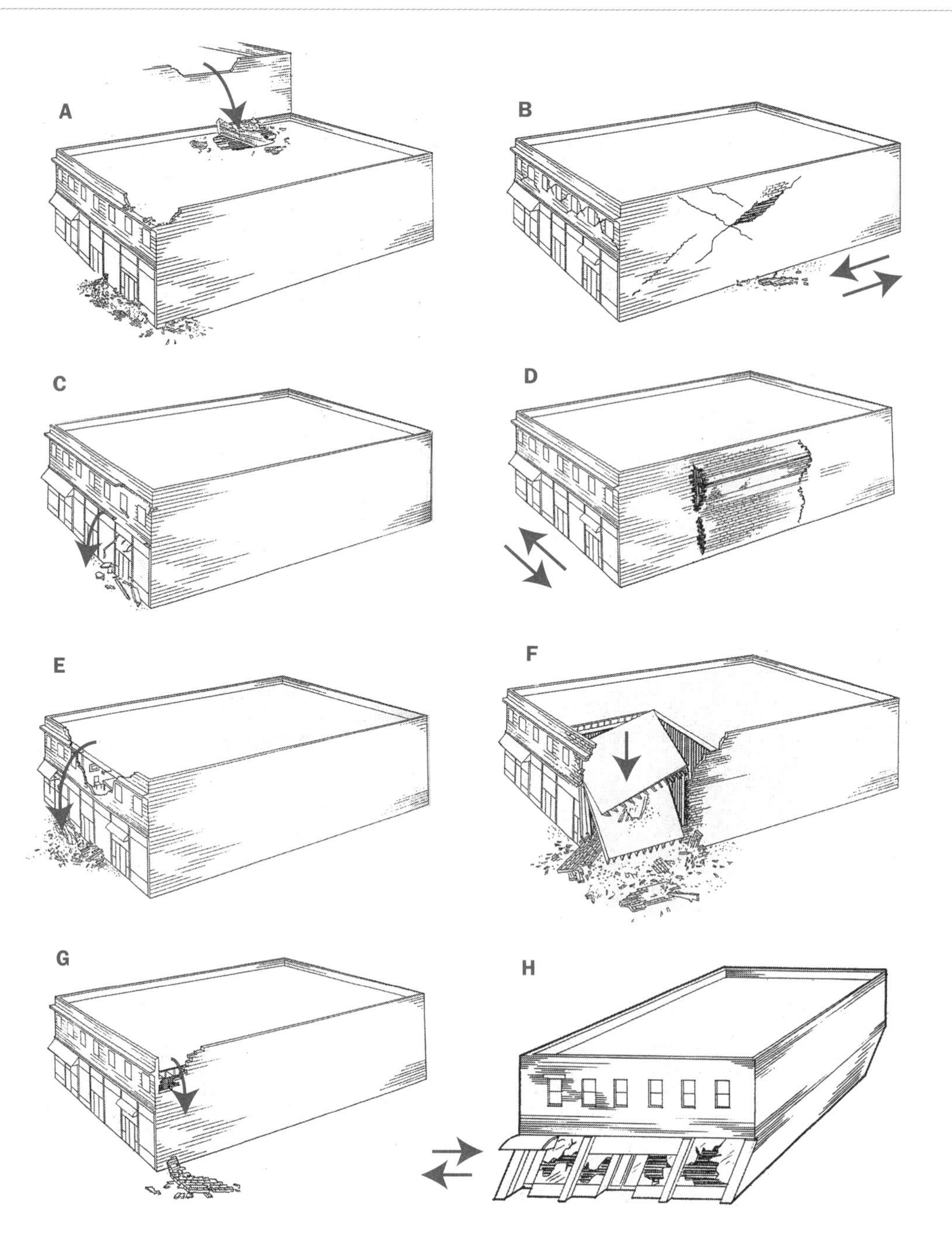

In the 1870s engineering was not, although it soon would be, exclusively responsible for structural systems. "Earthquake-proofing" was a free-for-all. Machinists and inventors who had applied their skills to supplying whatever was needed for excavating and processing Comstock silver now turned their attention to inventing earthquake-resistant systems. Amateurs, like George Gordon, and architects, like John Gaynor, also became involved. Scores of articles in local papers had suggestions for improved brick construction. For example, the author of an article titled "California Architecture" published on October 24, 1868, recommended the following strategies, certainly sound suggestions for their time:

1. Cease building heavy cornices, which invariably collapse in earthquakes.
2. Limit height to three stories.
3. Use wrought iron and cast iron, which have a "greater strength and reliability than wood."
4. Use mansard roofs, eliminating parapets and cornices, which collapse.
5. Provide independent support (in addition to the exterior walls) for floors and roof.[7]

Other articles urged better foundations and better attachment to foundations, and in one, the author advised constructing brick buildings with an "interior frame of wood for the support of floors and partitions."[8] In an article titled "Earthquake-Proof Buildings—How to Construct Them—Suggestions by a Professional Architect," the author, correctly noting that damage in brick walls sometimes occurs at abrupt changes in wall thickness, recommended remedying the problem by making such changes gradual.[9] Most of the innovations took for granted the utility of iron ties and bond iron.

Bond Iron

The use of bond iron (figures 4.6a and 4.6b) in San Francisco can be traced back to nineteenth-century England. In the 1830s, Marc Isambard Brunel, a famous nineteenth-century engineer, attempting to find solutions for his proposed Thames River Tunnel project, inserted ribbons of "hoop iron" in the brickwork to increase its strength.[10] These were flattened iron bands similar to those used in beer barrels. In 1835 he initiated the first test of reinforced brick construction with a twenty-five-foot brick beam laid seventeen courses deep with cement and reinforced with hoop iron in the lower five joints. He loaded this beam until it collapsed, bearing twelve tons.

Fascinated by the possibilities of reinforced brick, and backed by a cement manufacturing company, Brunel then decided to conduct an audacious test to prove its utility and economy in construction. In 1838, in the construction yard of his Thames River Tunnel, Brunel erected a scaled-down bridge pier that was intended to support twenty-five-foot arches extending out from the pier (figure 4.7). The arches were built without

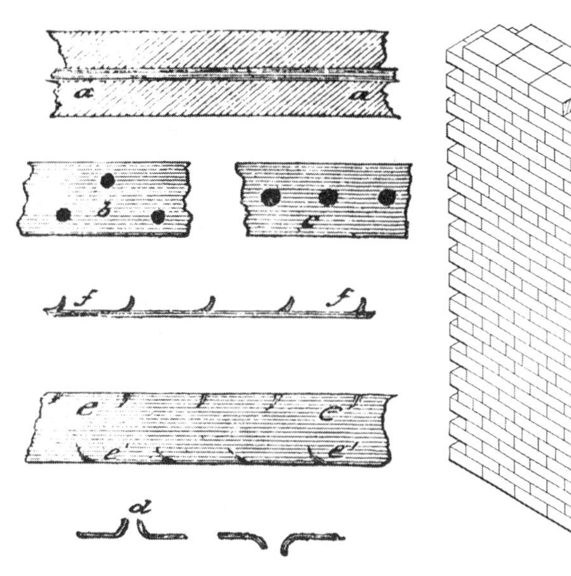

Figure 4.6a (below, left)
Bond, or hoop iron. English hoop iron is usually laid vertically in the center of the wall, as in a-a, between the bricks, to bind the wall together while being protected from rust. Holes can be drilled into it to improve the bond, as in b and c. A hoop-iron bond known as Tyermann's bond consists of bars with diagonal cuts on the edges (e-e), which are turned up, as in f, to more effectively catch mortar and bricks.

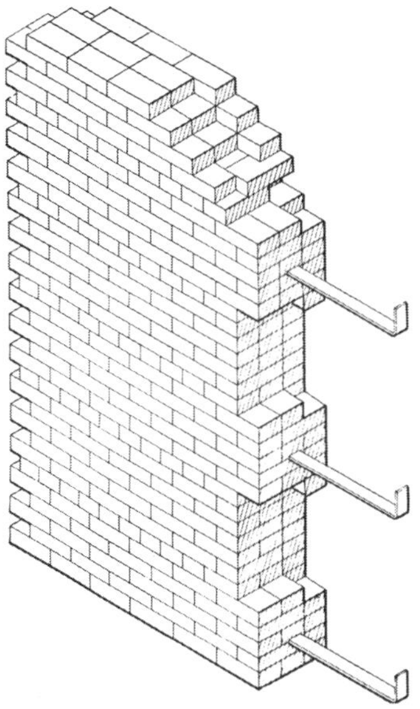

Figure 4.6b (below, right)
Bond iron in place. This diagram was by Frank Gilbreth, the efficiency expert. He labeled it "San Francisco bond."

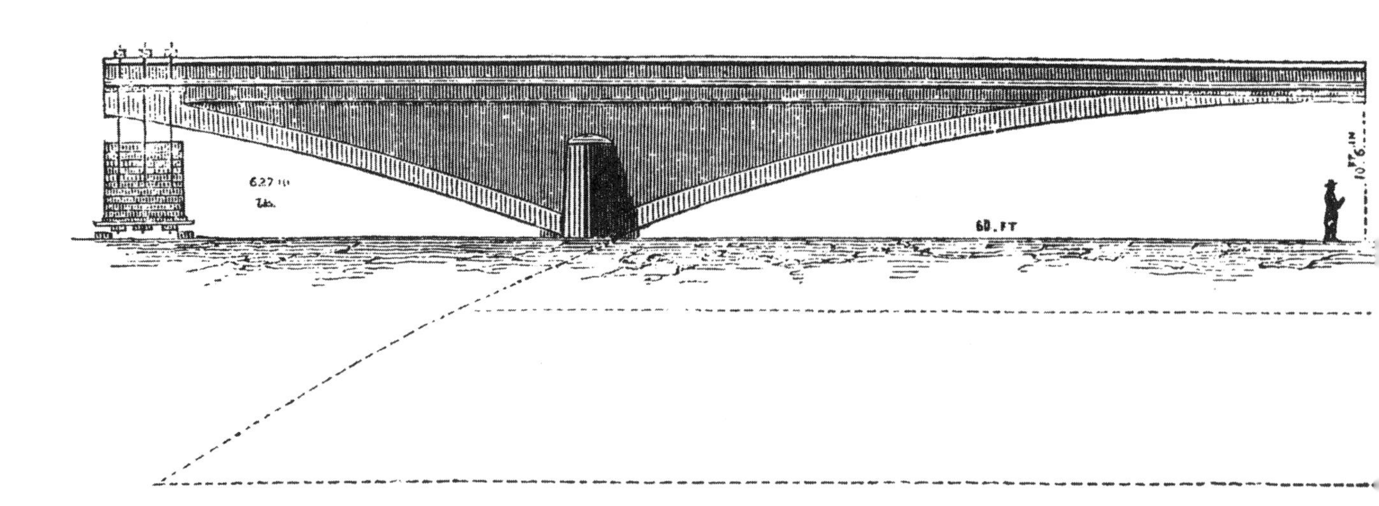

Figure 4.7
M. I. Brunel's bond-iron reinforced bridge pier experiment, 1838

costly centering or falsework usually necessary for their construction. (In order to build an arch, a semicircular falsework of wood is put in place and the masonry is rested on top of it. When the arch is complete and self-supporting, the falsework is removed.) By using the strength of the hoop iron and mortar, Brunel was able to create incomplete arches that supported themselves as they were built, which was a novelty. As the experiment progressed, Brunel decided to increase the span of the arches, but because of lack of space, he hung a counterbalance from one and extended the one opposite. On one side was the counterbalance, weighing 62,700 pounds, while on the other side Brunel was able to cantilever the arch for a length of sixty feet. This remarkable structure, a curiosity on par with the Thames River Tunnel itself, may have been responsible for the widespread adoption of hoop-iron bond in nineteenth-century masonry construction. Simultaneously, General C. W. Pasley was testing the strength of both reinforced and unreinforced brickwork. His 1838 book, revised in 1847, the standard in its day, discusses the advantages of hoop-iron reinforcement.[11]

Hoop-iron bond, as differentiated from Flemish bond or American bond, was known in the United States as "iron bonding" or "iron bond" (figure 4.6a) and was commonly used to strengthen brick walls here. Bond iron in lengths of six feet or more, two and a half inches wide by three-eighths of an inch deep, was laid transversely in exterior walls and bolted together so that continuous bands of bond iron ran through the walls of a building from one side to the other. A United States patent for using bond iron to reinforce seawalls, U.S. patent 74,547, was taken out by a San Franciscan named John Kelly in February 1868, nine months before the October 1868 earthquake.[12] Although "bond iron" is not explicitly named in the patent, it is described: "metallic strips are interposed between the stories in the outer course of masonry" in order to reinforce the wall.

Bond iron was commonly used in San Francisco at this time, though there is no record of payment to Kelly to honor his patent. It was not until 1901 that bond iron entered San Francisco codes, being required for all brick buildings.[13] Unfortunately, for reasons still unclear, the 1901 code required the bond iron to be positioned under joists, where it was easy to install but least effective.

The Success of Foye's Patent

William Foye, a little-known San Francisco machinist, patented a reinforcement system in 1869 that became famous, or infamous, in the city (figure 4.8).[14] Like Kelly before him, he may have been inspired by San Francisco's efforts to erect a huge seawall to stabilize its waterfront. He called his invention the "Improved Submarine Wall" and marketed it through his Pacific Submarine and Earthquake-Proof Wall Company. Although his method was not used for the seawall, it influenced masonry reinforcement in San Francisco.

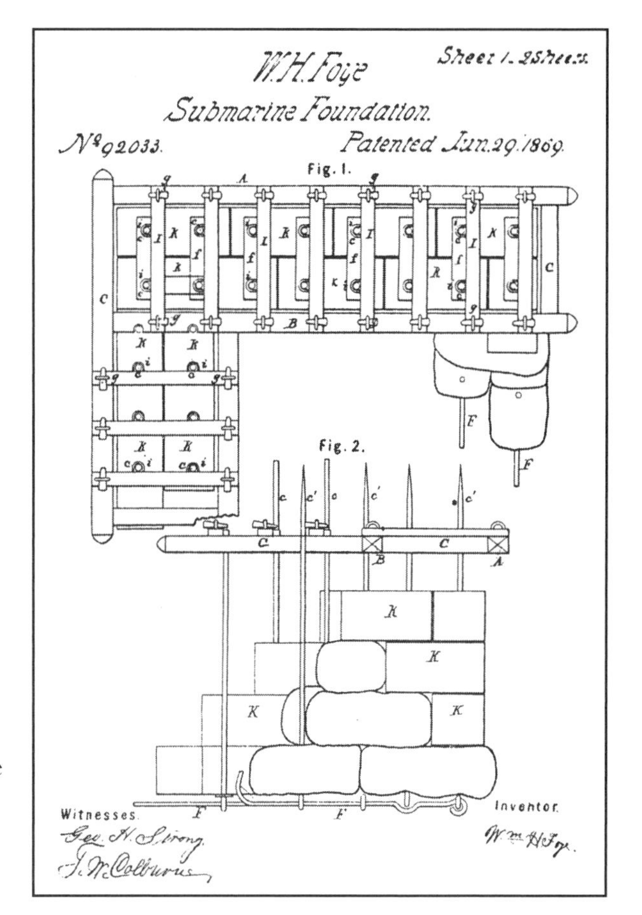

Figure 4.8
William Foye, U.S. Patent for an Improved Submarine Wall, 1869

Foye's invention depended on iron-bond technology. His idea was to sink vertical rods attached to an iron latticework into a foundation. After being drilled with holes corresponding to the positions of the vertical rods, the stones or bricks would be lowered into place. Each stone or brick would be held in place by this network of vertical iron rods and horizontal iron straps, and thus disparate pieces of brick or stone could be united into a single reinforced unit that could resist both in-plane and out-of-plane bending. This original contribution addressed the problem that had long haunted bond iron, that it tended to weaken the vertical continuity of masonry walls because it replaced a horizontal bond of mortar with iron. Foye described the usefulness of his invention:

> Walls for earthquake-proof houses can be built in this manner of concrete or artificial stone, binding them together with the guides and girders making them proof against the most violent convulsions and would be especially useful where earthquakes are common, as they cannot be thrown down unless bodily, and although the stones or blocks may become broken and thrown out of their proper line, they cannot fall and endanger the lives of persons who may be passing.[15]

Other builders copied Foye's idea. Federal and civic buildings used systems that were so similar that Foye took them to court. Though he freely admitted in court that he saw bond iron used in innovative ways in San Francisco before he invented his system, he insisted on the singularity of his idea.

> William H. Foye, the patentee, testified prior to his invention he had seen bond iron used in brick and stone walls in a great many different ways and that builders generally understood and were familiar with the use of bond irons in brick and stone walls in the construction of buildings. That the

> advantage of the method described in the patent over other methods was that an iron frame was constructed thereby, which held the wall together and in place, as a window sash holds the glass in it. That when an ordinary bond iron runs through the wall, if the wall cracked at all, the crack would follow the iron.[16]

Foye sued the federal government and the city when a version of his patent was used in the new 1871 City Hall. After three trials, in 1873, 1877, and 1878, he finally won in the United States Circuit Court.

Perhaps, as was claimed by Alfred Mullet, the architect of the San Francisco branch of the U.S. Mint, Foye simply appropriated what he saw on construction sites like the Mint's. But whether or not this is true, Foye gained from his notoriety. William Ralston approached him about using his construction method in the Palace Hotel. For the use of his patent on the Alameda County courthouse, Foye claimed he was given one-half of one percent of the gross cost of the building. He also claimed that his system was used in the Napa Lunatic Asylum. The popularity of Foye's system came from the fact that it did tie a building together. The *Morning Daily Call* praised Foye's submarine earthquake wall:

> It proposes an easy and apparently practicable mode of binding the walls of a building, making the whole structure an indissoluble unit, affording to each and every part the power of resistance and the weight of the whole, as a ship's keel is a unit, as a wheel is bound to its tire, or, rather, as we see the unit of strength in the construction of a basket.[17]

Foye's system was adopted by the San Francisco board of supervisors' Committee on Public Buildings for construction of all public buildings. Lieutenant Colonel B. S. Alexander of the United States Corps of Engineers, who had been a member of the Joint Committee on Earthquakes, wrote a letter supporting Foye, also using the analogy of a ship's keel, and warning that heavy bars of iron ought not be used.[18] "Their spring," he wrote, "or rebound, after sudden tension, would jar the walls and loosen the mortar during an earthquake." Instead, Alexander proposed the use of regular bond iron, presumably with the additional vertical rods, although he does not mention them. He concludes, "The value of such a building, if such iron ties were introduced according to the principles of the 'Foye patent,' would be enhanced more than the extra cost of the iron." A glowing article appeared in the *Mining and Scientific Press* in 1872 lauding the Foye invention because it would hold walls together not only in earthquakes but fires as well: "Firemen will feel a greater sense of security to fight fire in a brick building when it is known that walls cannot fall and crush them." The author did not address the problem of iron melting in fires. He recommended the system to the people of Chicago, who had just experienced the terrible fire of 1871, concluding:

> If this system of iron binding had been incorporated in the walls of the different factories in the East, which have fallen and killed so many...we should be spared the pain of such occurrences, and that constant dread and feeling of insecurity of those living in brick buildings in consequence of earthquakes which sometimes occur in California, would no longer be felt...No engineer, architect, builder, or practice man who pretends to a knowledge of the principles of construction has examined this system without endorsing it.[19]

Emerson's Iron Bracing

Another system, this one incorporating external bracing of masonry structures tied together by bond iron and fastened back to front to internal iron rods, was proposed by David L. Emerson of Oakland in 1872 (figure 4.9).

Figure 4.9
David Emerson's U.S. patent for iron reinforcement, 1872

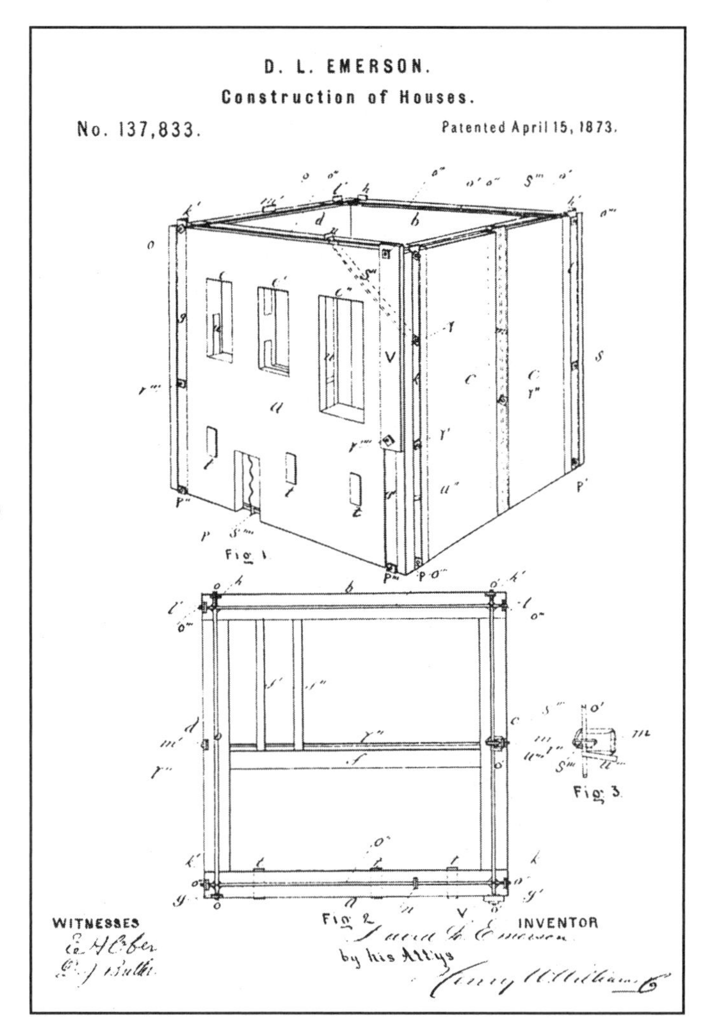

> The walls are each braced by vertical rods or plates and by anchors passing through them, the plates and anchors being connected by strap-iron [bond iron], which is also worked through the courses. The several walls are united and bound by horizontal tie-rods secured to vertical plates at the corners.[20]

A similar system had been developed for earthquake-resistant construction in southern Italy. The Bourbon government of the Two Sicilies developed a plan for iron reinforcement following strong temblors in 1854. This new system depended upon the introduction of iron rods into existing and new buildings. The installation of iron tie-bars both longitudinally and transversely, and putting them under tension by heating them and cinching them down, was meant to strengthen the floor, which it was hoped would work as a diaphragm. They were then connected to the story above or below by straps. At first the system was implanted in masonry walls, but the metal cage vibrated at a different rate than masonry, causing walls to collapse even if the building as a whole did not. Later the system was applied only to the exteriors of buildings. A typical retrofit in

PACIFIC STONE COMPANY. See Pages 96 & 97 and No. 671.

No. 1481. Concealed Window Spring, for preventing the shaking of windows and keeping out wind and dust. Exhibited by Lloyd & Mooney. A first-rate invention. The depot of these articles is at Rosecrans & Co's Hardware Store, Montgomery street.

No. 1482. Lifting Table. J. H. Swain & Co.
No. 1483. Bale of Burlaps. Oakland Cotton Mills.
No. 1484. Patent Wire Bale-tie. Lawton & Lenox.
No. 1485. Oil Painting. Barefoot Boy. A. Hart.
No. 1486. Patchwork. Quilt of 1,740 pieces. Mrs. Sarah A. Reed, Oakland.
No. 1487. Stereoscopic Goods and Big Tree Specimens. Thos. Houseworth & Co.
No. 1488. Photographs of California, Japanese and Chinese Scenery. Thomas Houseworth & Co.
No. 1489. Wreath of Tarleton Flowers. Emma Newbauer.
No. 1490. Case of Blue Vitrol (sulphate of copper). Crane & Brigham.
No. 1491. Samples of Hunter's Pruning Knives. H. J. Barnum.

I. TOUAILLON,

INVENTOR AND OWNER OF THE

MOST VALUABLE INVENTION

On the Pacific Coast, viz:

PATENT EARTHQUAKE-PROOF BUILDINGS.

PATENT RIGHTS FOR SALE.

Call at 427 GREEN STREET, and see the principle of this wonderful safety invention explained.

Examine the Model in the Fair.

See Description in Catalogue, No. 1226.

Nos. 1492, 1493, 1494, 1495, 1496, 1497 & 1498. Billiard Tables and Fancy Furniture, exhibited by Jacob Strahle & Co., No. 563 Market street, San Francisco. The largest and most important house on this coast engaged in manufacturing billiard tables is that of Jacob Strahle & Co. This firm has nearly $200,000 invested in various branches of the business, employ between 25 and 30 hands, and effect sales to the amount of about $100,000 annually. They use none but the best material, and turn out only

221

SKATERS, ATTENTION! Boone's C-Spring Roller Skate wants to Speak.

Mileto, Calabria, probably dating from the nineteenth century, looks very much like Emerson's solution.

Emerson's undecorated external iron braces never gained favor in San Francisco, probably because they looked too cumbersome. But his idea may have inspired more decorative versions, like the iron pilasters David Farquharson used at the University of California on Berkeley's South Hall.

Touaillon's Base Isolation System[21]

While Foye was battling for his share of the City Hall budget, other inventors rushed their ideas to the patent office and published commentaries in local newspapers. The most avant-garde proposal was Jules Touaillon's system of base isolation, a strategy so advanced that it was not widely adopted in the United States until the 1990s. Base isolation makes it possible to uncouple a structure from ground motion by introducing flexibility at its base. The ground moves in an earthquake, but due to inertial forces, the building moves far less, if at all. Today the most common means of isolating a structure is to support its superstructure on extremely hard and dense rubber bearings. This is the method that Forell/Elsesser Engineers used to retrofit both Oakland's City Hall and San Francisco's "new" 1912 City Hall after the earthquake of 1989.

As early as 1868, the Scottish engineer David Stevenson had proposed a system of balls and cups, which he called an "aseismic joint," to isolate lighthouses and lighthouse lamp tables from ground motion. In the 1880s, John Milne, one of the famous pioneers of seismology, sparred with Stevenson's son, David A. Stevenson, as to who first conceived of and successfully tested base isolation.

Touaillon, a San Francisco resident, patented his system in 1870 (figures 4.10 and 4.11). Who he was and what prompted his invention is still a mystery. Neither Milne nor Stevenson nor any of the scientific establishment seems to know him. Yet Touaillon's is one of the earliest extant descriptions of a base-isolated building system. He proposed that brick structures be built on platforms that rested on balls, each free to roll within its own space, which would be

Figure 4.10 (top)
Jules Touaillon, advertisement for base isolation, 1871

Figure 4.11 (bottom)
Jules Touaillon, U.S. patent for base isolation, 1870

shaped like an inverted cup (figure 4.11). As the ground moved back and forth, the building's movements would be far reduced. Touaillon realized that buildings might rock back and forth in large events and provided an interesting mechanical solution: "In cities where buildings are built with very little space between them, the walls may be provided with springs or bumpers, made of India rubber or other suitable material, to prevent injury or destruction from striking together." Touaillon's idea remained nothing more than a curiosity for decades. It would be revived by a German inventor after the 1906 earthquake and rejected again before being revived at the end of the twentieth century.

Timber Frames with Brick Walls

Despite the ingenious earthquake-resistant inventions patented after the 1868 earthquake, architects and builders continued to prefer cheap, dependable interior timber frames to provide flexibility for their brick buildings. The timber used, redwood beams and columns sometimes as large as twelve by twelve inches, was much more massive than the wood used in balloon-frame construction. These large timbers were bolted together, using pins and iron plates, and then attached to exterior brick walls, using iron anchors. The anchors, placed in horizontal rows at each floor diaphragm and the roof, linked the interior wooden frame to the exterior brick wall to ensure that the building could act as a unit in an earthquake. In most buildings the timber frame was completely self-supporting, so that if the walls cracked and broke apart, the floors of the building would not pancake—collapse one on top of another. In the late 1860s these timber-frame buildings were considered to be earthquake-proof by John Gaynor and scholar Thomas Rowlandson. An advertisement for John Gaynor's new Grand Hotel stated it was earthquake-resistant (figure 4.12). The *San Francisco Newsletter* provided an even more complete description:

Figure 4.12
The Grand Hotel, San Francisco, 1870—John Gaynor, architect: an example of the earthquake-resistant system of interior wooden framework and exterior brick facades

> Last of all and perhaps first in consequence... The hotel is a complete frame building, surrounded by brick walls. The frame is of heavy timber, bolted, braced, and strapped together with massive iron bolts, bars, and anchors, attaining a strength almost rivaling that of a ship. To this frame the brick

> walls are appropriately fastened. But, should the city ever be visited by an earthquake so destructive as to throw down these brick walls, that must need fall outward, and will leave standing the skeleton of the Grand Hotel, with its roof and floors unmoved. In this view its modest four story height, its floors spreading broadly over the ground instead of soaring in ambitious tiers towards the sky, possess for the San Franciscan eye a hidden beauty where the earthquake inexperienced eye might think to discern a defect.[22]

The directory of 1870 labeled many prominent new timber-and-brick structures as earthquake-proof. The Occidental Hotel (figure 4.13) is "iron braced and thoroughly anchored, and secured by heavy tie-rods in every story, thereby securing the building from earthquake shocks."[23] Pacific Bank is likewise earthquake-proof, "the outer walls of brick and iron being fastened by strong anchors to an inner structure of heavy timber, so as to prevent the falling of floors should the walls give way."[24] The British Bank of North America and the adjacent buildings, "like the Pacific Bank are secured by iron braces to an inner structure of wood to render them, as far as possible, earthquake-proof."[25] The new Girls' High School too was built to be earthquake-resistant: "To secure it as much as possible against earthshocks, it is built of heavy timbers, anchored and bolted together on a solid brick foundation."[26]

Figure 4.13
The Occidental Hotel, which weathered the earthquakes of 1865 and 1868 with no damage. The north (Bush Street) side was begun in 1861 with a stone foundation "laid in the most secure manner." Like the Lick house being built across the street, the 1862 addition on Montgomery Street was probably, as the 1862 *San Francisco Directory* said, "braced with strong iron rods to secure it against earthquakes." In 1869 another addition, on Montgomery and Battery Streets, was "iron braced and thoroughly anchored, and secured by heavy tie-rods in every story, thereby securing the building from earthquake shocks," according to the 1869 *Directory.*

John Gaynor's Palace Hotel

John Gaynor's four-story Grand Hotel may have been modest, but there was nothing ordinary about the Palace Hotel that he designed for William Ralston (figure 4.14). At the time of its construction, it was one of the largest hotels in the world. Ralston, a powerful San Francisco booster, was a reckless investor but insisted that the Palace Hotel be earthquake-proof. He is reported to have consulted with Foye before John Gaynor.[27]

Gaynor used brick cross walls strengthened by strong cement mortar and bond iron to provide a solid wall inside the shell of the building. According to reports, the entire building was belted together with courses of bond iron embedded every four feet in the brick walls.[28] Combined with the strong mortar and frequent cross walls, the result was an extremely innovative and strong structure, quite different from the interior wooden frame and brick of the Grand Hotel.

A run on the Bank of California, which Ralston had used to finance his empire, led to his apparent suicide in 1875. He did not live to see the earthquake-resistant Palace Hotel perform well in the earthquake of 1906, only to be destroyed by fire.

Figure 4.14
The Palace Hotel, San Francisco, 1875—John Gaynor, architect. The Palace was a reinforced iron and brick structure. The iron boxes visible at the corner brick walls and between the bay windows were part of the reinforcement system. The Palace was described as "Spotted like the sides of an iron-clad with bolt-heads that cinch the great rods running over and under and through-and-through the building, making it a kind of Cyclopean open-work iron safe, filled in and lined with fire-proof brick, where all treasure of human life and limb should be secure against fire or earthquake while the peninsula stands." *(Overland Monthly,* vol. 15, no. 3, Sept. 1875)

San Francisco's New Earthquake-Proof City Hall

When the City Hall on Portsmouth Square (figure 2. 11) was damaged even more severely in the earthquake of 1868 than it had been in 1865, a debate ensued about whether to build a new structure on a new location. As has happened many times since, seismic upgrading became the catalyst for rethinking the utility and aesthetic worth of the building. In 1852 the city had purchased the Jenny Lind Theater, a four-story brick building built in 1851 with a classical pedimented facade of sandstone, and converted it to City Hall. As the city and its offices grew, the El Dorado Hotel on the north and the Union Hotel on the south were purchased and incorporated.

By 1868 this ad hoc arrangement seemed inappropriate to some San Franciscans who had a grander vision of their city as the premier West Coast port and terminus of the new transcontinental railroad. They saw other cities, like Philadelphia, building huge city halls, and they wondered whether San Francisco shouldn't seize the opportunity presented by the earthquake to finally build a City Hall.[29] Three days after the 1868 earthquake, the supervisors voted to repair the old City Hall. A struggle ensued. Within a week, P. J. O'Connor, a local architect, proposed abandoning the old building in favor of a new one on a new site. A month later, Mayor Frank McCoppin vetoed the bill to repair the old City Hall. In defiance, the supervisors approved another bill to repair it, but the momentum for a new building had grown. On January 2, 1869, a grand jury proposed putting all new public buildings on public squares and mentioned Yerba Buena Park, a former cemetery on Market Street, as a possible site for the City Hall. On April 4, 1870, the state legislature passed a bill calling for this. Three City Hall commissioners were appointed to oversee the construction: P. H. Canavan, member of the board of supervisors; Joseph Eastland of the San Francisco Gas Works; and Charles McLane of Wells, Fargo & Company. The city hired the architectural firm of Williams and Wright to develop specifications for the building and staged a competition to choose an architect. The firm of Laver, Fuller and Company from New York State won the competition (figure 4.15), but their design was modified to reduce cost, and a competing team, Patton and Jordan, was chosen to design a separate Hall of Records.

The 1871 City Hall was supposed to be earthquake-proof, and in the wake of the great Chicago fire of October 9, 1871, the commissioners insisted on making the building fireproof as well.[30] It proved to be neither, but not for want of trying, particularly in those first years. The specifications that architects Stephen Williams and John Wright drew up required "all the walls throughout, both the exterior and interior, to be thoroughly bonded with wrought iron, and every possible precaution used to prevent injury by earthquakes."[31] They recommended brick rather than stone for the structure's walls because they believed that with the proper "appliances" (perhaps bond iron and wrought iron), brick had elasticity and tenacity, "qualities first to be considered in resistance against earthquakes."[32]

The construction of the new City Hall was plagued from the beginning by change orders, cost overruns, discontinuities in design, and disputes, including whether Foye should be paid for the use of his patent. It is clear from newspaper reports and contracts that the stone walls of the foundation and the brick walls above them were supposed to have been strengthened by a network of vertical iron rods and horizontal iron straps as prescribed in the Foye patent. However, in 1994 when the old foundations of City Hall were uncovered, it was revealed that the foundation walls of the subbasement were built of irregular blocks of Angel Island stone alternating with brick over a concrete base. Iron cross-plates four inches wide, three-fourths of an inch deep, and three feet long were laid in the brick and attached at intervals to vertical rods (figure 4.16).[33] The rods did not pass through the stones and bricks as specified in Foye's patent, nor were they joined to the horizontal bands. Nevertheless, it was a variation of Foye's method, and Foye's system had been specified explicitly in the plans for the building and discussed with the architect. Foye won his case. Looking at the foundation work from the perspective of 1994, it is apparent that the builders used a far weaker system than Foye's patent called for. Early in the course of construction, Augustus Laver was forced to resign. His assistant, E. A. Hatherton, took over. Hatherton believed the iron reinforcements were useless,

Figure 4.15 (above, top)
Augustus Laver's winning design for San Francisco City Hall, 1870

Figure 4.16 (above, bottom)
The foundation of San Francisco City Hall as uncovered in the 1990s excavations for the San Francisco Public Library. The first course of stones is tied by an iron bond which is visible in the center of the wall.

and much to the horror of a modern observer, cut off the remaining vertical rods, further compromising whatever strengths the system had.

Foye's patent aside, the 1871 City Hall had little to recommend it as an earthquake-resistant structure. The site was above a spring on a former marsh. Laver's plan, an irregular W shape, or configuration, attempted to adapt the building's footprint to its awkward triangular site (figure 4.17). A symmetrical building works as a unit in an earthquake, with no part weaker than the next. The irregular footprint of the new City Hall was very dangerous in earthquake country. Imagine shaking a building in the shape of a W: each of the horizontal sections would pound against the other, because each would oscillate at a different frequency. The walls forming the interior angles, called reentrant angles, would bang together because of their proximity. Joints would stretch and compress, pulling the W apart.

Not only the building's shape, but its multiple towers compromised it. Moreover, the decorative classical columns and cornices were never tied to the frame of the building. It was the proverbial disaster waiting to happen, and an exception to the real earthquake-resistant buildings of the time.

Figure 4.17
A view of the 1871 San Francisco City Hall as it was to look in the 1890s, before the design of the tower was changed. The irregular shape of Augustus Laver's design makes it hazardous in earthquakes.

Mullett, the Custom House, and the U.S. Mint

Alfred Mullett, an Englishman, was appointed supervising architect for the Treasury Department in 1866. He designed San Francisco's imposing Greek Revival United States Branch Mint on Fifth and Mission Streets. Unlike his colleague Gridley Bryant, also an easterner who worked for the federal government, Mullett was keenly aware of earthquake danger and the peculiarities of San Francisco's site. In 1868 Mullett had to decide what to do with Bryant's earthquake-damaged Custom House. Writing from Washington to J. F. Morse, superintendent of repairs in San Francisco, Mullett recalled bitterly that he knew the building was doomed from the beginning:

> I may say...that on my first inspection of the building in 1864 I predicted...early destruction...[Through] the nature and character of the damage, these predictions have been verified, though they were at the time the subject of much comment and ridicule. I did not and do not consider the site a fit or a suitable one for the erection of permanent buildings of the character needed for government use and I also consider the plan adapted vicious and founded upon an entire misapprehension of the problem to be solved.[34]

Having studied the earthquakes of 1865 and 1868, Mullett concluded that there were three possible soil conditions in San Francisco, and the Custom House was poorly built on the least favorable soil. The best building site was rock, the next sand, the worst fill. "In my opinion," he continues:

> The use of piles [as in the Custom House foundations] in soil subject to lateral movement is injudicial in the last degree, and must as in this case split the building...It has always appeared to me that the only chance of success lay in the construction of a timber foundation well framed and bolted together, forming in fact a raft of sufficient strength to sustain the foundation, when the building would probably have moved as a unit, if at all.[35]

When the time came to build the Mint, Mullett was unhappy that it was to be built not on the rock of Rincon Hill but on the sand of South of Market, but he committed himself to building a structure that would endure. His answer to the sandy site for the imposing brick-and-stone building was to propose a floating foundation, not of wood but of concrete deep enough to protect the Mint from the unstable soil and "sufficiently homogenous and firm to support and bind the building together at the base...supporting the building on sand as a floating dock is supported on the water.[36]

Foye and Mullett

Just as Foye was about to patent his system, Mullet wrote to his project manager, William Stebbins, changing his mind about the design specifications for the brick and stone walls of the Mint:

Figure 4.18 (top)
View of the U.S. Mint, San Francisco, in construction, showing iron reinforcement being laid in the brickwork

Figure 4.19 (middle)
The U.S. Mint, San Francisco, in the final phase of its construction. Note the vertical and horizontal iron bars in masonry on the roof above the column bases on the porch.

Figure 4.20 (bottom)
U.S. Appraiser's Building, San Francisco, after the earthquake of 1906

> I have decided to put iron rods (or hog chains) through the foundations of the building under your charge, and from end to end and from side to side of the building the rods to be 1½ to 2 inch round rods, as may to your opinion be required. These rods are to be placed between the concrete foundation now completed and the granite, and are to be secured to suitable anchors which must be sunk in the concrete and dowelled into granite in the best and most secure manner of which you will be the judge. I have explained my views to you fully and leave this and other questions of the same nature to your discretion.[38]

There is a possibility that Mullett changed his plans because he understood the importance of Foye's system. He also had horizontal (figure 4.18) and vertical rods placed throughout the building, inside or attached to external walls, to further strengthen it. (Figure 4.19 illustrates vertical and horizontal iron bars at the roof level over the incomplete entrance.) Mullett used a very similar system in the U.S. Appraiser's Building (figure 4.20), constructed adjacent to the old earthquake-damaged Custom House on Battery Street. Mullett resigned before the Appraiser's Building was completed, but it was constructed using his plan, if not under his supervision. Foye claimed that his system had been used in the Mint and discussed for the Appraiser's Building. He sued the federal government for patent infringement. There is no question that elements of Foye's system were present, but the court ruled against him on the grounds that the vertical reinforcement in the Mint was not tied to the horizontal reinforcement as consistently or in the same way as called for in Foye's patent.

South Hall, University of California at Berkeley

Meanwhile, the controversy over the construction of South Hall (figure 4.21) at the University of California at Berkeley continued. The regents were ultimately forced by public opinion to consider a masonry building. A competition was held to choose the architects who would be responsible for construction of all campus buildings. When the final competition winners, architects John Wright and George Sanders, presented their designs on July 5, 1869, they included

descriptions of seismic solutions for wooden, stone, and brick buildings to assuage the fears of the regents (as was the case in the 1870s, these architects engineered their own structures). Wright and Sanders described the three earthquake-resistant options. The first option, a wooden building, would have diagonally sheathed walls to limit lateral deformation and drift in an earthquake; diagonal sheaths running between the supporting members helped to counter the shearing forces in the wall.

Four years earlier, Wright and Sanders had used an interior timber framework and bond iron in the British Bank of North America in San Francisco, which they must have considered to be earthquake-resistant. They explained a similar strategy to the regents as an earthquake-resistant option for stone buildings:

> In case of stone being selected as the material of construction, the walls should be much heavier laid on broad and deep foundations of concrete, strengthened and tied by means of iron bars, laid continuously in through the length and breadth of the building above, through the walls, to be laid of fine tooled and coursed ashlar, faced on the inside, laid with strong iron bond at each story, and above and below all windows. In all cases heavy iron bolts to be continued through the joists at every floor, and the whole of the building to be thoroughly earthquake-proof, *having strong internal*

Figure 4.21
South Hall, University of California, Berkeley, 1873—Kenitzer and Farquharson, architects:
A. Bond-iron courses through masonry
B. Iron pilasters held in place by bond iron
C. The position of floor anchors in masonry
D. The position of internal iron girders

> *timber framing bolted to the walls, and affording independent support to the roofs and floors through.*

Wright and Sanders offered a similar strategy for brick. They described a dual system in which the internal wood framework and the exterior brick walls would be independent of each other, though attached. The exterior brick or stone walls would be additionally tied together as effectively as possible by iron rods and bars like those employed by John Gaynor on the Grand Hotel.

Wright and Sanders never got the chance to put their design ideas into effect, because the regents decided to employ someone else to construct the building, in order to avoid paying the usual five percent of the cost of the completed structure. Wright and Sanders contested this decision and resigned from the job by August 1869. Within days, David Farquharson of Kenitzer and Farquharson architects (and a member of San Francisco's Joint Committee on Earthquakes) had signed a contract for all architectural designs on the campus.

Enthusiasm for a wood building had faded, and in 1870 Farquharson began to build South Hall using a composite of stone, brick, iron, and wood. Describing the building in 1873, he wrote, "The building is bonded throughout in every direction with wrought iron, and the floors are supported by heavy wrought iron girders..." While he was aware of the possibility of using a wood frame on the inside surface of the brick, as Gaynor had in the Grand Hotel, Farquharson thought iron inserted directly in the wall was more appropriate than the dual wall system:

> Another system of construction with reference to earthquakes has occurred to me, which might be practiced in cases where expense was not the first consideration. The making use of cast-iron for the first story fronts of business buildings is now the general practice, but I propose to go a step farther, and make the bearings and supports throughout the building wholly of iron. There can be no doubt of the fitness of cast-iron for this purpose, its elasticity is well known, and there is no instance—that has come to my knowledge—of a cast-iron column, or a lintel, being broken by the earthquake. It is not proposed to dispense with the use of brickwork, but the use of iron and brick together, in such a manner that although the brick would form the bulk of the walls, yet the dependence for support would be wholly in the iron.[40]

In explaining his concept of iron columns and brick walls working together, Farquharson anticipated metal-frame construction, the basis of the modern skyscraper, though it did not become a reality in the United States until the 1880s. In Farquharson's system, the iron embedded in the wall kept the building upright while the brick load-bearing wall acted as a horizontal damper, resisting lateral forces in earthquakes by cracking and dissipating energy.[41]

Farquharson's earthquake-resistant system was an architectural composite; it depended upon a building's brick walls, wood supports, wood diagonal sheathing, wood floors, iron tie-bars, iron anchors, and iron columns working together. He considered how every part of the structure, from its foundation to its chimneys, could be tied together. He believed that a building's structure as well as its decoration could aid in its seismic resistance. His use of earthquake-resistant ornament heralded a new style of architecture that was beautiful because of its frank expression of purpose.

South Hall is bound together by bond iron (figure 4.22), and the brickwork and lime mortar are exceptionally strong, even by modern standards.[42] Pieces of bond iron measuring two and a half inches by three-eighths of an inch were worked through the brick above and below the apertures on each story and at the joist level. These pieces of iron were spliced together with two bolts at each joint to form a continuous belt around the whole structure. As each belt of bond iron approached an end wall of the structure, it was forged into a threaded rod. Depending upon their position, these rods either entered heavy corner impost blocks or went directly through the wall of the building to be bolted to iron pilasters on the exterior. This network was clearly intended to hold the whole structure together should the bricks begin to fail.

A second line of defense can be seen on the building's exterior, which is decorated with vertical ornamental panels made of cast iron. They appear at the corners and sides of the building, often with the threaded rods of the bond iron protruding through them. The rods are secured to the panels by decorative bolts that form a regular pattern, appearing even where no rods are present. Rather than securing the panels (which are held in place by special iron hooks), these bolts unite the bond iron from one side of the building to the other. This linking suggests that Farquharson hoped to form a sort of exoskeleton.

Farquharson seems to have taken great care to make sure the floors functioned as diaphragms, tying the exterior walls to them and thus helping the building move as a unit. South Hall is an I-shaped building with a corridor running down the middle. Farquharson lapped every other four-by-sixteen-inch joist over the top of the corridor, effectively tying the building together. Every joist in the structure was either nailed to a hanger or extended out into the brick walls (figure 4.24). Large, round iron anchors are buried three widths into the brick exterior walls, bolted to the end of huge iron angles attached to the joists.

If the brickwork began to break up, vertical iron T's implanted in the north and south walls of each of the large lecture halls on the wings of South Hall provided support. Two great iron

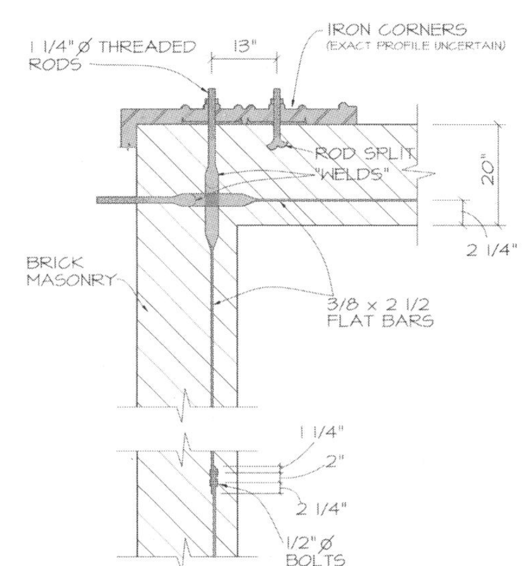

Figure 4.22 (top)
South Hall, University of California, Berkeley: diagram of bond iron in walls and attachment to external iron pilasters

Figure 4.23 (bottom)
South Hall, University of California, Berkeley, wall in demolition: iron T-shaped columns hold up girders with bond iron in outside walls. Note one row of bricks has been removed

Figure 4.24
South Hall, University of California, Berkeley, internal walls in demolition: iron anchor and hanger in brick wall, which is attached to floor joist—Farquharson's way of linking wall and diaphragm

girders spanned the north and the south lecture halls, supporting four-by-sixteen-inch wooden joists. The vertical iron T's supported the iron girders on each side of the room, creating a redundant brick and iron wall support; the iron would probably have buckled without brick around it. Farquharson's construction points to the significance of redundancy, another important idea in earthquake-resistant construction.

South Hall's construction proceeded in fits and starts, and Farquharson's seismically resistant design remained intact. The first stone was laid on July 17, 1870, but work was suspended for lack of funds on January 28, 1871. Construction began again in July 1872, and by March 15, 1873, the structure was roofed. Ironically, North Hall, started later and constructed of wood for the sake of economy, was already complete. At the university's first commencement exercises, held at North Hall on July 17, 1873, University President Daniel C. Gilman, said:

> Incomplete as are the surroundings, the plan of these edifices is obvious; the one a costly, massive, and enduring hall, proof, it is hoped, against the quakings of the earth and the inroads of time; the other spacious, economical, and in a high degree convenient, but possibly liable at some future day to yield its place to a more solid structure.[43]

Wood had yielded its place as an earthquake-resistant structural material to innovations in brick and iron. With the right foundations and devices to ensure the integrity of walls, brick regained its preeminence as the safest building material, not only in fires but in earthquakes.

CHAPTER 5

Earthquake-Resistant Construction, 1889–1905

Civic-mindedness in San Francisco declined in the 1880s and 1890s as bureaucracy, graft, and machine politics increased.[1] City politics were heavily influenced by Southern Pacific Railroad executives. Nevertheless, Blind Boss Buckley's machine, supported by labor, won the mayor's office in 1885 with the promise of keeping taxes to one dollar per person, to the detriment of city parks, streets, schools, and infrastructure. The Comstock Lode boom had dwindled by the 1870s, but the economy improved after the depression of 1873 and through the 1880s. Another crippling depression occurred in 1893, and the economy again rebounded as San Francisco outfitted prospectors for the Yukon gold rush of 1897 and provided arms, munitions, and staging facilities for the United States war against Spain in 1898. Labor and business fought bitterly, particularly along the waterfront.

When the conservative business candidate James D. Phelan won the mayoral election in 1897, he began a campaign to beautify the city by applying Beaux-Arts aesthetics. His attempts at beautification did nothing to satisfy labor's demands for better wages, and in 1901 a huge citywide strike erupted. In its wake a unionist mayor, Eugene E. Schmitz, was swept into office.

With Schmitz came yet another city boss, Abe Ruef, who made an art form of graft and corruption. Some departments within the city government managed to function, but others did not. San Francisco's fire chief, Dennis Sullivan, felt that the quantity of water supplied by the privately owned Spring Valley Water Company was inadequate for firefighting. In 1904 he crusaded for an auxiliary water system, to be used only for fighting fires. But laissez-faire capitalism and corruption boded ill for any improvement in the infrastructure.

The city continued to grow: newspapers, insurance companies, banks, railroads, steamship lines, newly formed corporations, and privately held retail, manufacturing, and warehousing companies all required office space.[2] Competition and interdependency dictated that these offices be concentrated in one district of the city, the "downtown." The safety catch, invented in 1854, made elevators, previously used only for freight, acceptable for transporting people. The first passenger elevator was built in New York in 1857; the first in Chicago was an Otis Patent Steam Passenger Elevator installed in 1870. Following the lead of Chicago and New York, the San Francisco business community embraced the image of the vertical city, offices stacked one above the other in multistory towers. Three- to six-story fireproof brick buildings had been the norm, and brick buildings were still prevalent, but now taller iron and steel buildings changed the architectural paradigm. Between 1889 and 1905, tall office buildings transformed San Francisco's skyline and introduced a new building technology to earthquake country.

Architects and Engineers in the 1890s

As one technology replaced another, questions of earthquake safety had to encompass new parameters. New experts were educated who made the domain of iron and steel buildings their own. Prior to the late 1880s, architects assumed responsibility for both form and structure. Now all steel-frame systems would require an engineer trained to calculate the loads and to determine how the frames should be connected for what appeared at the time to be tremendously tall structures. An unintended cloak of mystery began to descend over the rationale for the decisions these professionals might make. Their work, hidden by the architect-designed facade, was unseen and difficult, if not impossible, for a nonengineer to understand, as an article in *Engineering News* of 1896 explains:

> The construction of the exceptionally high buildings with steel framework which now form such a notable feature of the larger American cities has led not only to the development of a new and large market for structural steel and iron work, but also to new problem in structural design...These difficulties are probably appreciated not at all by the owner, and only to a limited extent by the architect (unless he is himself an engineer experienced in structural work), and the designing engineer can expect but little credit for the successful results of his labor, since the work is all built in and enclosed as soon as erected, so that the casual observer never gives it a thought, and even if anyone were interested in it, it is hidden from examination.[3]

The professionalization of both architecture and engineering occurred simultaneously in the United States between the 1830s and 1900s.[4] Engineers trained in mathematics, physics, and mechanics were indispensable to armies, which had to have roads, fortifications, and accurate ordnance.[5] The title "engineer" first appeared in modern usage in the Italian Renaissance, and it was increasingly associated with warfare technology in the sixteenth century after the French first effectively used moveable cannons to destroy Italian fortifications.[6] The Italians responded by inventing a new system of low walls guarded by equally low, heart-shaped bastions on each corner. Architects who specialized in these fortifications were often called engineers. The French in the eighteenth century realized that a standing army required its own specially trained corps of engineers. Further, such a corps could also be employed by the state to service its roads, bridges and canals. Hence the French Corps de Ponts et Chaussée, founded in 1716, and the first state school of engineering, the Ecole Nationale des Ponts et Chaussée, founded in 1747.

Engineers were naturally included in the establishment of the military in the new United States. The Continental Congress legislated the appointment of engineer officers in the army, most of the positions being filled by Europeans. When the United States Military Academy was created in 1802 it was intended to educate military engineers, and it did so until the Civil War. In early nineteenth-century America, all military engineers were trained at West Point and formed part of the United States Corps of Engineering. Henry W. Halleck, who designed the foundations for the Montgomery Building in San Francisco, was a graduate of this program. But some engineers in the early nineteenth century were people with little or no formal education who had acquired their expertise through self-study or apprenticeship. When Congress enacted legislation for surveys of major roads and canals in 1821, it distinguished between "engineering officers" and "civil engineers."

With the increasing need for major public works—bridges, railroads, water and sewer systems—and the more extensive use of iron in construction, the United States needed more civil engineers. In 1821 the American Literary, Scientific and Military Academy (later Lewis College and, later still, Norwich University) offered the first civil engineering course outside West Point. The first civil engineering degree was conferred by Rensselaer Polytechnic Institute in 1835. By the mid-nineteenth century, engineering courses were being offered by Union College (1845), Harvard College (1846), and Yale College (1846). In 1869 Massachusetts Institute of Technology began its program in civil engineering. The American Society of Civil Engineers was founded in 1852 to better regulate and represent the profession.

Architects, who had either been educated with engineers or come up through the building trades, claimed their own professional identity.[7] They organized their own clubs, wrote codes of ethics, and pressed for professional licensing, which would first occur in Illinois in 1897. Like engineers, architects in the early 1800s could be self-taught; anyone who had acquired the expertise to build might call himself an architect. Schools were formally established late in the century at the Massachusetts Institute of Technology (1868), Cornell University (1871), and the University of Illinois (1873). To further the profession, Richard Morris Hunt, a powerful New York architect, founded the American Institute of Architects (the AIA), which first met in 1837. It took until the late 1890s for architects to win public visibility, partly through the triumph of the 1893 Columbian Exposition,

The architectural aesthetic of the exposition, so shaped by the French Ecole des Beaux-Arts, demonstrated that American architects could produce elegant Continental architecture in an American context. Richard Morris Hunt himself had been the first Paris-trained American architect. He was followed by scores of other aspiring Americans who wanted to be educated in the most famous and influential architectural school of the day.

By the 1880s and 1890s, architects were viewed as artists and creators, and engineers and builders were seen more as technical wizards whose expertise lay in bridges, dams, and railways. Architects needed to have engineering training or to employ engineers as consultants on tall buildings. John Wellborn Root was the engineer/architect/designer in the famous Chicago firm of Burnham and Root, which designed the first earthquake-resistant tall building in San Francisco, the Chronicle Building, a hybrid constructed of iron, steel, brick, and wood. The Reid brothers, architects, collaborated with Charles Strobel, an engineer, to build an equally innovative building to house the *Call* newspaper. Looking at these and several other major buildings will lead us to the eve of 1906.

The First Earthquake-Resistant Skyscraper

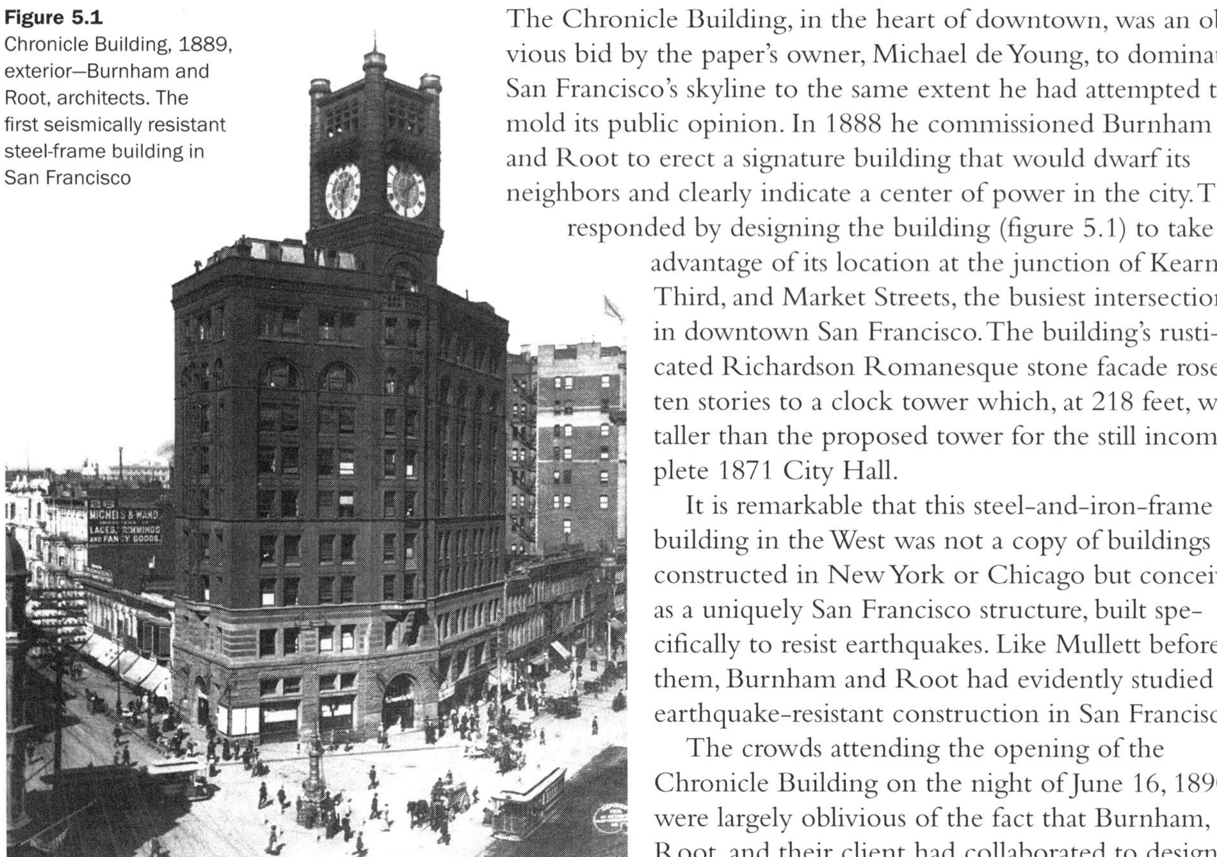

Figure 5.1
Chronicle Building, 1889, exterior—Burnham and Root, architects. The first seismically resistant steel-frame building in San Francisco

The Chronicle Building, in the heart of downtown, was an obvious bid by the paper's owner, Michael de Young, to dominate San Francisco's skyline to the same extent he had attempted to mold its public opinion. In 1888 he commissioned Burnham and Root to erect a signature building that would dwarf its neighbors and clearly indicate a center of power in the city. They responded by designing the building (figure 5.1) to take advantage of its location at the junction of Kearny, Third, and Market Streets, the busiest intersection in downtown San Francisco. The building's rusticated Richardson Romanesque stone facade rose ten stories to a clock tower which, at 218 feet, was taller than the proposed tower for the still incomplete 1871 City Hall.

It is remarkable that this steel-and-iron-frame building in the West was not a copy of buildings constructed in New York or Chicago but conceived as a uniquely San Francisco structure, built specifically to resist earthquakes. Like Mullett before them, Burnham and Root had evidently studied earthquake-resistant construction in San Francisco.

The crowds attending the opening of the Chronicle Building on the night of June 16, 1890, were largely oblivious of the fact that Burnham, Root, and their client had collaborated to design the world's first earthquake-resistant building with a skeleton of steel and iron. The *Chronicle* explained it to its readers:

> With the knowledge...that earthquakes of more or less severity may shake the city, Messrs. Burnham and Root have been obliged to make special provisions in the lofty *Chronicle* structure, and have thus added to the strength such features of construction as are not found in Eastern buildings of the same class.[8]

Like the experimental buildings going up in New York and Chicago in the late 1880s, the Chronicle Building was unique; it pushed the technological envelope. Load-bearing masonry walls are extremely heavy, and wary of adding too much weight to the ten-story building, Burnham and Root had framed the clock tower in wood. The combination of load-bearing walls and an iron-and-steel support cage was touted by the *Chronicle* article as particularly effective:

> The stonework and brickwork carry but little load other than that of the walls in the construction of which they are used. The labor is divided. The steel cage, which the walls surround, carries the weight of floors and roofs, and the walls in turn afford support and protection to the upright columns of the cage, and in this way there is mutual support and protection to the upright columns of the cage, and also to an extent independence.

In the transitional moment in which the Chronicle Building was created, Burnham and Root chose the most appropriate technological solution at their disposal. How could they build the Chronicle Building to resist the lateral forces of earthquakes, which up to this time had not been calculated? They looked to the weight of the exterior brick shell of the building as previous architects and engineers had, but they also saw that the bricks, if not bound together, would fail. This took them back to the earthquake-resistant solutions of the 1870s. They seized upon bond iron and straps, both of which had proven effective in San Francisco, wrapping them around the entire exterior of the building:

> To further stiffen and strengthen the walls, there are four flat bands of quarter-inch steel built inside the exterior walls around the entire structure. One of these bands is five inches wide and the other three are each two inches wide. They are placed at varying heights between floor and ceiling and at varying distances horizontally from the outer surfaces of the walls, no two straps being directly over one another.[9]

Having attempted to stabilize the exterior masonry wall, Burnham and Root turned their attention to the interior skeleton. The Chronicle Building's footprint looked like an irregular, bent "I." They were particularly concerned that the floors of the building would lose their integrity as planes or diaphragms, and distort and collapse the skeleton. They understood, as many architects of our own

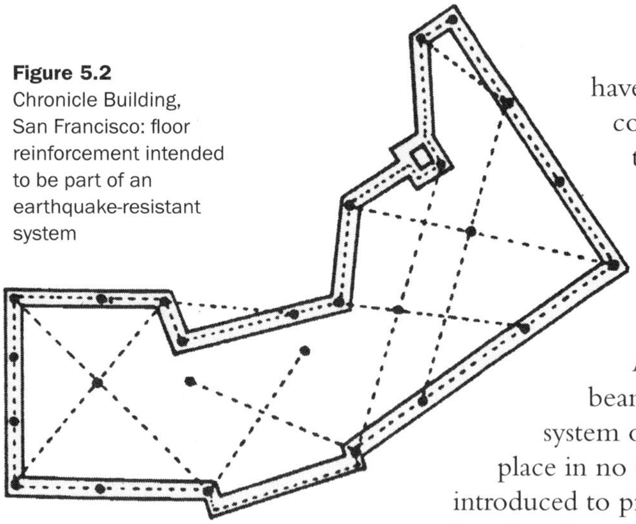

Figure 5.2
Chronicle Building, San Francisco: floor reinforcement intended to be part of an earthquake-resistant system

have forgotten, that asymmetrical, irregular shapes or configurations are dangerous in earthquakes because they complicate load paths and introduce torsion, or twisting. They chose to stabilize the interior frame by using a system of diagonal steel straps to help the building work as a unit if swayed in an earthquake (figure 5.2):

> At every floor...there is, in addition to all the steel beams and the rods necessary to this construction, a system of diagonal bracing with steel straps which find place in no other building of its class. These braces have been introduced to prevent the lateral racking and dislocation which an earthquake of great force might bring...At each intersection of one strap with another a bolt is passed through them, and at every place where a beam is crossed by a strap they are bolted together. On each floor there is thus arranged a steel net.[10]

The idea behind these experimental earthquake-resistant features, to tie the building together so it would move as a unit, was sound. But the irregular configuration of the Chronicle Building introduced a level of complexity into the forces affecting the building that could not be corrected through strapping. Within three years, Burnham and Root themselves, along with their colleagues, would invent more effective methods for combating the problems associated with lateral forces.

The *Call* Trumps the *Chronicle*

The Chronicle's hegemony over the intersection, the downtown skyline, and earthquake-resistant construction did not long go unchallenged. In 1881 the *Chronicle* began a crusade against Claus Spreckels, who had amassed a fortune by developing sugar plantations in Hawaii and shipping sugar to the San Francisco Bay Area to be refined and distributed. The paper alleged, probably correctly, that the Hawaiian plantations were virtual slave camps, and went on to claim that Spreckels had swindled the stockholders of the Hawaiian Commercial and Sugar Company. So incensed was Claus Spreckels' son Adolph that he walked into Michael de Young's office and shot him.[11] De Young survived and Adolph Spreckels was acquitted, but a deep and enduring enmity divided the two families. In retribution Claus Spreckels bought the *San Francisco Call* in 1895 and set out to beat the de Youngs at their own game.

Spreckels commissioned the construction of the headquarters for the *Call* at Market and Third, across the street from the *Chronicle*. At 310 feet, the Call Building (figure 5.3) would be 102 feet taller than the Chronicle Building, and it had a more advanced design, which must have warmed the Spreckels' hearts. They had asked a local architectural firm, James and Merritt Reid, to design their building, but for their engineer they looked farther afield and chose Charles Strobel of Chicago, one of the most famous engineers in the country. Strobel was a recognized bridge engineer and scholar, and inventor of the Z-bar. The widely used steel Z-bars could be riveted together or used independently to create extremely strong columns in buildings and bridges.[12] Together the Reid brothers and Strobel created an outstanding building that would outperform the Chronicle Building in the earthquake of 1906.

Strobel's approach to the Call Building was probably informed by the ASCE meeting of 1892, at which members discussed the effects of wind and earthquakes on tall buildings. The sixteen-story Call Building would be a narrow tower just 75 feet by 75 feet, topped by a gracious baroque dome. A tall, narrow building like this, even though symmetrical, presented specific and difficult problems to its engineer: how could it be braced to resist the wind that blew upon its surface, or earthquakes, which might whip it back and forth from below?

To solve the problem of overturning, Strobel used a foundation of steel grillage and concrete that had already been pioneered in Chicago.[13] He extended the two-foot-thick concrete foundation to the end of the lot lines, an area of 96 by 100 feet, even though the structure itself was only 75 by 75.[14] The foundation was laid nearly twenty-five feet, nine inches below street level, on wet, compact, hard sand. Next, two layers of fifteen-inch steel I beams were placed at right angles to one another and covered in concrete. On top of this, the engineer positioned twenty-eight sets of twenty-inch steel I beams to carry the structure's columns and pedestals. The major support columns were held by cast-plate pedestals attached to the very lowest of these I beams. The whole foundation, from concrete to grillage to column supports, acted as a unit, distributing the building's weight over the widest possible area. As a 1906 observer reported:

Figure 5.3
Call Building, San Francisco, 1898, exterior—James and Merrit Reid, architects; Charles Strobel, engineer

> It can be seen that not only was the weight of the superstructure, 12,000 tons, and the wind force allowed for, but that the tipping action from the earthquake wave would be restrained by the entire weight of the wide, broad, flat reinforced concrete table and its adhesion to the soil.[15]

Strobel attacked the problem of wind or earthquake forces with enormous ingenuity. The building is a tower, its floors a series of radially symmetrical polygons within the square exterior (figure 5.4). This costly solution, which makes for a jigsaw plan of bays of different dimensions, was obviously included to provide a very stiff diaphragm for the tower. It could be pushed from any side and the floor planes would resist distortion—looking at the complexity of the steelwork, it is hard not to think of a bridge engineer. Strobel went on to brace his steel frame

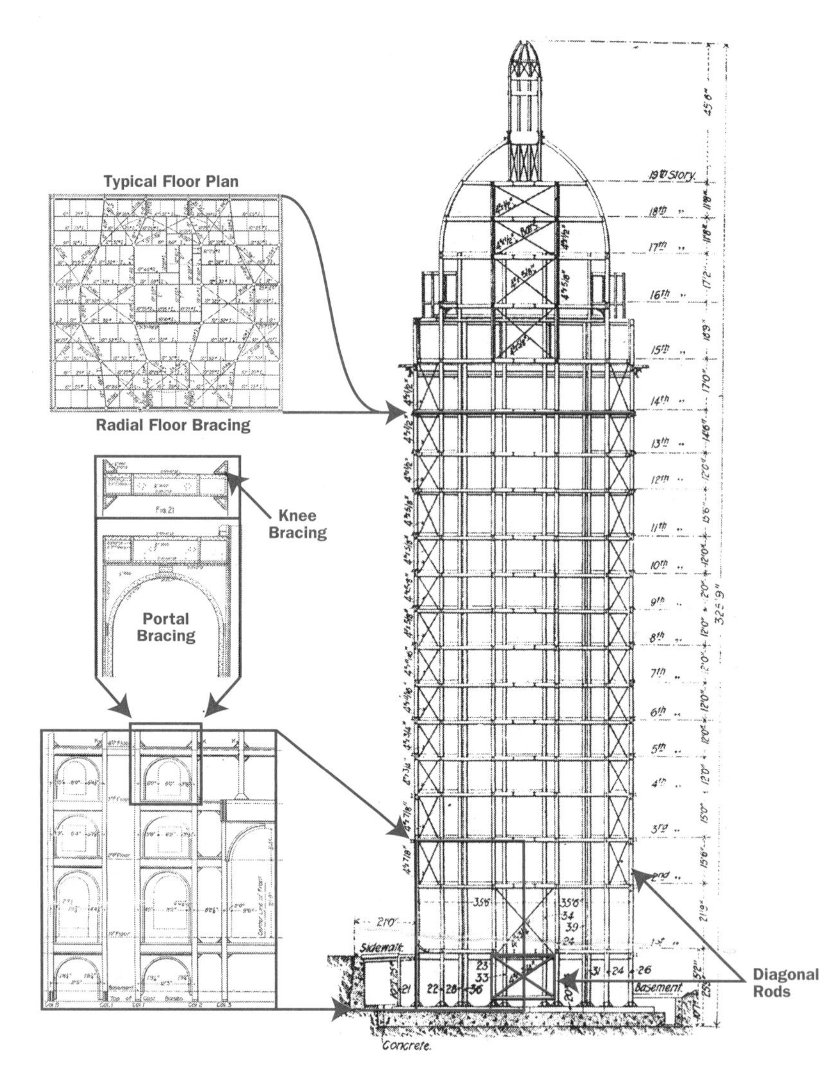

Figure 5.4
The Call Building, San Francisco, 1898. The floors are unusual, in that they are radially braced and stiffened with X-braces to ensure they work as stiff, integral diaphragms. The tower's lateral strength is assured by a robust steel skeleton reinforced with diagonal braces on each floor and in the central core, deep spandrels, knee braces, and portal braces.

with every stiffener known at the time. From the first to the third floors, he linked columns at the corners with deeply rounded fillets. These rigid portal frames (figure 5.4), as they are called, effectively resisted lateral forces. A second series of braces, perhaps more adapted to bridges than buildings, consists of eight bays of diagonal bracing per floor, from the foundation to the sixteenth floor, and three sets in the dome, in the form of adjustable rods. The rods, with adjustable turnbuckles, run from the base of one column to the top of the next, forming a series of X's. This bracing method is cheaper and lighter than portal bracing but has major drawbacks. Rods cannot counter lateral forces as effectively as portal braces because they are vulnerable to bending and distorting in shear. Even if they are effective initially, after some time, or perhaps a major earthquake, they will distort and loosen, making it necessary to retune the turnbuckle. For example, the lowermost rods which can be seen today in the basement of the Call Building are somewhat distorted—they worked hard to restrain the building in the earthquake of 1906, and perhaps in 1989.

In addition to portal bracing and diagonal bracing, the main beams of the Call Building are connected to the columns with knee braces. These are diagonal webs riveted to columns and beams, creating a very rigid connection. Last, Strobel stabilized the center bays of the building with spandrel girders—deep girders that are designed to stiffen the columns of a building.

The Call Building survived the 1906 earthquake and fire, while the Chronicle Building survived the earthquake but not the fire. In 1938 engineers calculated that the Call Building was constructed to be very stiff, with a wind resistance of 50 pounds per square foot (psf), phenomenal for its time.[16] B. J. S. Cahill, a local critic, called it the "handsomest tall office building in the world," which it probably was, until the owners decided, because of its strength, to remove the dome and add six extra stories in the 1930s, destroying its architectural integrity.[17]

A Short, Steel-Frame Building

While the Call Building was in construction, the federal government was erecting an important building on one of the most dangerous sites in the city, the former marsh at the northeast corner of Seventh and Mission Streets.[18] Why the government was building on this site we do not know; civic corruption may explain it. Whatever the reason, the architect, James Knox Taylor, understood the problem of filled land and designed the new United States Court of Appeals and Post Office with an unusual structural system. Although it is only a three-story building, its brick-and-granite exterior wall is supported by a steel frame with deep knee braces that could have been used as wind bracing on tall buildings (figures 5.5a and 5.5b). Here the knee braces are used to stiffen a relatively low building to restrain its sideways movement. These knee braces are repeated in each bay of the building and inverted at ground level. They are securely attached to a deep foundation to further control lateral drift. To keep the masonry facade intact during an earthquake, each block of stone is attached to adjacent blocks by shear keys, and straps embedded in mortar beds are tied to each column. Opulently ornamented, perhaps in competition with the 1871 City Hall, this federal building was constructed with surprising diligence in relationship to seismic safety, much like the U.S. Mint before it. Unfortunately, there were weaknesses too, which would become apparent in 1906.

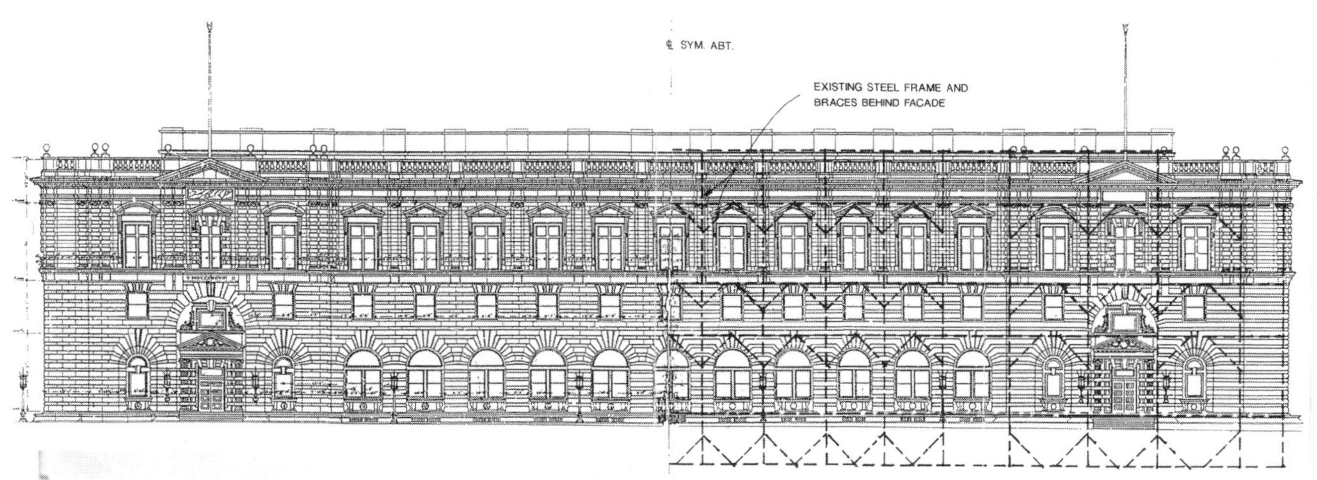

Figure 5.5a
Post Office and Court of Appeals Building, 1897, James Knox Taylor, architect: exterior facade with diagram of diagonal bracing

Figure 5.5b
Post Office and Court of Appeals Building, after earthquake and fire of 1906: although the building suffered extensive interior damage, it did not collapse.

During the same period, Burnham and Root continued to design with seismic problems in mind, as a description of the Mills Building (figure 5.6) reveals. Designed for silver baron Darius O. Mills and constructed just after the Chronicle Building, in 1891, it was also ten stories but with a completely steel frame and load-bearing walls, erected under the direction of project architect Willis Polk and advertised as being earthquake-resistant:

> The internal pinning principle employed in the Palace Hotel [was used] in the erection of a modern skeleton frame building at the corner of Montgomery and Bush Streets. [Mills] chose for his architects the Chicago firm of Burnham and Root, who designed a building particularly square...which in spite of its ten stories was believed to possess far more quality of resistance against earthquake shock than any two-story construction of redwood.[19]

Root died the year the Mills Building was finished. Burnham continued his practice. His next San Francisco building, the Merchant's Exchange (figure 5.7), was completed in 1905. Willis Polk was credited with the building's design. Burnham and Company used an all steel frame from which curtain walls hung, thus making the transition from a hybrid structure like the Chronicle Building to the modern steel-frame skyscraper.

Figure 5.6 (left) Mills Building, San Francisco, 1891, exterior, Burnham and Root

Figure 5.7 (right) Merchant's Exchange Building, San Francisco, 1903, exterior, Burnham and Co. The Safe Deposit Building (1875) is in the foreground.

Steel Frames Hit the Heights

After the unique Chronicle and Call Buildings, a flood of steel-frame structures was built without mention of earthquake concerns. These were standard issue, formulaic designs, the same as their counterparts in Chicago and New York. Some of the scores of steel-frame buildings that rose in San Francisco each year survive today: the Alvinza Hayward (now Kohl) Building (1901), the Rialto Building (1902), the Mutual Saving Bank (now Citizens Savings, 1902), the Bullock and Jones Building (1902), the Aronson Building (1903), the Atlas Building (1904), the St. Francis Hotel (1904), the Grant Building (1905), the Pacific States Telephone and Telegraph Co. Building (1905), the Shreve Building (1905), and the J. E. Adams (now Alto) Building (1905). By 1906, forty-two of these tall, "fireproof," metal-frame structures stood in San Francisco's downtown, and several others, like the fourteen-story Whittell Building, were in construction. Someone was making decisions about earthquake safety, and that expert, the quiet member of the design team, was the civil engineer. Because no documents survive to describe the decision-making process, and no critiques appeared, we must try to reconstruct what these silent engineer-designers were thinking.

In the case of the Whittell Building (figure 5.8), begun before the earthquake of 1906 and completed afterward, there is little doubt that the local architects Frank and William Shea and an unknown engineer designed it to be seismically resistant. According to structural engineers George A. Bruch and J. B. C. Lockwood, analyzing the Whittell Building after the 1906 earthquake, its major strength was its foundation.[20] Engineers knew that a seventeen-story building, over two hundred feet tall, with a small base (forty-five feet frontage by sixty feet) needed maximum depth, mass, and integrity in its foundation. Twenty-four feet deep, it consisted of a solid bed of concrete five feet six inches thick topped by a grillage of I beams, representing about 150 tons of steel. Bruch and Lockwood spoke of the building and its foundation as having the buoyancy of those small toys "which appear top-heavy but which always return to an upright position when tilted, because of the mass at the base." [21] The Whittell Building's steel frame was also heavily braced against lateral movement, with deep spandrel plate girders on the first six floors, because the maximum stress would be exerted on the stories closest to the ground. The architects and engineers of the time did not see the danger of a connection that strengthened the beam while weakening the column.

In contrast to the Whittell Building, the Flood Building (figures 5.9a and 5.9b) appears to depend

Figure 5.8
Whittell Building, San Francisco, 1906, steel frame, Frank and William Shea, architects

solely on a standard steel frame for seismic resistance.[22] Designed by Albert Pissis for a triangular lot at Market and Powell Streets, the building was started in 1901, but labor disputes delayed construction until 1904. The twelve-story triangular steel skeleton supported facades of brick and gray Colusa sandstone. Figure 5.9a shows the skeleton in construction, but no evidence of strong lateral bracing, no portal braces, knee braces, or deep lattice or spandrel girders appear in the photograph. Nothing distinguishes this from similar buildings elsewhere in the United States. The Flood Building survived the 1906 earthquake because of the good design of its massive steel frame. Most of the rivets on its first floor sheared, which easily could have meant total collapse.[23] But because the building remained standing, the rivets were replaced; it could be quickly repaired and put into service.

Figure 5.9a
Flood Building, San Francisco, 1901–1904, steel frame, Albert Pissis, architect. Although the building lacks earthquake-resistant features, the frame is very heavy and strong.

Figure 5.9b
Flood Building, exterior

The Continuing Story of City Hall

The new steel-frame buildings had an indirect but critical effect on the seismic integrity of San Francisco's most important structure in progress, the 1871 City Hall. Plagued by delays and problems, the initial "earthquake-proof" brick structure was now in danger of being dwarfed by the skyscrapers downtown.[24] Eighteen years after construction had begun, the tower had yet to be built. Officials pleaded for approval to update its old-fashioned design, despite the building's history of huge cost overruns and delays:

> The building of the mansard roof raises the City Hall over 40 feet, or one-third higher, making an imposing and massive structure...all fair persons must demand that the building with...a suitable main tower be finished, if San Francisco is to keep pace with the new era of high buildings, running from ten stories, like the "Chronicle" Building, to sixteen stories high, as in New York, Chicago, and elsewhere.[25]

New plans for a square, domed tower rising 335 feet from the ground were challenged on both structural and aesthetic grounds.[26] When critics attacked the proposed new height, city commissioners asked the local chapter of the American Institute of Architects (AIA), as well as six engineers, to discuss the merits and debits of the high square tower versus a high drum and dome. The AIA liked the idea of a high tower but decided it would look better if it were round (figure 5.10). Engineers testified that the high drum and dome would have greater structural integrity. Calvin Brown, a civil engineer, reported that it was "more than ample in strength to sustain the weight which could be placed upon it, as also to resist the force of winds and earthquakes."[27]

Figure 5.10
San Francisco City Hall, begun 1871, as it appeared in 1906, Frank and William Shea, architects: an altered photograph that illustrates both the new dome and drum from the exterior, and an interior section showing the steelwork

It is rarely advisable to change a building's plans in mid-construction and, indeed, we can see a number of flaws in the 1871 City Hall's new high drum and dome—starting with the fact that the foundation was designed to support the smaller, square tower. Architects and engineers had to design an entirely new iron frame for the dome, within strict shape and size limitations. Not only did this produce a structure less heavily reinforced than comparable domes;[28] it also appears that certain connections between the original building's load-bearing masonry and the new iron dome were badly engineered. Finally, the new design called for diagonal bracing that was, inexplicably, never installed. Even with this bracing, the structure might not have withstood earthquakes. Without it, disaster was almost inevitable. The dome even "foretold" its own demise during the 1898 earthquake.[29] The quake weakened its masonry cladding, a large number of the terra-cotta plates forming the dome's outer wall fell about one hundred feet to the pavement, and a well-defined crack extending about one-third of the way around the dome indicated that its lateral bracing was deficient. Yet the city architect failed to raise the alarm, perhaps because of the rampant graft of the era, or perhaps because he misunderstood what the cracking might mean.

What Engineers Knew

By the 1880s, engineers were expected to specify the construction material, the construction method, and the required strength of the members and connections in the steel frames they were designing. To do this, they relied on recent advances in structural understanding and in mathematical modeling that greatly simplified the process of calculating loads and stresses on bridges and tall buildings.[30] Using Navier's 1826 solution to the problem of bending—the elastic theory of structural design—it was possible to use data from laboratory tests to calculate the maximum permissible or working stress of steel members. Safety factors against failures could then be stipulated numerically in building codes.

The first law regulating the working stress for wrought iron on railroad bridges was fixed by England's Board of Trade in 1840. It was W. J. Macquorn Rankine who in 1866 further defined the safety factor as "the ratio of the ultimate strength of the material to the maximum stress permissible under the action of the actual or working loads acting on the structure." In his calculations Rankine for the first time differentiated between live and dead loads (defined in chapter 4). In 1858 Thomas Tredgold published solutions to column design. In 1875 Otto Mohr published a new way of calculating the strain on rigid frames, using an abstract concept from an 1864 paper written by Clerk Maxwell which Mohr applied to a structural problem. With the Maxwell-Mohr method and the method proposed by Alberto Castigliano in 1873, engineers had at their disposal the equations to calculate how much force it took to deform a steel frame.

After the 1890s, the topic of earthquake-resistant construction in San Francisco disappears from architectural records and from the press releases of well-known architectural firms. Newspaper notices of construction starts mention nothing about earthquake-resistant construction. Engineers are rarely mentioned. Yet dozens of engineers, either steel company consultants or the partners of architects, were deciding just how safe the new tall buildings were going to be. How were they deciding? Wind force calculations provide the most parsimonious explanation.

As engineers designed ever taller frames, and as these frames came to support exterior walls as well, calculation of lateral loads became mandatory. Because earthquakes were so peculiar to San Francisco, they were initially disregarded in national discussions, but engineers knew that tall buildings were affected by wind and had to be built to resist and dissipate its force (figure 5.11).[31] This lesson had been hard-learned from a number of catastrophic bridge failures partially caused by underestimating the force of winds. In 1759 John Smeaton, engineer of a successful lighthouse in stormy Eddystone, England, proposed to the Royal Society that the maximum lateral pressure on a building would be 6 pounds per square foot for high winds, 9 psf for very high winds, and 12 psf for storms. Sir George Ayrey, the royal astronomer, was more cautious, stating that sometimes the wind pressure might reach 40 psf, but he then diminished this figure, deciding that a planar surface like that of a bridge might be subjected to only 10 psf. These low figures led Sir Thomas Bouch to underestimate the wind forces that would act on his Tay Bridge in Scotland, the longest in the world at the time of its completion in 1878. During gale-force winds in 1879, a three-thousand-foot section of the bridge collapsed, dumping a passenger train into the waters below. The resulting inquiry in 1880 revealed deficiencies in workmanship and materials, but more importantly ignited a controversy over wind forces.

The first paper discussing wind bracing for tall buildings in the United States appeared in the *Transactions* of the American Society of Civil Engineers (ASCE) in 1892. It ushered into scholarly debate a question that was already

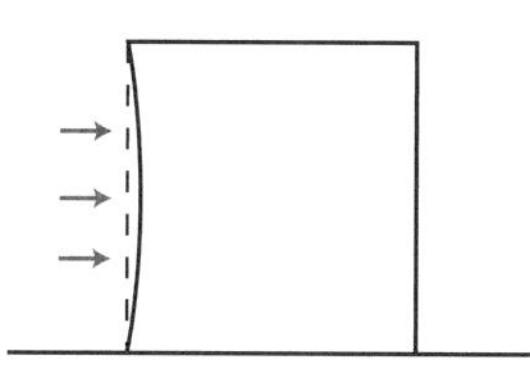

Direct Pressure

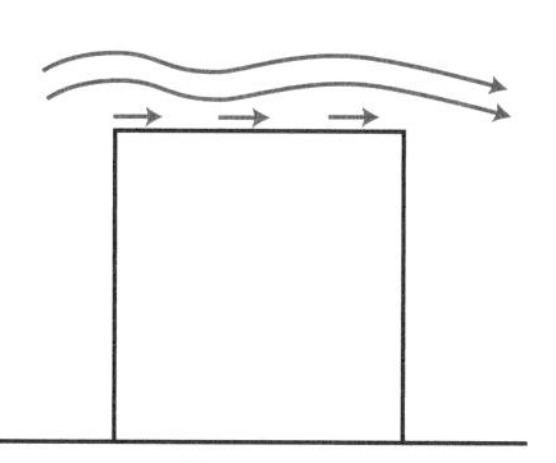

Drag

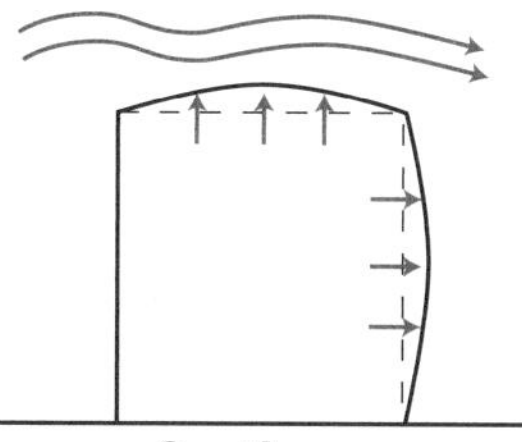

Suction

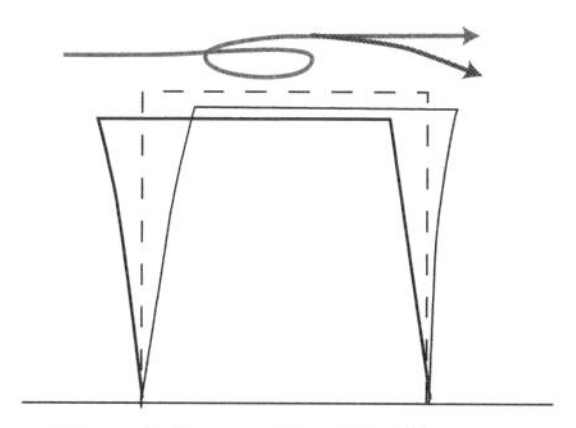

Rocking, Buffeting

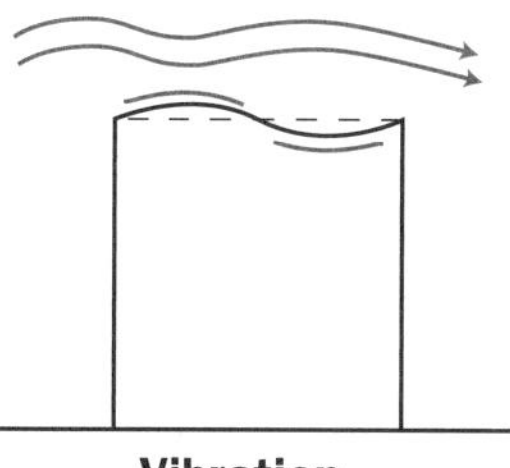

Vibration

Figure 5.11: Wind Effects on Buildings

Wind is air that is moving, that has a particular velocity and moves in a given direction. When wind encounters a stationary object, it exerts force on that object. The simple measurement of that force, adopted in nineteenth-century building codes, was the static force of wind blowing upon a surface: pounds per square foot (psf). This is still the standard today.

High wind velocity is associated with windstorms that can damage buildings—tornadoes, hurricanes, and local conditions like the Santa Ana winds of southern California. Since high wind velocity was common in windstorms in New York and Chicago, 30 psf became the standard wind force for tall buildings. In San Francisco, where there are no such high wind velocities, the allowable modern wind pressure is 20 psf; building to pressures above 20 psf is unnecessary for Bay Area wind conditions.

Wind damages buildings because it exerts direct positive pressure on surfaces facing the wind (direct pressure diagram). Wind does not stop when it strikes an object, but flows around it, causing aerodynamic drag on the surfaces parallel to the direction of the wind (drag diagram). As the wind passes over and around surfaces it can create a suction effect, pulling surfaces outward on the leeward side, that is, opposite from the direction of the wind (suction diagram). These effects combine to create rocking, or buffeting (rocking, buffeting diagram). If the period of vibration of the building as a whole or any of its parts matches the velocity of the wind, harmonic effects can result, causing vibration (vibration diagram). When wind blows over a building surface quickly and encounters an obstruction like a chimney or canopy, these can be swept away in a "clean-off" effect. Diagrams and phraseology based on James Ambrose and Dimitry Vergun's *Simplified Building Design for Wind and Earthquake Forces* (New York: J. Wiley, 1997) being discussed within the profession: how many pounds of lateral pressure per square foot should a building be constructed to withstand? Somewhere between 30 and 40 psf were generally considered sufficient. One engineer, Henry H. Quimby, wrote:

> The forces which we have to deal with here are not, as in other departments, limited and known, and the practice must be, to some extent, empirical. We do know that the wind sometimes develops an energy which must be far beyond what has ever been measured; the efforts of investigators by means of artificial and confined currents of air have failed to obtain velocities and pressure sufficient to account for some of the feats of the natural article. In view of the constant liability of any locality to the occurrence of wind storms of destructive violence and unknown force, every tall building should be assumed to be subject to a wind pressure of 40 pounds per square foot of exterior wall surface, and be braced to resist this with iron or steel rods or stiff braces.[32]

George A. Just, representing those who thought 40 psf "hardly necessary," commented that structures designed with no thought for wind pressure had survived:

> Buildings have been erected in this city [New York]—and, no doubt, elsewhere—with such a disregard to resistance to wind that the conclusion seems inevitable that they remained intact only by the operation of forces that are ordinarily ignored by engineers in their calculations.[33]

He then concluded that 30 psf "will answer the requirements of safety."[34] A majority of engineers agreed, and 30 psf became the standard code specification at the turn of the century throughout the United States.

Wind and Earthquake Forces Erroneously Equated

Today engineers design differently for wind than they do for earthquakes, taking into account factors that were barely considered in the nineteenth and early twentieth centuries. For instance, the fact that earthquakes exert movements and therefore forces from the foundation up, rather than—as wind does—from the top down, calls for significantly different compensations (figure 5.12).[35] Increasing the building's mass will help it resist the wind but make it more dangerous in an earthquake. Still, because both wind and earthquakes exert lateral forces on buildings, some reinforcement strategies are helpful in both situations. In fact, a major topic of discussion at the 1892 ASCE meeting was the wisdom of earthquake-proofing a building with special bracing designed for wind pressure. Quimby's discussion of the problem in 1892 illustrates a profound understanding of earthquake-resistant methods that still seems valid today. He proposes that some form of bracing be added to the steel frame:

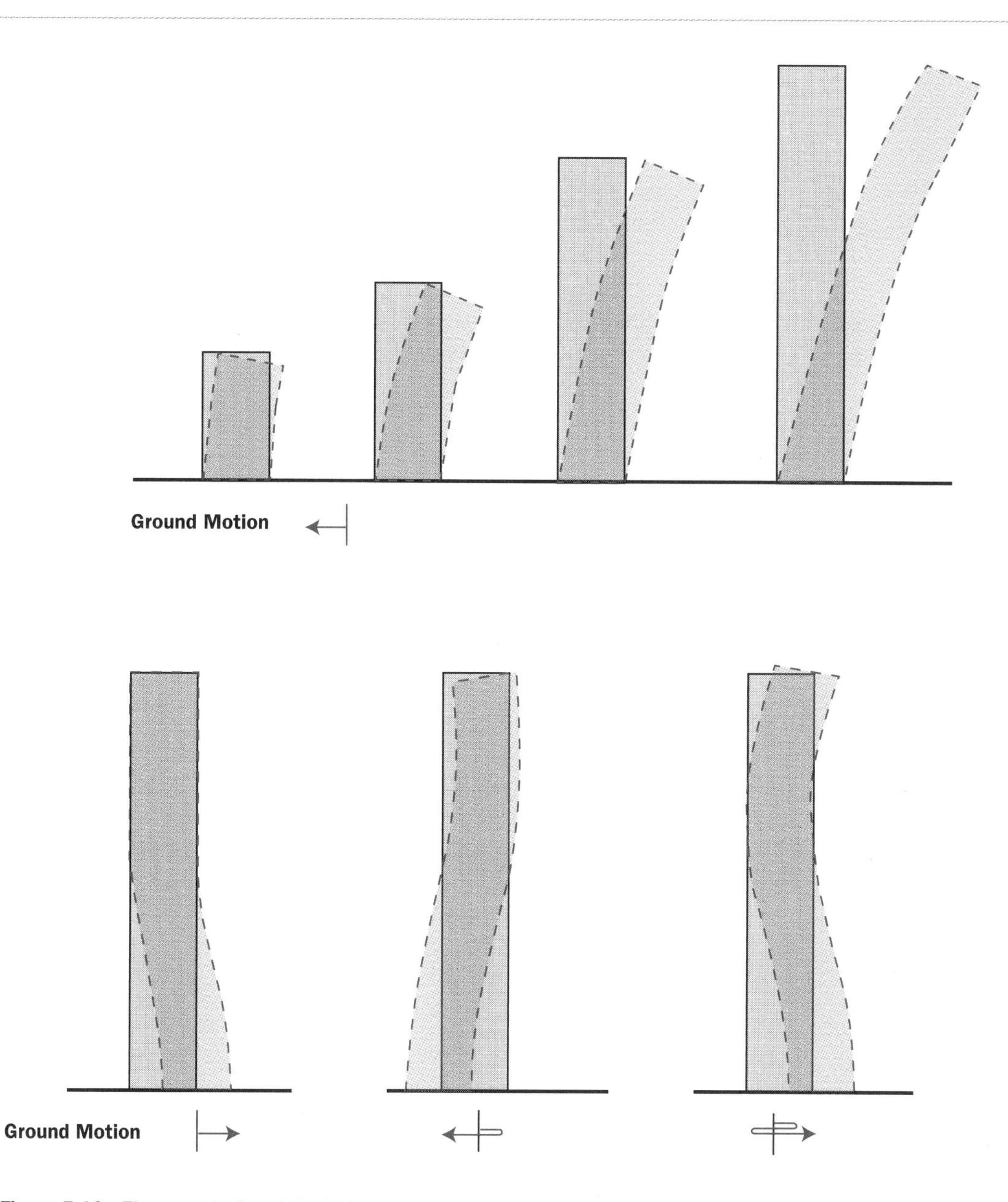

Figure 5.12 : The complexity of designing tall buildings to survive earthquake ground motion.

Fundamental Period (top): Unlike desiging low and midsize buildings for wind, designing for such buildings in earthquake country requires engineers to consider, among many factors, the building's flexibility or rigidity in relation to ground motion, and its natural period of vibration. The period of vibration of a building is the time it takes to oscillate back and forth to the postion it started from. Low masonry buildings can be very stiff and have short periods, while tall flexible steel frame buildings have long periods. A four-story building can sway in a half-second period, whereas a ten- to twenty-story building's period might be one to two seconds. As the building sways back and forth, it must be strong enough and stiff enough to not deflect or drift so far sideways that its structural members, or nonstructural components like cladding and windows, fail. Yet it must be flexible or elastic enough to bend and dissipate energy and return to its original form without structural damage. Further, it must be designed so that its natural period does not match the period of the earthquake, which would cause destructive resonance (see Sather Tower discussion in chapter 13).

Higher Modes (bottom): In a tall building, the fundamental period may be so long that the ground motion may excite the building to higher modes of vibration in addition to the fundamental mode, causing the building to sway back and forth in a complex whiplash effect, with different parts of the building moving in opposite directions at the same time.

> The only absolute security against danger from this source [earthquakes] is in a system of bracing with some elastic material of positive strength that will so unify a structure that it will hold together even to the point of being overturned bodily...A reasonable degree of security against injury from any of the causes mentioned can be obtained with the...provision for wind strains, care always taken that all details, fastening of bracing, etc., are equal in strength to the main members and competent to properly transmit components of stress.[36]

Another civil engineer, J. P. Snow, thought less emphasis on bracing was appropriate and advocated steel frames alone. If the steel frames were properly designed and all parts were well connected, they would be sufficient in even the tallest buildings, as long as there was "a systematic reduction of weight from the bottom to the top; care being taken to avoid all heavy cornices and projecting front ornaments at the top of the building."[37]

The simplest and most persuasive, though ultimately erroneous, argument was that of J. Foster Crowell:

> Speaking, of course, of buildings in this part of the world...if a building is secure against the combined action of gravity and of the high-wind forces of this vicinity, there need be no apprehension that it will not withstand earthquakes that are likely to occur in these latitudes...I think [earthquake danger] is a good deal of a bugbear, a view which I think is borne out by the experience of the San Francisco builders, whose more modern constructions resemble those in other localities, with rarely any special consideration for earthquakes in their design.[38]

Low-Risk Low Buildings?

It was the practice of engineers at the time not to brace buildings under four stories tall for wind. Architect William Birkmire published this rule of thumb, typical of the time: "A building whose height does not exceed three times its base, and which has a well-constructed base, scarcely needs a special system of wind bracing to secure it."[39] Because earthquake forces were equated with wind forces, buildings of this size were no longer included in seismic efforts.

While San Francisco engineers did not all design similarly, it is likely that they based the calculations they gave architects on two premises: that when they talked about wind forces they included earthquake forces; and that these forces should follow the national standard, 30 psf. They no longer considered San Francisco's earthquake-prone site to be important, because the engineering profession as a whole did not identify it as a problem. Standardized steel frames could therefore be used in a formulaic manner that was assumed to cover seismic concerns. In fact, there no longer seemed to be any need to talk about earthquake forces even on the tallest buildings, because calculations for wind forces, including lateral drift and resistance, seemed to include all the necessary precautions.

Engineers were not concerned with vernacular architecture, which was seen as the province of architects and builders. They were not interested in the old brick technology. The cutting edge was steel-frame construction: the tall, expensive buildings that were redefining downtown. As if by sleight of hand, "wind forces" silenced the earthquake debate among engineers, and earlier experiments with earthquake-resistant design in brick seemed so antiquated as to be irrelevant.

The Late-Nineteenth-Century Earthquakes

On March 26, 1884, San Francisco experienced two shocks, the most violent since 1868.[40] The first was the worst, consisting of three tremors: "an upheaval, and settling back, and wavy vibration."[41] It was felt throughout the city, although its intensity varied according to location. On the hills the motion was slow and rocking, but on the "made ground" from Montgomery Street to the bay it was strong, probably MMI VII. Reporters who described the shock on made ground wrote that it made the telegraph wires slap together down Montgomery Street, nearly overturned acid batteries in the Western Union Building near the old City Hall, and severely rattled the windows in the old City Hall itself. The heaviest damage, including a few cracked brick walls, occurred at the foot of Market Street and in the South of Market, on made ground.

The next quake, on February 29, 1888, was very small. Two months later, on April 28, a heavy earthquake hit Colusa County, in the Central Valley.[42] A year later, on July 31, 1889, a strong shock hit San Francisco, and a similar temblor occurred the following year, on April 24, 1890. Neither quake did much damage; an 1890 *Chronicle* headline summarizes the common feeling: "Earth shaken—but as usual no one hurt." [43]

The next earthquake, on April 21, 1892, had its epicenter in the northern portion of the Sacramento Valley. Many buildings failed in the small towns of Winters and Vacaville, probably because they were badly constructed, with none of the earthquake-resistant features used in San Francisco. *The Morning Call* reported:

> The fact is...that brick buildings in the interior are, as a rule, very flimsy affairs. Their foundations rest almost on the surface: There is an absence of iron rods to hold them together, such as are common everywhere else.[45]

The Chronicle Building and other steel-frame structures did well in this earthquake. One insurance expert wrote an article contrasting San Francisco's tall buildings with some allegedly "flimsy" structures in the state's interior:

> Here we have been erecting lofty buildings and several of these are still incomplete. What was our experience during recent quakes? Not one of these high buildings was in the least affected. They are built in the modern

fashion, braced as strongly against the elements as are the large business structures of the leading Eastern cities like Chicago and New York, but more strongly.[46]

The earthquake of April 1, 1898, provided another opportunity to assess the newest tall buildings. According to an editorial in the *Call,* their own building rode out the earthquake quite well. Tall buildings were again wholeheartedly endorsed as the answer to earthquake problems:

As a matter of fact, the new style of architecture is about the safest that could possibly be adopted to resist the shocks of earthquake. A one-story adobe house would go to ruins before one of the skyscrapers...It has been years since we had a shock equal to that of Wednesday night...Even if such shocks came every year they would do no harm to modern structures. The steel skyscraper is proof against any danger of that kind, and hereafter there will be hardly a doubt of it in the minds of even the most timid on the subject.[47]

Newspapers and Boosterism

By 1892, newspapers had become increasingly dismissive of seismic danger. Reporting ranged from wishful thinking to the absurd. Even the more sober pieces managed to minimize earthquake danger while simultaneously listing earthquake damage. On April 23, 1892, the *Chronicle*'s headline read, "Damage from Earthquakes Was Exaggerated," with this lead sentence: "No Californian proud of his state likes to hear it spoken of as famous for earthquakes."[48] In a similar vein, the *Call* ran an editorial titled "The Earthquake Bogie":

Newspapers which are filling their columns with exaggerated accounts of the recent earthquakes are engaged in poor business...The simple truth is that the temblor of Wednesday and Thursday was no worse than the gentle earth waves of which hundreds or more are recorded by seismometers every year. It is very thoughtless to describe it in the language which was fitting to paint the Japan earthquake of last October. Such accounts do the State no good in the East and create an uneasy feeling at home. Winters and Vacaville owe no thanks to the newspapers which represent them as scenes of desolation and ruin, when in fact, a few carpenters and glaziers will efface all traces of the catastrophe in a couple of weeks.[49]

In fact, local newspapers featured few such "exaggerated accounts," choosing to protect California's image. A *Chronicle* editorial illustrates the tone found in most papers of the time:

> When there comes to be an assessment of damages caused by the "quake" in Vacaville, Winters, Woodland, and Dixon and other places, it is safe to say that the first estimate will have to be divided by two to reach the exact results and if what had occurred shall teach the interior to build substantially for the future the lesson will not be a very dear one. Nobody was killed and very few persons even bruised.
>
> It is bad enough, no doubt, but has anyone stopped to reflect that every week in one or more of the great cities of the United State there is a fire that causes more loss of property by far than this earthquake, and not infrequently loss of life as well, yet people do not flee from the cities nor give up business for fear of a fire.[50]

The same piece concludes with dizzying logic and not a little ambivalence that although earthquakes are frequent, we should not think of them as characteristic of California; we can't ignore them, but we can't accept them either:

> There have been, says an old resident, 250 shocks of earthquake felt in this city during the past forty years, and out of all those there has been but one that did any material damage, and that was nearly twenty-four years ago. It is folly to think of California as an earthquake country because we have a shock occasionally...We...need to exercise a little common sense after the earthquake is over...and while we cannot and should not ignore such occurrences...we certainly need not plume ourselves on them as though they were as much a part of California as our climate.

Even when damage really did occur in San Francisco, the papers did not sound a warning. On June 20, 1897, nine years before the 1906 earthquake, a strong MMI VI quake hit San Francisco, cracking tiles on the 1871 City Hall dome and severely shaking St. Ignatius Church, the Palace Hotel, and the Occidental Hotel.[51] The April 1, 1898, earthquake caused major destruction at the Mare Island Station in Vallejo. Probably as high as MMI VIII, this was in some ways a rehearsal for the earthquake of 1906. The shock damaged many structures in San Francisco that would be damaged again in 1906 and set off several fires.[52] The *Call,* however, took a jocular, almost giddy tone in reporting these events: "Everyone in San Francisco talked about the earthquake yesterday. Laughable were some of the incidents of the shocks, which seem to have caused more merriment than damage in this city."[53]

On the Eve of Disaster

Although the newspapers reported on earthquakes in detail, not a single editorial advocated inspections, better construction practices, or the enactment of new codes. The New Washington Hotel at Fourth and Harrison Streets, built on fill, suffered a shock in the 1898 quake that jammed the main doors. The Palace, also near the border of the old bay marsh, was so badly shaken that many eastern guests wanted to leave, and the Folsom Street wharf, built on mud in the bay, fared little better. Hundreds of dishes broke at Deiffenbacher's Store on Front Street, constructed on the Yerba Buena cove fill. The same building had also done poorly in 1868. Across town, the Concordia Club, at the corner of Van Ness Avenue and Post Street, "sustained as much damage as any building in town." All of its chimneys needed to be replaced; one was "twisted nearly at right angles from its original position," wrenching out a heavy granite cornice.[54] This structure would be ruined in 1906.

Ironically, a *Call* editorial after the 1898 earthquake stated, "California is particularly blessed... Its earthquakes are not frequent... And a violent one, such as that of Wednesday night, can be classed as nothing more than an interesting experience, doing less damage than the sudden April shower."[55] Such boosterism may have been disingenuous, but it is not entirely unexpected. No one had been killed in San Francisco earthquakes in the 1890s, damage was minimal, and engineers were confident about the strength of reinforced brick buildings and the new steel-frame skyscrapers. Editorial writers were not prescient; they clearly could not have divined that catastrophe loomed eight years ahead.

At this point in a morality play, the flaws of the main characters would have been drawn in order to implicate them in the catastrophe to follow. Could San Francisco's citizens and professionals be held responsible for their fate in April 1906? Perhaps they could, to the extent that the water system suffered from government neglect and the 1871 City Hall showed signs of weakness; but given the technology of the day, they could not be accused of ignoring earthquake and fire dangers in building construction. Idiosyncratic buildings like the Post Office and the Chronicle and Call Buildings, as well as the formulaic steel-frame structures, had all done well in the earthquakes of the 1890s. Engineers had been reassured that building to a wind force of 30 psf, a standard practice, was sufficient to address earthquake forces. As far as they knew, well-built, tied brick buildings would perform satisfactorily, as would the new, preferred steel-frame technology.

CHAPTER 6

What Really Happened in the Great Earthquake of April 18, 1906

At approximately 5:12 a.m. on Wednesday, April 18, 1906, the tectonic plates along the San Andreas Fault slipped, releasing a surge of energy from a hypocenter close to San Francisco.[1] The exact origin of this famous earthquake is still somewhat uncertain, but new studies have placed the epicenter out to sea, just south of the Golden Gate near the coast of Daly City.[2] Seismic waves radiated from the hypocenter, as great masses of rock on the Pacific Plate moved northward in relation to the North American Plate in what geologists call a "right-lateral strike-slip motion," causing a 270-mile section of the San Andreas Fault to rip apart at about 1.9 miles per second.

From the hypocenter the fault tore north through Stinson Beach and Bolinas up to Olema and on to Shelter Cove in Humboldt County, causing offsets as much as sixteen to twenty-one feet. Simultaneously, it ripped south through San Francisco to San Juan Bautista, and close to Hollister and Salinas. Because the earthquake had a right lateral displacement and not an upthrust, it did not generate a significant tsunami. However, Captain Svenson of the schooner *John A. Campbell*, sailing one hundred and fifty miles west of the Golden Gate, felt the ship shudder as if it had hit a shoal.[3]

The duration of the shaking varied from place to place. According to investigator Harry F. Reid, the earthquake began with a fairly strong movement that continued with increasing strength. Very violent shocks followed this, and quiet was restored about three minutes later. One reliable witness, Professor A. O. Leuschner, director of the observatory at the University of California, Berkeley, estimated that the shaking gradually became more violent for forty seconds, stopped for ten seconds, and then resumed more violently than before for another twenty-five seconds.[4] The Modified Mercalli Intensity in San Francisco varied

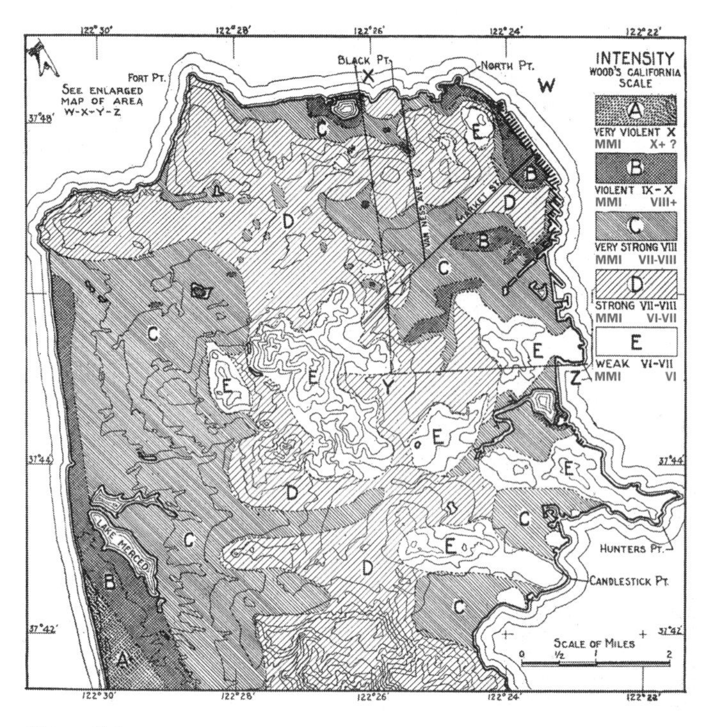

Figure 6.1
Map of San Francisco illustrating the intensity of ground shaking on April 18, 1906, drawn by Harry O. Wood and redrawn in black and white by John R. Freeman in *Earthquake Damage and Earthquake Insurance.* Wood invented his own San Francisco intensity scale, which he used for this map drawn for the 1908 report of the State Earthquake Investigation Commission. The equivalent Modified Mercalli Scale (MMI) appears below Wood's scale. This map was important to the history of seismology because it correlated shaking intensities with geological conditions. Violent shaking (B) occurred on fill, whereas weak shaking (E) occurred on rocks. The commission estimated that very violent shaking (A) only occurred south of Lake Merced (lower left).

from MMI VI on hills to MMI VIII+ on filled ground.[5] Present-day seismologists, trying to reconstruct the earthquake, have downgraded it from a Richter magnitude of 8.25 to a lesser magnitude between 7.7 and 7.9.[6]

The earthquake of 1906 has been viewed as a catastrophe of mythic proportions. Coupled with the ensuing fire, the earthquake has been variously seen as cleansing San Francisco of graft, corruption, and immorality, as in the 1936 MGM epic *San Francisco*, starring Clark Gable and Jeanette MacDonald, or unleashing the forces of racism and bigotry, as in Philip Fradkin's book *The Great San Francisco Earthquake and Firestorms of 1906: How San Francisco Nearly Destroyed itself* (2005).[7] Estimates of the destruction wrought by the earthquake in San Francisco and the loss of life it caused have climbed and continue to do so without an end in sight. While earlier estimates placed earthquake damage at around 5 percent of the total, compared to 95 percent from fire, the late Karl Steinbrugge, who tracked earthquakes worldwide, raised the percentage to 20 percent.[8] Gladys Hansen, every year accumulating more data on anyone who in any way was impacted by the earthquake or fire, in 1987 hypothesized that more than fifteen hundred people had died in San Francisco because of the combined catastrophe.[9] With no corroborating published documentation of any kind, various writers have now set the number of casualties at "perhaps" ten thousand.[10]

In order to place the damage to San Francisco in context, let us consider another earthquake in which catastrophic damage and loss of life occurred. On December 28, 1908, a major earthquake, estimated to have been magnitude 7.5, struck southern Italy, destroying the twin cities of Reggio Calabria, on the Italian mainland, and Messina, across the Straits of Messina on the coast of Sicily.[11] These two cities were, like San Francisco, seaports poised on the edge of continental plates. At two hundred thousand, Messina and Reggio's populations were half that of San Francisco. The people of Messina and Reggio, like those of San Francisco, were sleeping when the earthquake hit. The difference is that San Franciscans were sleeping in wooden houses and buildings of brick and reinforced brick. The Italians in Messina were sleeping in unreinforced stone masonry buildings. Looking over the hills of Messina after the earthquake, we see a city in which 98 percent of the dwellings were destroyed (figure 6.2). Looking at a similar view of San Francisco after the 1906 earthquake, it is hard to see the damage. It is estimated that more than seventy-two thousand people died in the rubble of the unreinforced masonry buildings in Messina, Reggio and their hinterlands—sixty thousand in Messina itself, about 42 percent of its urban population.[12]

Figure 6.2
The city of Messina after the earthquake of 1908

The official San Francisco Board of Supervisors' death toll for both earthquake and fire in San Francisco was 478 fatalities, which may have been significantly less than the actual deaths. In 1987 Hansen estimated that 1,498 people were killed in the earthquake and fire.[13] Of the total she tallied 69 people officially listed as earthquake deaths and an additional 427 people crushed to death. There may be victims in other categories, like asphyxiation, but these cannot be proven to have been caused by the earthquake, as opposed to the fire. Even if we double Hansen's estimate to a thousand dead, San Francisco's population of four hundred thousand suffered an earthquake mortality rate of less than 1 percent, in comparison to Messina's 42 percent.

Any loss of life is regrettable, but in relation to the buildings of Messina, a majority of the buildings of San Francisco survived the earthquake well, and for the most part did not kill San Franciscans as they slept. In other words, because most people were indoors, they survived—which is remarkable, given the high casualty rates caused by collapsing buildings in recent earthquakes around the world. By saving more than 99 percent of San Franciscans from death, most of San Francisco's buildings can be said to have performed satisfactorily in relation to present-day codes. Current California law, Section 1626 of the Building Code of 2001, states, "The purpose of the earthquake provisions herein is primarily to safeguard against major structural failures and loss of life, not to limit damage or maintain function."[14] Even if a building is unsalvageable after an earthquake, it will have performed satisfactorily according to present-day code if it has not killed its inhabitants or by presumption anyone in the street.

After the 1989 Loma Prieta earthquake, engineers swarmed over San Francisco. They used a standardized system to indicate building damage: green tags meant the building could be entered, yellow meant it could be entered with caution because it was hazardous, and red meant that the building could not be entered, because of past or impending structural failure. No one tagged the buildings in 1906. No quantitative survey of damage exists. How can we establish some measure of success or failure from the distance of one hundred years? We have to rely on external visual assessment of gross damage. Has the building collapsed or fallen in on itself? Even this kind of assessment is limited, because external visual assessment cannot, obviously, measure internal damage. Inhabitants can be killed when internal walls collapse or furniture falls. Yet in order to assess earthquake damage, collapse is our measure of total failure.

Building damage or collapse in the San Francisco earthquake can only provide an estimate. Neither local nor out-of-state engineers could get to the scene of the earthquake promptly enough to study single buildings in detail. Within hours, fires blazed through the downtown and South of Market, burning evidence as they went. Photographers scurried out to take pictures as fast as they could, but in those early hours they saw "disaster," not careful documentation, as their subject. Only by chance are their photographs useful for studying structural issues. But by using a combination of documentary evidence—eyewitness accounts, experts' analyses after the fire, and contemporary photographs—it is possible to evaluate how buildings of various types and materials performed, and to reconstruct damage patterns throughout the city.

Three Eyewitness Reports from the Street

Eyewitness accounts are anecdotal and idiosyncratic, but nonetheless a priceless source.[15] What follow are several accounts of the earthquake of 1906. The first three are from people who were standing on streets in San Francisco. The next four were inside buildings (see figure 6.3 for their locations). After the eyewitnesses give us their description of what happened, we will see for ourselves by taking a virtual walk through San Francisco and using photographs taken before fire obscured the earthquake damage. The italics in the following descriptions are mine.

The first of the eyewitnesses is Jesse B. Cook, a patrol sergeant with the police department who was in the wholesale district in the middle of downtown San Francisco, on the filled ground of Yerba Buena cove, when the earthquake struck.[16] Cook was standing in front of A. Levy and Company's store at the northwest corner of Washington and Davis Streets, talking to Sidney Levy, when a nearby horse (hearing the sound of P waves not yet audible to humans) began to whinny and paw the ground. Cook told Levy to talk to the horse and quiet him down. Just then, the men sensed the approaching earthquake. As a deep and terrible rumbling noise grew nearer, they understood immediately what was happening: "Gee!" Cook said to Levy, "It's an earthquake, and it's a dandy! Listen to it!" Cook then looked up Washington Street toward Russian Hill, ten blocks to the west. He could actually see waves approaching.

> The whole street was *undulating*. It was as if waves of the ocean were coming toward me, and billowing as they came. The houses were nodding and bowing to each other across the thoroughfare, and these antics were approaching more quickly than you could run. The houses were taking up the motion like cards in a children's house of cards when you spill them from one end to the other.
>
> I knew then that the place I was then *standing upon was made ground and therefore dangerous in an earthquake*. It was all made ground thereabouts, and the high Levy Building under which we stood was likely to collapse if the quake was a severe one.
>
> Therefore I hurriedly glanced about for a safer refuge. Diagonally across the street, on the southwest corner of Washington and Davis, was a one-story building, another produce merchant's place, which I felt would be safer. I darted for it on the run, but before I reached the middle of Davis Street the earthquake overtook me.
>
> Davis Street split right open in front of me. A gaping trench that I think was about six feet deep and half full of water had suddenly yawned

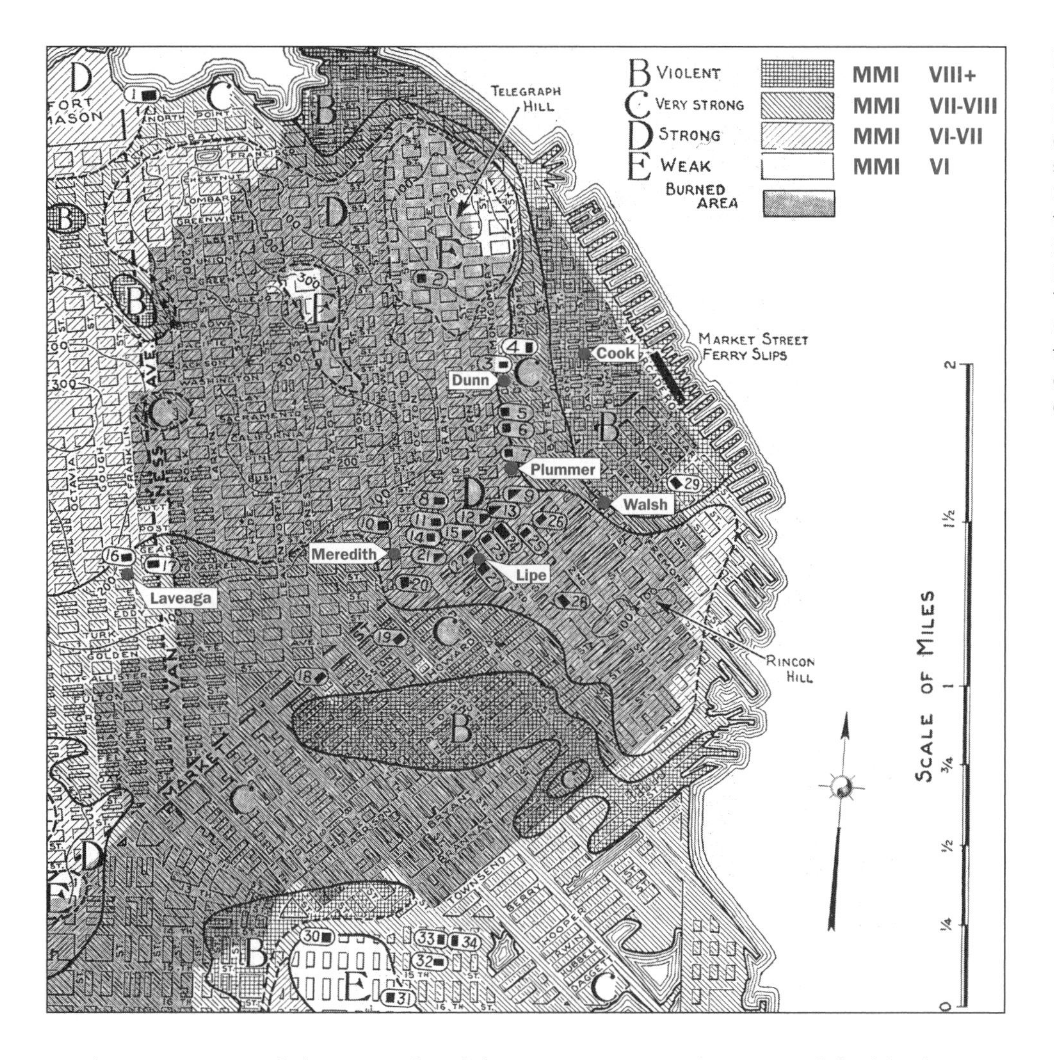

Figure 6.3
Map of eyewitness locations in San Francisco on April 18, 1906, superimposed on Wood's map. Wood's intensities from B to E are expressed in the Modified Mercalli Intensity Scale (MMI). The numbers refer to buildings that, according to Freeman, survived the earthquake with little or no damage.

between me and the east side of the street. It seemed to extend for blocks along the thoroughfare north and south. But I had no time to judge how far it ran, nor how deep it was. I simply took it at a running jump, and sprang up on the sidewalk at the southeast corner while the walls of the building I had marked for my asylum began tottering. *Before I could get in the shelter of the doorway those walls had actually fallen inward.* But the stack of produce that filled the place prevented them from wholly collapsing. They just seemed to lean in, and the structure did not fall.

[The second shock] came in a couple of instants. It was wholly unlike the first shock, fiercer and sharper. The ground seemed to twist under us like a top while it jerked this way and that and up and down and every way.

Opposite, on the southwest corner of Davis and Washington Streets, we could see the building that stood there shaking and crumbling like a house of sand. In front of the place, which was occupied by Bodwell Brothers commission merchants, cases of oranges or some other fruit were piled four tiers high. These blocked the way from the sidewalk to the thoroughfare. As we watched the front wall toppling forward, we also saw Mr. Frank Bodwell and his entry clerk, Nelson, rush out from the inner part of the store to the front. With them was the porter. It was all over in an instant. *The wall cracked as they topped the cases and buried them utterly under six feet or more of bricks and mortar.*

As I stepped down from the sidewalk, I looked for the trench I had been obliged to jump over. It was gone; closed up again; and I had never noticed it closing. But a man beside me pointed up along the street to the north, and called my attention to the fact that the center of *the thoroughfare between Washington and Jackson had sagged in about six feet deep and the hollow was filled with water.* Nevertheless, the street where I had jumped the gap seemed quite solid, though later, when real traffic was resumed, the surface collapsed under the first teams that tried to pass.

Patrolman Harry F. Walsh was also on the downtown landfill, and his experience was similar to Cook's.[17] Walsh was in the back of Meese & Gottfried, a machine shop located at 169 Fremont Street. The forceful shaking sent heavy machines on the second floor tumbling down to the first. When Officer Walsh opened the building door, he saw the earthquake as Cook had:

Fremont Street was then *rolling in waves*. First it opened and closed in big cracks. At least I first noticed *big cracks opening and shutting* as I looked at them. Then I noticed that the street was moving in waves like the sea, and was lifting under me like the sea.

I ran for the alarm box on the corner, and before I reached it, the Meese & Gottfried Building and the *power company's building crumbled up and fell flat*...When I tried to pull the alarm in the call box near Howard Street, I found that the *wires were all down, and no signals were working.*

I then hurried up Fremont Street to Mission...the second shock seemed to me to feel more like the pull-back in the snap of a cracking whip than anything else. It just jerked the ground from under everything...the *cornices and chimneys were still dropping* into the street every now and then from the places where they had been loosened.

Police Officer E. W. Meredith was on the east side of Powell Street between Ellis and O'Farrell, walking toward Union Square.[18] While this area is not on fill, the shaking Meredith describes seems as strong as that experienced by Cook and Walsh:

When I noticed that the sidewalk had begun to tremble under me...I ran to the center of the street for safety, and tried to stand there. But the first big shock threw me over to the west side of the street. As I was trying to recover my balance the big shock came, threw me back toward the east side again, and shot me just far enough to get me out of the way of the falling debris of the Colonial Club. The *whole front of this building was shaken out by the twister quake, and the masonry was flung halfway across the street* right where I had first been thrown. *No other building had been crushed in that collapse.* So I ran south down Powell Street toward Market, where I could see that the Columbia Theater wall had crashed down through the roof of the lodging house, and I feared there had been a terrible disaster.

It was astonishing that none of the people in the lodging house had been killed. And we were all amazed when we found [that] *men and women who had been driven through one or two floors into the Oriental Cafe below were all able to walk when we fished them out of the debris.*

These descriptions make several important points. First, the earthquake was felt as two distinct shocks, the second of which had a component of rotation. Next, surface waves of the earthquake could be seen and felt, reminiscent of surface waves reporters like Mark Twain saw in the 1865 earthquake. In 1906, as in 1865 and 1868, these waves were powerful enough to leave their silhouettes frozen on street surfaces. Cook's comment that he knew he was on filled ground and that it would be hazardous in an earthquake indicates that this was probably common knowledge. Still, he was not prepared for the dramatic results of liquefaction along Davis Street. The earthquake caused several different kinds of failure in brick buildings. Cook saw two brick structures respond to the same earthquake in opposite ways: one fell in on itself, allowing its contents to hold up the roof and floors, while the other building fell outward, killing people on the street. Walsh had firsthand experience of how heavy equipment could destroy a building's interior. Meredith saw that a brick building need not collapse despite losing its facade, but also that falling walls could destroy a neighboring structure. An 1870 advertisement for the Grand Hotel had proudly explained that its walls would fall outward in an earthquake, protecting the building's inhabitants. Luckily, when the 1906 earthquake struck, the walls of the Grand Hotel did not fall at all. However, many earthquake fatalities occurred when brick walls and cornices fell outward, the most common failure in brick buildings.

Four Eyewitness Accounts from Inside

The shaking was more pronounced inside structures, especially on higher floors. The following accounts were written by people who had been sleeping when the earthquake struck. Their experiences cover an array of building types: an earthquake-resistant brick building, an unreinforced brick building, a brick building with bond-iron reinforcement, and a wooden frame house.

Although most steel-frame office buildings in San Francisco were empty at 5:12 a.m., one was occupied, the ten-story Mills Building, which had load-bearing exterior brick walls and an interior steel frame. The building's superintendent, who was living in the penthouse with his family, reported that the earthquake knocked crockery off shelves but left the penthouse otherwise undamaged.[19] Having worked his way through the building and having found no major damage, he fired up the boilers and waited for the business day to begin. The boilers continued functioning until two in the afternoon, when he shut down the building because of the approaching fire.

Fred G. Plummer, a civil engineer working for the Forest Service at the time of the earthquake, had no idea he might be saving his own life when he chose the seismically strengthened brick and timber Occidental Hotel for his sojourn in San Francisco.[20] Plummer's room was on the north side of the hotel's third floor, facing Bush Street (figure 6.4). He later wrote to Grove Karl Gilbert, the famous geologist, that he had acquired the habit of keeping a watch, pencil, and paper handy, night or day, to record meteoric or seismic activity:

> Being awakened by the violent movement, my first thought was to turn on the electric light and note the time. Owing to the rapid motion of my bed, from which I tried to reach the overhanging light, it was some six or eight seconds before this was accomplished. I then *noted sixteen seconds till the end of the temblor, and that the time of greatest intensity was about ten seconds before the end.* I noted the position in which the bed and bureau were left and the tracks made by the castors on the plaster-strewn floor [figure 6.5]. I then opened the door to the hall and placed a chair against it that it might not jam if the building settled, and then called to Mr. Bradley [his assistant in the adjoining room] to do the same to his door...After dressing, which required four minutes, Mr. Bradley and I made our way to the street without difficulty except that the *air was dense with plaster dust and the floor and stairways were somewhat littered with debris.* The night clerk of the hotel was at his post behind the counter, and after seeing that there was no mail for me, he remarked that he thought this was the worst shock ever experienced in California. I assured him that it was not the most severe, but agreed with him that it was worse than that of 1868.

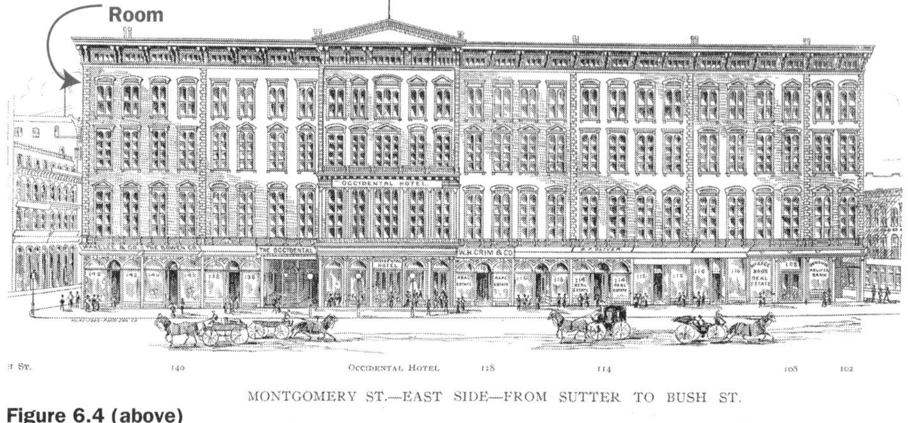

Figure 6.4 (above) The location of Fred G. Plummer's room on the north side of the Occidental Hotel, April 18, 1906

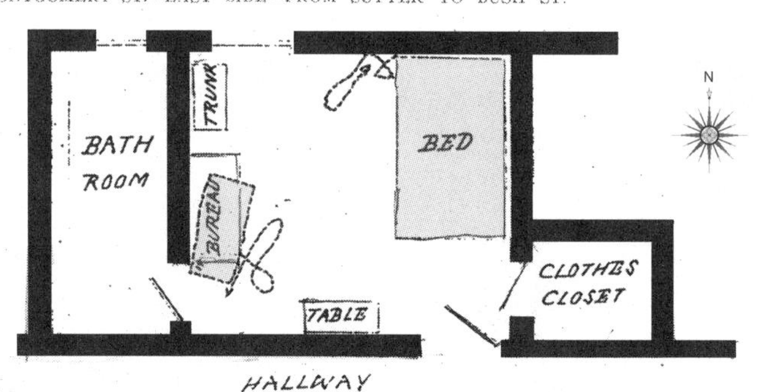

Figure 6.5 (right) Plummer's sketch of the movement of his furniture in the earthquake, April 18, 1906

Figure 6.6
The location of Horace D. Dunn's room at 610 Montgomery, April 18, 1906

Horace D. Dunn was standing up alongside his bed, facing south, on the fourth floor of an unreinforced brick building at 610 Montgomery Street on the corner of Merchant Street (figure 6.6), in the filled area of San Francisco, when the shaking started.[21] Although he was just a few blocks north of the Occidental Hotel, his experience was more harrowing than Plummer's:

> The first vibrations were light and apparently were from the West-North-West or the reverse and did no damage. With but a slight interval there came a series of *violent shocks* from the north. These shocks—i.e. the first one—*threw me to the bed* to the south, where I remained on my hands with feet on floor until the earthquake ceased. The sensation was as if I was pushed at my back without, however, the feeling of hands on my shoulders. *At about the fourth shock, or rather thrust, the chimneys and south wall down to the ceiling of my room went crashing into the court below.* How many thrusts came, I do not know, but I think there were at least six and probably more. I did not feel any return motion to the north, which probably escaped my attention as I raised myself on my hands from the bed after each thrust. I did not notice any easterly or westerly shocks except those at the commencement as above stated.
>
> The building I was in *also lost its north wall down to the ceilings near the fourth floor on the easterly half of Merchant Street, where horses and men were killed. Having a hallway in the middle and girders in the room on both sides,* i.e. four interior supports, in my opinion *saved the building from a collapse,* though it would have to be completely rebuilt from the foundation. I had lived in the building for a period of eleven years and knew its inferior condition.

DeWitt J. Lipe watched the quake from his room on the fifth floor on the west side (the back) of the Winchester Hotel, located on the west side of Third Street between Market and Mission (figure 6.7), about one hundred feet south of the Call Building, the city's tallest.[22] He was awake when the first shocks began and evidently felt no fear that his building would collapse. He describes what he saw as he looked out his west-facing window:

> As I looked out on this vista, *everything was swaying*; and the first among these swaying structures to attract my particular attention was *the tall steel smokestack over the* Call *power plant, which toppled down within the first few seconds*, and the simultaneous *disappearance of the big iron water tank* from the roof of the same building. That happened during the first few seconds that I was watching the effects of the quake.
>
> Almost immediately after these incidents, *the great chimney of the San Francisco Gas and Electric Company's powerhouse*, which had been waving like a whip, broke off about the middle, and the top *went crashing down on the powerhouse itself,* which caused a terrific explosion as the debris smashed the steam pipes or boilers beneath.
>
> About this time, *the plastering in my room began to fall in*, and the earthquake seemed to be pumping up and down with a vertical motion. It was time for me to be doing something to save myself, so I tried to dress while the place was shaking and showering plaster. While I was so engaged I saw the swaying *steeple of St. Patrick's church give way and fall down en masse* toward Mission Street. I then heard a tremendous noise of some other falling building. Thinking it was the Call Building, I looked to my right toward Market Street; but that structure was still standing and apparently uninjured.
>
> Then I noticed that other guests from the Winchester were crowding out on the roof of the annex, which was about on a level with the fourth-floor windows of the rear of the main building underneath me. To get out on that roof they had to jump the three-foot light well, or alley, that separated the main building from the annex. While I was watching these people *they were running away from the rear wall of the main building*, and were looking up apprehensively toward it, and me, as though momentarily expecting the wall to fall out and crush them.
>
> By this time I was partly dressed. The shocks had ceased, and it seemed as if the earthquake was over. *Except for fallen plaster and some disturbance of the furniture nothing seemed to be much the matter* with my room. The door was not jammed or anything. When I had hurriedly dressed myself, I went out...I ran down the stairs and saw that *the damage was far more extensive than I imagined,* and the plastering was down everywhere.
>
> [Later] I went upstairs to my room...I made a point to see what I could of the damage to the hotel. *The rear walls*

Figure 6.7
DeWitt J. Lipe's view on the morning of April 18, 1906:
A. Examiner Building
B. Winchester Hotel with approximate position of Lipe's room
C. Spire of St. Patrick's
D. Power station chimney
E. Call Building power station chimney
F. Call Building

and the wall of the "L" were badly cracked and fissured, the largest cracks being along the line of the iron rods running through the brick walls. About three feet from the fire wall along the roof was shaken off; the arches over the windows were badly cracked and bricks had fallen out of them, and the walls of the two courts or air shafts were in a similar plight. *A chimney or part of a wall had fallen on the roof of the annex and had passed clean through, leaving a large hole.*

On the morning of April 18, Edward I. de Laveaga was jarred awake with the rest of his family at 5:12 as the earthquake shook his wood-frame house at Geary near Franklin:[23]

> Every one of us fully believed that our last hour had come—*the house was wrenched back and forth and seemed as though any minute it would fall* on our heads. The big, heavy bookcase in my room was thrown down; the clock on the mantel was broken into three pieces and lay on the floor, the strike going at a rate of about 50 miles an hour. When the quake was over, I got out of bed, lit the lights and picked my way over the fallen bric-a-brac into father's room. He was fumbling at the button in a vain endeavor to light the lights. Good reason why he could not—the chandelier was in a heap on the floor. After finding out that father was all right, I hurriedly dressed and went out on the street to see what damage had been done. The first thing that caught my eye was the church on the corner of Geary and Franklin [the Unitarian Church]. *Two of the corner supports of the steeple had fallen away and stones from these two had gone through the roof of the church* proper. It certainly looked like a picturesque wreck. Bricks covered the streets from the fallen chimneys, which reminded me that I had better go on the roof and examine ours. I found *every one of our six brick chimneys down*, and for that matter, so were everyone else's as far as you could see.

From the eyewitness accounts we learn how intense the shaking was throughout the city in every kind of building. Although some witnesses thought their buildings would collapse, none of these did. But all suffered some kind of damage. Well-built and reinforced structures like the Occidental Hotel suffered minor damage, including the wreckage of internal plaster and the loss of one chimney. The Mills Building performed even better. But poorly built, untied buildings (Dunn's rooming house) lost whole walls and major parts of cornices. Though a damaged structure might not collapse, its falling parapets and walls could destroy neighboring structures and kill people in the street. The shaking and internal damage in the Laveaga house were severe, but the wooden house held together. None of the house's six brick chimneys crashed through the roof, but the potential for massive chimneys falling through the light roofs of dwellings was great. Two witnesses said they could turn on the lights; from other accounts we know that electricity still flowed through selected lines after the earthquake.

A Virtual Tour[24]

How widespread was the partial failure or complete collapse of buildings in the 1906 earthquake? To answer this question, let us take a virtual tour of San Francisco immediately afterward, using photographs of the time. There have been allegations that photographs were retouched to minimize earthquake damage. The photographs used here have not been altered as far I can determine (see note 26). I have made one composite photograph of the Chronicle Building to suggest how it looked before the fire. Throughout I alert the reader if I edit photographs to illustrate structural practices.

Figure 6.8
Map of virtual tour of San Francisco, morning of April 18, 1906. The base map was drawn in 1905. The dotted line indicates the direction of the walk, and the arrows indicate figures in this chapter and chapters 5 and 8.

In the 1950s a group of San Francisco structural engineers led by Karl Steinbrugge and later Henry Degenkolb began traveling to earthquake sites as soon as permitted so that they could examine the evidence of structural damage before it was cleared away. They investigated the issues that nonengineers might miss and personally assured the accuracy of data they collected. This kind of investigation became a mainstay of the Earthquake Engineering Research Institute (EERI). I am going to use the 1906 photographs as if they were evidence for an EERI report. This kind of report is usually divided into three sections: Seismology, devoted to ground shaking and soil response; Building Damage, evaluating damage by building type and configuration; and Lifelines, which considers damage to necessary services like electricity, water, gas, and communication. The report that follows focuses on building damage.

I was the leader of an EERI reconnaissance team investigating the Umbria-Marche earthquakes in Italy in 1997, and I have been a member of similar teams investigating earthquakes in Kobe, Japan (1995) and Northridge, California (1994). I also made individual reconnaissance trips to Ismit, Turkey (1999) and Puebla and Oaxaca, Mexico (1999). If I had been in San Francisco on the morning of April 18 with an EERI reconnaissance team, having just a few hours to investigate before the city was engulfed by fire, I would have divided the team into four groups: two would study damage due to liquefaction and violent shaking, one in the Mission District and one at the waterfront; the third group would go to the Western Addition, and I would lead a fourth through the fire district.

Touring the Fire District

I now imagine that my reconnaissance group and I are in one of the turrets of the Hopkins Art Institute, housed in the former Mark Hopkins mansion on Nob Hill (on the site of the present-day Mark Hopkins Hotel). The Hopkins mansion, designed and built by the society architect John Wright in a mélange of Queen Anne and Eastlake styles, was undamaged by the earthquake. For all the excess of its decoration, the building, built on rock, is sound. Wright, who had designed earthquake-resistant structures for the University of California at Berkeley and others in San Francisco, may have included iron bars and braces to tie the building together.[25]

This is a key vantage point for observing the city: our group gathers here several hours after the earthquake. A number of photographers have dragged their equipment here to take shots as the downtown begins to burn. One of the photographers with R. J. Waters & Company sets up his tripod for a panoramic shot stretching from the cornice of the Fairmont Hotel with Telegraph Hill in the background to the northeast, to the ruins of San Francisco City Hall, in the southwest (figure 6.9).[26] Looking at his work and that of the other photographers, we see much evidence of the earthquake in the foreground: piles of bricks on the roofs, where chimneys once stood, and some downed cornices. But the city looks intact. The major buildings of downtown San Francisco—the Merchant's

Figure 6.9 (above, pages 120-123)
"The City Burning, 10:00 a.m., April 18, 1906," R. J. Waters Co., 1906. This famous panorama was taken from the Hopkins mansion on Nob Hill. A close examination of the panorama in relation to photographs taken by others (figure 6.10) reveals that the damage recorded in the photographs is identical. These photographs give us a trustworthy impression of the impact of the earthquake on the city of San Francisco. Circles indicate apparent earthquake damage.

Figure 6.10 (middle) and 6.11 (bottom)
Detail of damage from Nob Hill, April 18, 1906

the roofs, where chimneys once stood, and some downed cornices. But the city looks intact. The major buildings of downtown San Francisco—the Merchant's Exchange Building, the Mills Building, the Chronicle Building, the Shreve Co. Building, the Mutual Savings Bank Building, and the Call Building—appear, from this vantage point, undamaged. Even buildings in construction appear unhurt. The earthquake-resistant buildings of the 1870s, like the Palace Hotel, the Occidental Hotel, and the Appraiser's Building, appear intact. Towers of religious buildings around us—Old St. Mary's and Grace Church, on California Street, and the First Congregational Church, at Mason and Sutter—also appear undamaged. Several other towers, like the Hall of Justice and Selby's shot tower, are still standing. While the 1871 City Hall dome stands, we can see right through it: the masonry of the drum has collapsed, leaving just the steel frame.

As we look out, we realize we cannot see a whole block, just the side closest to us and the roofs, so damage may elude us.[27] The chimney that fell on Fire Chief Dennis Sullivan's firehouse from the east side of the cupola of the California Hotel is absent; only from the east would we be able to see the gaping hole in the cupola and wall where the chimney had been (see figure 8.16). In front of us are the more than 2,086 separate buildings in the fire district of downtown San Francisco. Just 2 percent are fireproof with steel frames (Class A and B). Brick buildings with brick bearing walls (Class C) constitute 68 percent, and 30 percent are frame buildings (which are not included in the classification system). The overall impression from our vantage point is that most buildings, including the overwhelming majority of brick buildings, survived the earthquake with little exterior damage. If we look carefully, there is plenty of evidence of fallen walls, chimneys, and cornices. Collapses appear to have been rare, but overviews can be deceptive. We must walk through the city to get a better impression of the damage.

We walk down the stairs of the Hopkins mansion, turn right, and look over the east garden wall to the undamaged, wooden Leland Stanford house (figure 6.12),

Figure 6.9 (continued)

and down California Street toward the bay. Across California Street is the unfinished Fairmont Hotel (a portion of it appears on the left of figure 6.12).

We walk to Powell Street and turn north to Sacramento Street. Walking west down the slope of Sacramento, we join the crowd looking down at the downtown fires (figure 6.13). This is the location from which Arnold Genthe takes his famous view of San Francisco burning.

What is most arresting about this view of Sacramento Street is that the wooden houses around us are intact, but a brick facade has broken away from its interior wooden frame to fall into the street, exposing the inside of the building. The interior floors and furniture remain, and it looks for all the world like a huge doll house. This kind of failure has occurred frequently, due to one or more omissions in construction. First, the facades themselves were not thick enough or well enough bonded; that is, they didn't have enough rows of bricks, and these rows didn't adhere to one another, because the mortar wasn't strong enough or because there were insufficient headers to link one row to the next. Second, the bricks were not strapped with metal ties to the wood behind, nor were the facade walls tied to the side walls with bond iron.

Turning around and walking to the top of Nob Hill, we retrace our steps up Powell Street, looking at the east side of the Fairmont Hotel, filled with unopened crates containing new furniture. The exterior granite-and-terra-cotta facade is undamaged. We pause at the cable car tracks at California and Powell, looking south to Union Square (figure 6.14). On the right is the stone retaining wall of the Stanford house's garden. The stones are the same as those used for the Southern Pacific tunnels and they were assembled by the same construction crews. The wall is intact. Looking down Powell, we note how little damage there is to the brick and wood buildings on either side of the street. Walking down Powell to Sutter and turning left, we notice that the two brick towers of Temple Emanu-El are still standing. This is not surprising, since the architect, John Patton, who also designed

Figure 6.12
Looking northeast down California Street from the Hopkins mansion on Nob Hill, April 18, 1906

Figure 6.13 (right)
Looking down Sacramento Street from Nob Hill, April 18, 1906, Arnold Genthe

Figure 6.14 (above) Looking down Powell Street to Union Square, April 18, 1906

Figure 6.15 (right) View from Geary and Powell, April 18, 1906. Union Square is to the left. In its center is the Dewey memorial column. The unfinished Whittell Building is in front of us. The St. Francis Hotel is immediately to our left.

the city's Hall of Records, may have ensured that the building was earthquake-resistant by inserting iron in the brickwork to hold it together.[28]

We continue south to Union Square. Standing at Powell and Geary, near the corner of the St. Francis Hotel (figure 6.15), we notice the granite Corinthian column, designed by architect Newton Tharp, commemorating Dewey's triumph over the Spanish fleet in Manila Bay. The Victory, modeled after the statuesque Alma de Bretteville, who will later marry Claus Spreckels, stands undamaged at the top of the column.[29] Tharp had inserted a medal rod through the length of the column to ensure its stability. On the west side of the square is the new steel frame for the extension of the Class A St. Francis Hotel. The hotel, with its granite base and Colusa sandstone walls, was built in 1904, designed by the architects Bliss and Faville and engineered by John D. Galloway. It survived the earthquake with little damage.[30] Examining the entire perimeter of Union Square, we see damage on the south side only. The debris from the Heine Piano building cascades across Geary Street (figure 6.15). Otherwise, the buildings here, including the incomplete Whittell Building, appear undamaged.

Walking south on Powell toward Market Street, we can see that debris from the Columbia Theater has crushed a neighboring building, just as Officer Meredith described. Turning left down O'Farrell Street, we notice the same kind of damage we saw from the top of Nob Hill: a piece of a brick cornice and two brick facades have fallen out into the street, but the damaged buildings have not collapsed (figures 6.16 and 6.17); their wooden partitions, not meant to be bearing walls, are holding up the wooden floor and roof joists. Because wood and brick vibrate at different rates, a flexible wood interior can knock a brittle facade into the street. The parapet failures are similar to those we have seen in earlier earthquakes, caused when an improperly bonded or improperly tied brick fire wall or parapet is thrown off its building as it oscillates.

We continue to walk down O'Farrell toward Market Street until we reach Spreckels' Call Building, which seems

Figure 6.16 (above)
Looking down O'Farrell toward the Call Building, April 18, 1906

Figure 6.17 (left)
The collapse of the facade of the Shiels apartment building, 32 O'Farrell Street: detail of view from the middle of the block toward the Call Buidling

to be in excellent condition (figures 6.18 and 6.19). The building is locked up tight because the employees are producing the *Call* from the Chronicle Building across the street, due to the encroaching fires. The rival papers and their owners buried the hatchet in 1905, when the Chronicle Building's tower burned.[31] Fireworks set off to celebrate Schmitz's mayoral triumph sparked a blaze that destroyed the tower but left the building otherwise undamaged. The *Call* offered to help bring out the *Chronicle* newspaper, and now the *Chronicle* is returning the favor. Some people say the Call Building facade is out of plumb, leaning toward the intersection of Market and Second, but it doesn't look distorted.[32] Although stones from its upper floors may have been loosened, as some investigators claimed, they have not fallen from the facade. The elegant building has no stress cracks and appears to be completely intact.

Diagonally across the street, the oldest tall building in the city, the steel-frame, earthquake-resistant Chronicle Building (figure 6.20), has survived the quake—but a few hours after our visit, its iron-and-steel skeleton will collapse in the fire.[33] The upper corner of the ninth-floor wall was knocked out when heavy linotype machines smashed against it, but one witness reports that "the old building was not seriously hurt, and one elevator

Figure 6.18 (above)
The Call Building, Third and Market Streets, before the earthquake

Figure 6.19 (left)
The Call Building burning, April 18, 1906. The fire had already burned east from Fourth Street (far right), weakening the internal structures of buildings along Market Street. The seven-story building to the right of the Call, marked for its cornice failure, is the Kamm Building, whose steel skeleton nearly collapsed due to the intensity of the fire. Its cornice and entire lateral wall collapsed as the steel gave way. The next cornice failure to the right is the Bancroft Building, which seems to have suffered severe damage in the earthquake and fire. To the right of the Bancroft, there are several cornice failures and major collapses.

when workers showed up to produce the morning paper.[34]

The newspaper has already outgrown this building. It is slated for renovation: three floors are to be added to its facade, and its clock tower is to be reerected above the addition. The new, taller annex is under construction to the north. The annex seems to have suffered an unusually large lateral movement in the earthquake (figure 6.21): notice the X-shaped cracks between the windows from the seventh to the thirteenth floors.[35] As the steel frames oscillated in the earthquake, they pushed against the brick walls that they supported. In the case of the annex, the moving frames were stiffened by extensive spandrel braces with gusset plates said to be earthquake-resistant. But the braces could not ensure that the building would be absolutely rigid. The building swayed back and forth, dissipating energy as it oscillated. As the steel moved, it in turn pressed against the adjacent walls of terra-cotta tiles and brick. These walls, stiffer, massive, and brittle, resisted the movement of the steel, cracking to release the energy. Looking up at the damage, we can that see this structure drifted (or bent sideways), despite the heavy bracing, and its steel frame or rivets probably suffered some damage that we cannot see.

To the west of both the Chronicle and the Call Buildings, on Market Street, is the Mutual Savings Bank Building (now Citizens Savings), designed by William Curlett and erected in 1902 (figure 6.22).[36] This steel-frame, terra-cotta-clad building completes the ring of handsome facades

Figure 6.20 (above)
The Chronicle Building with its annex in construction just before the 1906 earthquake. The Chronicle Building has already lost its tower to the fire of 1905; only a stub remains. This image is composed of two photographs altered and blended to give the impression of the building just before it burned.

Figure 6.21 (right)
The Chronicle Annex with X cracks, after April 21, 1906, published by Himmelwright. The old Chronicle Building was gutted by the fire and collapsed.

Figure 6.22
Newspaper corner from Kearny Street, April 18, 1906. Intact buildings will soon be ruined by the fire. Call Building begins to burn.

encircling Lotta's Fountain, given to the city in 1875 by the opera signer Lotta Crabtree in commemoration of her outdoor concert here. The Mutual Savings Bank, with its carefully detailed Beaux Arts Baroque facade topped by a three-story, Dutch-style gabled roof, is an elegant foil for both the Romanesque Chronicle Building and the Baroque dome of the Call Building. The bank will burn in the fire. Continuing east on Market Street, toward the San Francisco Bay, we pass the incomplete Monadnock Building on our right.[37] Market Street is surprisingly free of rubble. Our destination is the Palace Hotel (figure 6.23), in front of us and on the right.

The Palace Hotel was designed specifically to resist earthquakes and fire (see chapter 4); its many interior cross walls were tied to the exterior walls by brick and iron bond, while the interior floors and some partitions were constructed of wood. According to the accounts of many guests, the building rocked violently during the earthquake, shaking down plaster and large chandeliers.[38] Some doors jammed when the steel pins in their hinges snapped, and a number of plate-glass windows blew out, although the main glassed court apparently lost only a few panels. Some people said that the elevators did not work; others (including one guest on the top floor) claimed they did.[39] The telephones kept working, some gas lines broke, and at least one sink was out of order, although no flooding or dripping water was reported. The hotel's standpipes remained essentially intact, and motors in the basement generated enough pressure to lift water all the way to the roof for fire-fighting. These mechanical systems survived because none of the hotel's walls, ceilings, or floors gave way. The stairs were passable, and though the rooms were full of plaster, the building's exterior was intact; surprisingly, no windows were broken on Market Street. The bar was still serving drinks, and water was still available from the hotel's cistern.

The Grand Hotel, also designed by John Gaynor, was no Palace. It was much lighter and cheaper. It too featured a brick exterior, but its internal walls were made from huge wooden timbers tied together. This gave the structure a degree of flexibility that did not work well in relation to its less flexible parts. The Grand moved a good deal more than the stiffer Palace, and its guests gave correspondingly more panic-stricken accounts of the quake. The Grand's elevators did not work—the shafts were full of debris. Huge pieces of plaster fell in the interior, the stairs collapsed in places, and one account mentions a timber pushing into a room, which suggests a rather serious structural failure.[40] Despite all this, the exterior brick walls, which (as mentioned before) had been designed to fall outward in an earthquake, held fast, and windows in its Market Street facade also were unbroken. An aftershock knocked one of the Grand's chimneys into a room, an indication that although the brickwork remained standing, it was damaged in places.

We continue east down Market toward the Ferry Building, which is already shrouded in smoke. Looking from Second Street east to the corner of Bush and Battery, we see that an 1860s building on the north has lost its cornice, but there are no failures in the 1860s facades to the south (figure 6.24). On the south side of First Street and Market (figure 6.25), we can see the failure of the east part of the Sheldon Block (built in 1895), perhaps due to liquefaction, and the probable failure

Figure 6.23 (below, top) The Palace Hotel and Grand Hotel from Lotta's Fountain, April 18, 1906, just before the fire. All the buildings in the photograph are intact.

Figure 6.24 (below, bottom) Looking east on Market Street in the direction of the Ferry Building, toward Battery and First Streets, April 18, 1906. The buildings in the photograph are as depicted in the *Directory* of 1895. They appear to date from the 1870s and 1880s. Of eleven brick buildings on fill, all but the two on the right appear undamaged.

Figure 6.25a (right) and Figure 6.25b (below) captions:

Figure 6.25a (right)
The south side of Market Street between Fremont and First Streets, April 18, 1906. The photograph shows that the Sheldon Block, built in 1895, was badly damaged by settlement or shaking. The Lachman Block may have also lost its cornice.

Figure 6.25b (below)
The 1895 *Directory* gives a clear idea of the elevations of the south side of Market Street between Fremont and First.

of the cornice of the Lachman Block (1887). As we turn around and look north (figure 6.26) down Battery, every brick building we see looks intact. The Italianate building on the corner, designed by Edward R. Swain for the Clark Estate in 1896 and constructed of load-bearing brick, seems to be in excellent condition.[41] We turn around and make our way back to Second Street.

Figure 6.26
Market Street looking north to Battery. The Mechanics Monument is on the right, where it still stands today. Seven major brick Class C buildings built on fill are shown completely intact after the 1906 earthquake.

After purchasing bicycles from bystanders in front of the Palace Hotel, we make our way past the Grand Hotel, and taking a right, bicycle south on Second Street. We stop at the Jessie Street intersection (figure 6.27). In front of us is a scene of total collapse and loss of life. The unreinforced brick walls of the Wilcox House, a modest boarding house, have crumbled and fallen outward, no longer supporting the wooden floor joists. One floor has collapsed, burying inhabitants below.[42] The external walls were simply too thin and too weak. The scene is surreal because no other building in sight is damaged.

As we continue down Second Street past the intact Wells, Fargo & Co. Building, we turn west on Mission. On the corner is the small, unreinforced, two-story,

Figure 6.27 (above)
The Wilcox House collapse, Second and Jessie Streets, as seen from the Wells, Fargo & Co. Building directly across Second Street. This corner building has totally collapsed in the earthquake. The other buildings on Jessie Street are apparently undamaged. The undamaged rear facade of the Palace Hotel can be seen at right rear.

Figure 6.28 (right)
The Burdette Building, on the northwest corner of Second and Mission Streets, was completely intact after the fire, with its merchandise undisturbed. Although the Atlas Building, on the left, was wracked by the earthquake (see the X cracks), it was quickly rehabilitated.

brick Burdette Building. It must have been constructed on a good foundation with rich mortar to have survived the shaking so well (figure 6.28). Protected by the taller buildings around it, it will be the sole building in the fire district to survive with its contents intact and windows unbroken, without people on the inside or outside fighting to save it from the flames. According to a newspaper report, the owner of the building had wanted to erect an eleven-story building on the site and had therefore built a deep foundation before deciding on only two stories.[43]

Next to the Burdette Building is the Atlas Building, at 602–606 Mission, designed by Frank S. Van Trees in 1904 but not completed until 1906.[44] The earthquake struck as tenants were moving into the building. This is a steel-frame building with brick infill, the only lateral bracing other than the column girder being knee braces at each floor on the south Mission Street facade. It is hard to know whether the architect or construction engineers were thinking about earthquake resistance when they designed the building, but the knee bracing has served to limit drift east to west. A. L. A. Himmelwright, who examined the state of the building after the fire, wrote, "The steel frame is uninjured... With the exception of the walls which are badly cracked, this building can be repaired without serious difficulty."[45]

Figure 6.29
The Atlas Building after the 1989 Loma Prieta earthquake. X cracks appeared on the facade as the building oscillated from east to west.

The 1906 earthquake shook the Atlas Building north to south, causing the X-shaped cracks on the east side of the building above the Burdette Building. If we travel forward in time, the Atlas Building will illustrate the problem of excessive drift. The Loma Prieta earthquake of 1989 shook it from east to west, causing the X-shaped cracks that can still be seen on the south Mission Street facade, and damage to walls on the north (figure 6.29). The worst cracks in the 1989 earthquake were on the seventh floor, the height of the cornice of the building next door; the Atlas Building was breaking itself up as it pounded its eastern neighbor.[46]

Riding our bicycles, we skirt the fires encroaching on Mission Street and zigzag west to examine the old United States Mint (figure 6.30). Superficially, it looks like an ordinary, unreinforced masonry building, which would do poorly in earthquakes—but of course we know better. Alfred Mullett reinforced the Mint by building it on a massive concrete foundation. It has done extremely well.[47] Even the two smokestacks survived with little damage. None of the heavy cornices fell, and all the columns supporting the front portico remain in place. The Mint is a success. We hear from other investigators that the United States Appraiser's Building on Battery Street (figure 4.20), also built under Mullett's supervision, survived with little or no damage. Like the Mint, the Appraiser's

Figure 6.30
The United States Mint at Fifth and Mission Streets, having survived the quake and fire with little damage

Building incorporated special iron reinforcement and a deep concrete foundation. Unlike the Mint, it was located on filled ground, yet it still did extraordinarily well in the earthquake. Mullett did not live to see how well his buildings performed; he died in 1890.

Riding west on Mission, still skirting fires, we make our final stop, at the United States Court of Appeals and Post Office, at the northeast corner of Seventh and Mission Streets (figures 5.5a and 5.5b). This is the building designed by James Knox Taylor and finished in 1905 that includes an elaborate system of lateral earthquake-resistant knee bracing (see chapter 5). Significantly, its southwest corner is on the edge of the old Mission Bay swamp, where we see that shaking has loosened blocks of granite cladding.[48] We know the building has its deficits: a floor system of shallow, unreinforced concrete arches, hollow clay tiles (which can crack in earthquakes) used for fire protection and partition walls, and a U-shaped plan. But the foundation was built well enough so that while the structure has sustained some exterior damage, it has not collapsed. There is no doubt that the interior was damaged, with decorative features falling away from the wall. Misled by smashed and fallen pieces of the highly decorated stone, plaster, terra-cotta, and mosaics, a custodian wired Washington that the post office might collapse at any time, but that seems unlikely. On Mission Street, trolley tracks are buckled, and the street and sidewalk on the southwest corner of the post office have subsided several feet, making the building's performance all the more impressive.

Fire is threatening the downtown fire district from several directions. I can see the ruins of the 1871 City Hall dome and I want to investigate, but soldiers are urging evacuation. It is time for my reconnaissance group to leave. I will write a short EERI-style report that will form the core of the assessment in chapter 8. Each of the four reconnaissance groups is like a blind man trying to describe an elephant. I trust that their reports will be as reliable as possible, because we must rely on each other to describe the elephant as accurately as possible. As I am writing these notes, the fire is erasing the evidence upon which the report will be based; the elephant we are trying to describe is suddenly disappearing like Alice's cheshire cat.

CHAPTER 7

The Fire: April 18–21, 1906

Like most San Franciscans, my grandfather, Dr. Oscar Tobriner, was jolted awake by the earthquake. He was staying with his father-in-law in a flat on the steep slope of Jackson Street at the corner of Presidio Avenue, in Pacific Heights. Fallen chimneys abounded, but otherwise damage seemed minimal. His wife, Maude, was recovering from surgery at the old Mt. Zion Hospital on Sutter and Steiner.[1] Walking quickly across the Western Addition, Oscar was relieved to find the hospital damaged but intact. Not far away, Beth Israel Synagogue and the Pike Memorial were in ruins. Borrowing a wheelchair, Oscar pushed his wife northwest, up toward the Laurel Hill Cemetery, and back to Jackson Street. Smoke from the downtown fires filled the sky. His brother, Ike, brought a horse and wagon to carry the family to the Hyde Street pier, where a flotilla of ferries and other craft were waiting to take them to Oakland.

Oscar never considered leaving the city permanently and always claimed he was among the first to build a new house in Pacific Heights. My father, Mathew, only two years old at the time, later insisted that he remembered the city burning. My family's experience was just one variation of what San Franciscans faced during and after the largest urban fire ever to occur in the United States.[2]

Fire Breaks Out

There is no doubt that the earthquake started fires on the morning of April 18, 1906; in the first fifteen minutes after the quake, scores of small fires broke out (figure 7.1). Having climbed to the top of the California Hotel after its chimney had collapsed into his firehouse on Bush Street, Fire Battalion Chief Cook reported

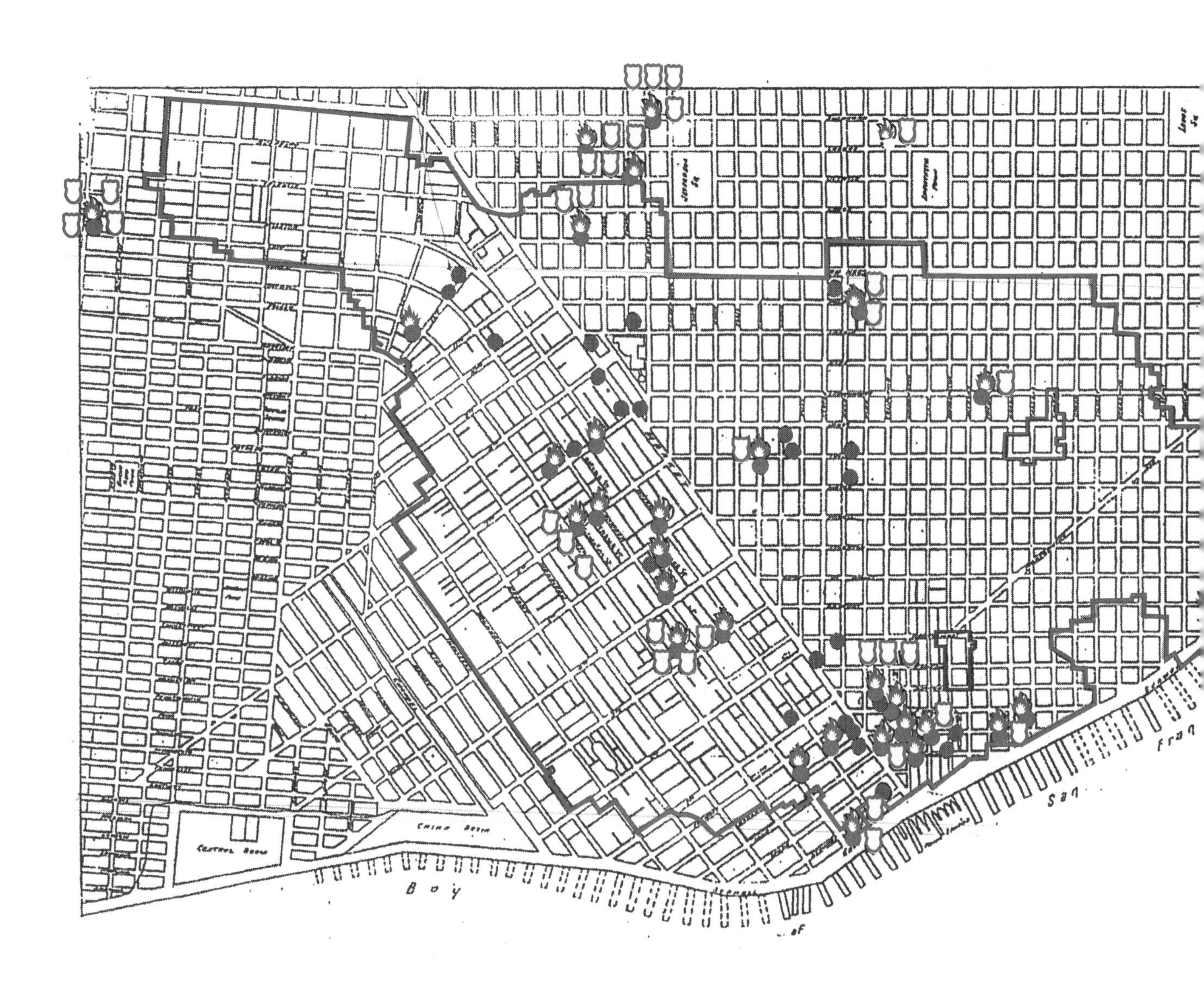

Figure 7.1
A map illustrating the fires that started immediately after the earthquake of April 18, 1906, based on an anonymous map from The Bancroft Library, probably originally part of the Bowlen papers. The dots represent unverified fires; the dots with flames represent fires corroborated in other sources. The shields represent the engine, truck, and chemical companies reported to have been fighting the fires (not all companies are represented because documents are incomplete). The fire at the upper left corner of the page is the Lippman fire. The map illustrates how difficult it would have been to form a single defensive line at Market Street as Sullivan had planned during the first day of the fire.

that "from the eighth floor I saw fires in every direction."[3] Captain J. R. Mitchell of Chemical Company Number Two stated that immediately after the quake, "fires started in different parts of the city" and that his crew "worked for seventy-five hours immediately after the quake and helped extinguish twenty-three fires."[4]

The fires had diverse origins.[5] Lanterns fell. Boilers ignited in damaged industrial buildings. Huge power-station chimneys collapsed onto their own facilities and their neighbors' structures. Ninety-five percent of residential chimneys suffered quake damage; cooking fires ignited timbers. Fuel and other flammable substances spilled and exploded in warehouses, manufacturing plants, and drugstores. Live electrical wires fell, igniting what they touched, and gas exploded in the mains. According to A. M. Hunt, secretary of the Citizens' Committee on the Restoration of the Light and Power Service, gas explosions continued for hours:

> Along Van Ness Avenue there were a number of explosions in the 30-inch main on that street which split lengths of the pipe and burst the bituminous pavement above them. As soon as the earthquake took place all valves at the works shutting gas off the city mains were closed. The mains themselves contained several hundred thousand cubic feet of gas, and it was some time before this gas was entirely expended; in fact, several persons living at elevated points in the residence district state that they cooked with gas as late as Thursday evening after the [Wednesday] shake. It is more than probable that some of these parties were responsible for the explosions in the mains, for as the pressure weakened, the flame from any lighted burner would flash back through the main to a point of explosive mixture of gas and air with disastrous results.[6]

Hunt's statement about gas left in the mains might explain the experience of F. Ernest Edwards, a printer. He walked downtown to check on his shop in the Benson Building, on Leidesdorff and Pine Streets, and reported finding the building (later destroyed in the fire) "in very good order." In his fourth-floor office he found the gas still lit under the linotype machines, and nothing disturbed.[7] Whether the gas was officially turned off or not, there seem to have been ample supplies in the lines, ready to blow.

Electrical lines, which according to Hunt had been turned off directly after the earthquake, were as live as gas was plentiful. Standing outside the Palace Hotel, the dramatic if somewhat dubious eyewitness Frank Louis Ames reported that "electric wires were on the pavements, spitting blue flames and writhing like snakes."[8]

Fighting the Fires

How was the fire department able to fight so many fires simultaneously? The fledgling volunteer outfit established in the 1850s was now a professional force. In the 1860s, the first steam engines arrived in San Francisco by ship from the East

Coast, creating a revolution in firefighting and departmental organization.[9] These steam boilers on wheels could produce constant pressure to propel water through the hoses. Unlike firemen "manning the brakes," the steam engines never tired. The days when scores of men pulled and pumped fire engines were over. Horses could haul the engines. Fewer firemen were needed, but they needed technical expertise in order to handle the new engines.

During the Civil War years, volunteer fire departments across the country often joined warring factions of northern and southern sympathizers.[10] In San Francisco the antagonism sparked a riot between engine companies on the way to a fire in 1865, an incident that contributed to the abolition of the volunteer department in favor of a paid municipal force. At that point there were still only fourteen engine companies, three hook-and-ladder companies, and five hose companies in the city. Four of the engine companies had steam engines; only one was horse-drawn. The city had 448 fire hydrants, 38 cisterns in good condition, and a steady water supply from the newly formed Spring Valley Water Association.[11]

The large fires of the late nineteenth century led to further improvements in equipment and firefighting. An 1871 conflagration in the Harpending Business Block, a group of three-story brick buildings, revealed that the department's ladders would not reach above three stories.[12] San Francisco fireman David Hayes had already invented a ladder that met this need. The Hayes ladder, mounted on a turntable and operated by a crank, was added to the department's equipment.[13]

By 1873 the city had installed 940 fire hydrants.[14] A fireboat was added to the department's equipment in 1878, withdrawn from service for lack of funds, and then reinstated in 1880.[15] The number of hydrants now numbered 1,352, and the department had seventeen steam engines, five hook-and-ladders, six wheeled hose carriages, twenty-three tenders, and several buggies, with a complement of 303 firemen.[16] In 1890 a chemical company was added and in 1891 the department acquired a water tower (a hose on top of a long boom). By the turn of the century, the fire hydrants numbered.3,721.[17]

At the outbreak of the 1906 fire, the department had thirty-eight engine companies, ten hook-and-ladder companies, seven chemical companies, one hose company, and one hose tower company, with 585 firemen in all.[18] The firehouses were well spaced throughout the city, and the equipment was up to date.[19] The National Board of Fire Underwriters visited San Francisco in 1905 and gave the department a high rating, though the city did receive a warning because of its inadequate water system and flammable wooden buildings.[20]

The outbreak of multiple fires after the earthquake strained the resources of the department. Ten of the thirty-eight firehouses reported significant damage and three reported slight damage.[21] Deployment of engines was delayed by jammed doors and stampeding horses.[22] Nevertheless, nearly all the engines and ladder trucks made it out of their firehouses, unimpeded by rubble as they made their separate ways to fight the small fires.

After the disastrous Baltimore fire of 1904, San Francisco's charismatic fire chief, Dennis Sullivan, had decided that if a conflagration broke out in San Francisco, a single fire line should be established along the broad expanses of either Market Street or Van Ness Avenue to stop the flames. Unfortunately, Sullivan was one of the first casualties of the earthquake.[23] The chief's quarters were located in the same Bush Street firehouse that housed Chemical Company Number Three. When the California Hotel's chimney collapsed into it, parts of the firehouse's roof and the second and third floors collapsed. The chief fell several stories and was badly burned by scalding water from an engine boiler or a heater. He was taken first to the Southern Pacific hospital at Fourteenth and Mission Streets and later to the Presidio hospital, where he struggled for his life for several days, finally succumbing on April 21.

With fires breaking out in every direction, Sullivan's plan for a single fire line was unfeasible; buildings were burning all over San Francisco, and each company was on its own. There was no plan for this situation. Communication between the firehouses was impossible. The electric alarm system was completely disrupted when the earthquake upset the system's batteries. (Despite a similar disruption to Western Union batteries in the 1889 earthquake, no bracing had been added to hold the fire department batteries in place.[24]) Downed telephone wires further isolated firehouses. But the lack of alarms didn't hamper the response of individual companies; fires were so widespread that each company simply looked for smoke or asked passersby.

In firefighting, every minute counts. Crucial minutes were lost extricating equipment from damaged firehouses and finding viable water sources from which to pump. Many citizens had been, like Chief Sullivan, injured or trapped by earthquake collapses. In one small area south of Market, several rooming houses caved in, trapping their occupants. These included the Valencia Hotel, the Portland House, the Brunswick, the Cosmopolitan House, the Wilson House, and a large structure at 119 Fifth Street.[25] Firemen stopped to aid in rescue efforts. In 1906 there were no cranes or other mechanical means of lifting heavy beams or quickly moving debris; everything had to be done by hand. Sometimes the helpless police and firemen could only offer the trapped victims swigs of liquor as the flames approached.[26]

Lack of Water

Without water, fires cannot be fought. During the course of the earthquake and fire, city water mains and house connections broke in twenty-three thousand places, draining much of the system.[27] Due to local failures, many hydrants, especially in the South of Market, had no water. Chief Sullivan, languishing near death in the Presidio, had foreseen this problem. Cities like New York had already installed auxiliary water systems dedicated to fighting fires. The problem with depending on only one system was that as the fire progressed and individual buildings collapsed, the lines for those buildings ruptured, spilling precious water from the system and causing a severe drop in pressure. The typical auxiliary system had its own water supply, stronger pipes, and fewer connections, and was maintained at greater pressure.

After studying the Baltimore fire of 1904, Sullivan felt that the quantity of water supplied to the city by the private Spring Valley Company was inadequate. (Recall that the National Board of Fire Underwriters had praised the fire department during their 1905 visit, but with respect to water supply they were highly critical, citing the friction between civic leadership and the Spring Valley Water Company.[28]) Sullivan crusaded for an auxiliary water system, advocating that the supply reservoir be situated on Twin Peaks and that the system be able to pump salt water from the bay if necessary. Sullivan ended his report by saying, "I would therefore again urge that there be no delay in commencing this undertaking, and sincerely trust that the close of the fiscal year we are now entering upon will find it fairly under construction."[29] It would be ten years before an auxiliary water system was in place.

South of the city, the Spring Valley Water Company was holding water in dams in the rift valley formed by the San Andreas Fault itself: the Crystal Springs Reservoir made of concrete, and Pilarcitos and San Andreas Reservoirs made of earth.[30] Miraculously, all three reservoirs remained intact after the earthquake. Pipes from the San Andreas Reservoir carried water to the College Hill Reservoir inside the city, continuing through the Mission District to Market Street and the central business district. Breaks in the main pipe, like the one at Seventeenth and Valencia, quickly emptied the College Hill Reservoir.

Similarly, water from Crystal Springs Reservoir was carried to the University Mound Reservoir and then piped to the South of Market. Liquefaction in the Mission and South of Market caused major lines from the University Mound Reservoir to break, reducing water flow from the reservoir to a trickle. The main pipe of the Pilarcitos Reservoir was far to the west of the other two, conveying water to Lake Honda, from where it was distributed to smaller reservoirs and tanks in the Western Addition. Lake Honda lost two-thirds of its water through leaks but was replenished by emergency water pumped from Lake Merced. Because this water was used to supply the Western Addition, all the fires in the outlying western part of the city were extinguished.

Chief Sullivan had noted wryly that though the city was surrounded by water, it had no municipal saltwater system. The Olympic Salt Water Company operated a private pumping station that brought sea water to a reservoir in the San Bruno Hills. Hoping to interest the city in using their resources, they had installed a demonstration hydrant at Fifth and Market.[31] There were drawbacks; the salt water eroded iron pipes and could rot materials it was sprayed on. However, this hydrant provided an excellent supply of water throughout the fire; it was used to spray the U.S. Mint, which also had a private water supply.[32]

Private sources of water were also crucial to other buildings. The private reservoir, in the Mission District, of wealthy San Franciscan John Center helped stop the fire in many locations.[33] The large factory of the California Electric Company had a private reservoir and hydrants at Third and Folsom.[34] Although the company's electric pumping system didn't work, firemen hooked into the supply at special hydrants and saved the building.

The hunt for water was maddening for the harried fire companies. After hooking up to a hydrant and discovering that it had no water or too little pressure, they could use other engines to relay water from distant hydrants. Water pressure in some hydrants was far below forty pounds per square inch (barely adequate); in some it was as low as fifteen pounds. The city's abandoned cistern system, which still held runoff water, was used after the hydrants emptied. Sometimes the cisterns' dirty water stopped up the boilers of the steam-operated fire engines. However, the cisterns were crucial to the fight in the Mission District, where the southern front of the fire was finally halted. Only 6 percent of the expected flow of water was available to fight fires in the fire district itself.[35] For lack of water, the fire companies within that district were in constant retreat.

Coordinated Firefighting

Outside the fire district, one specific fire illustrates the battles that firefighters were facing. A fire broke out in Lippman's Dry Goods Store, at the corner of 22nd and Mission Streets, threatening the large surrounding residential area.[36] Firemen from Engine 37 saw the flames from their undamaged frame firehouse at Utah between 24th and 25th Streets and rushed to the Lippman store. Although no alarm could sound, because there was no electricity, the fire must have been visible to other firehouses in the Mission. Engine 18 soon arrived from its seriously damaged brick firehouse at Dolores and Church. Engine 13, Engine 24, and perhaps Engine 25 responded as well. These engines worked together as a team. Finding the closest hydrants, at 22nd and Howard Streets, to be empty, Engine 37 found water at an old cistern on 22nd Street near Shotwell and relayed it to Engine 18, at the fire. Engine 13 used water from 22nd and Valencia. The crew of Engine 24 came later, having helped with rescue work at the Valencia Hotel. Engine 18 then hooked up to a hydrant on Hill and Valencia and worked the fire from 22nd and Bartell Streets with its own hose. Crews went to Engine House 25, stripped its extra hose, and relayed with Engine 37 to the Lippman building. It was crucial that they stop the fire from crossing to the unburned side of Mission. Five engine companies were engaged from immediately after the earthquake until noon. They validated the high rating given to them by the insurance underwriters, stopping the fire at Mission and 22nd.

Many fires were successfully extinguished by single companies, and teams of three engine companies working together had put out large fires in Hayes Valley and the Western Addition by ten in the morning.[37] By noon all the fires in outlying districts were out. But inside the fire district, at least four fires had combined to outflank the fire department by late morning. Uncontrolled fires burned on the south side of Market Street from approximately in front of the Ferry Building to Fourth Street, burning the Call Building among others. In the South of Market, fires burned from the waterfront on the east to Folsom on the south and Sixth

Street on the west. North of Market, several fires burned, the major front being at Sansome Street, with fires burning east to the waterfront.[38]

Another fire, which turned out to be the worst of the day, started in Hayes Valley when a woman on Hayes Street near Gough started to make breakfast in a stove with a damaged chimney, around ten o'clock.[39] Dubbed the "ham and eggs fire," it spread quickly. All the crews were busy with other fires. It was finally stopped twenty-four hours later at Octavia Street, after it had spread west with no opposition and burned the Mechanics Pavilion, where a temporary emergency hospital had been set up, and destroyed the 1871 City Hall.

The Fire District

In 1906 the highest ranked, most fire-resistant buildings were Class A: these were fireproof, iron- or steel-frame construction like the Call Building, in which the frame structure bears the entire weight of the building, as in modern steel-frame skyscrapers. In Class B buildings, like the Chronicle Building, the exterior walls carry their own weight, but an interior skeleton of iron, steel, or fireproof wood holds up the interior floors. The fire resistance of Class A and Class B was improved with interior metal lath and plaster partitions and exterior surfaces, other than masonry, covered in nonflammable materials. Class C buildings, like the Montgomery Block, were those similar to the best brick buildings of the 1850s. Outside they were brick with fire-resistant roofs, but inside they had wood or iron frames, without interior plaster lath or any other fire-resistant material.

The downtown fire district, with its fireproof buildings, might have been expected to retard the advance of the fires that had started there. But looking at the history of the fire district, we can see why it did not. Beginning with the first fire codes of the 1850s, an area of downtown San Francisco had been designated as a fire district, in which building exteriors were required to have fire-resistant brick facades. The first fire district restricting wooden buildings, in 1853, was bounded

Figure 7.2
The fire limits of San Francisco, 1866 to 1904

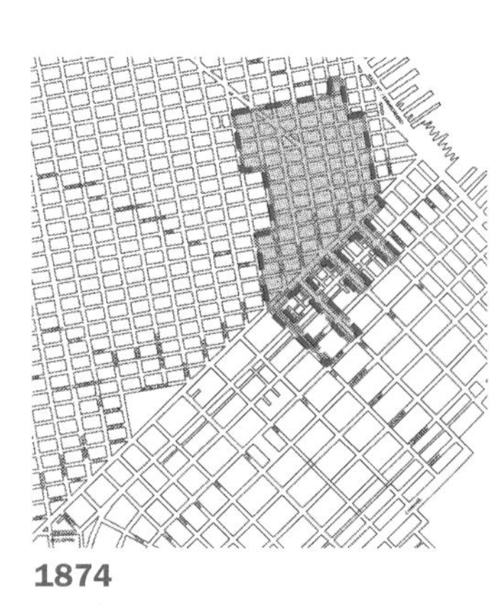
1874

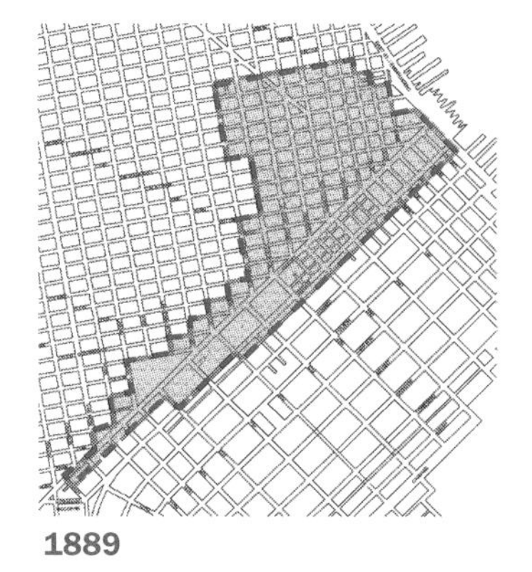
1889

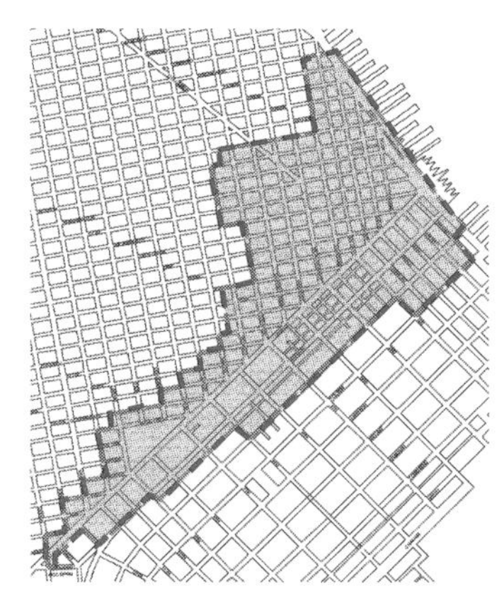
1904

by Dupont, Pacific, Pine, and Front Streets. By 1874 the district was extended to include lower Market Street and a portion of the South of Market, but the city grew far faster than the fire limits (figure 7.2).[40] The fire district was weakened in 1881 when new wooden buildings were allowed in the area in which they were supposed to be forbidden.[41] The limits were strengthened again in 1883, when wooden structures were excluded anew; however, existing wooden structures remained. The 1883 ordinance extended the fire limits from the waterfront to Union Square below Pacific, and up Market and Mission Streets to the 1871 City Hall.[42] In 1887, at the fire department's insistence, the limits were extended a few blocks farther to the south and west.[43] In 1889 Fire Chief Scannell predicted that without improvements, "a fire that would wipe the whole southern part of the city out of existence" was imminent, and he sponsored a proposal to extend the fire limits into Hayes Valley.[44] This was rejected at first because the extension would be a hardship for the people living in Hayes Valley, but a year later, in 1890, the extension was approved. By 1901 the fire limits of San Francisco were enlarged to include the area between the waterfront and Telegraph Hill, an additional block west of Union Square and north of Market to Gough and Valencia Streets, and an additional block along the boundary of Howard Street in the South of Market.[45] No further extensions occurred before 1906 and many old wooden buildings remained in the fire district.

The Baltimore fire of 1904 had already proven that fire districts and "fireproof" buildings could not stop a fire if buildings were breached. The Baltimore conflagration shocked the nation and served to focus attention on the folly of depending upon "fireproof" structures. The National Fire Association published a report in 1904 stating, "The old term 'fireproof,' so obviously a misnomer when applied to any sort of building, is herein discarded in favor of 'fire-resistive,' a name which, even when applied to the very best building yet devised, more correctly defines the character of the structure."[46]

The Baltimore blaze, which burned one hundred and forty acres comprising eighty city blocks, was concentrated in an area composed of substantial brick buildings, including twenty-seven "fireproof" structures which were not expected to burn.[47] The roofs of the buildings were for the most part made of fire-resistant tin or slate, with more combustible gravel and shingle interspersed. The fire started in a single structure, the brick Hurst Building, and was fought by a fire department with a good water supply and satisfactory equipment. But the interior of the building exploded. This explosion blew out the windows in the Hurst Building and those of many neighboring structures. Although the brick facades could stop fire, each one of the broken windows offered an unprotected entrance. Unfortunately, the wind was strong. The fire quickly became a conflagration. Firebrands were blown into the air, igniting buildings nearly. Once a fire becomes a conflagration, it is practically impossible to extinguish.

San Francisco's fire district was nowhere as secure as Baltimore's. The earthquake itself and the explosions which followed broke windows, providing many opportunities for fire to enter buildings designed to be "fireproof." The demise of the Call Building illustrates the vulnerability of the best buildings of the day. Across the street from the block on which the Call Building stood were two power plants, one belonging to the *Call*, the other to the city's gas and electric company. During the earthquake the smokestacks of both power plants fell in on themselves, hitting the steam pipes and causing huge explosions. Some investigators thought fire gained access to the Call Building through a basement tunnel from the power station.[48] Others blamed the advancing fire started by the power station collapses.[49] Either way, one spark getting into a single room of the building through a broken window and igniting wood furniture, carpets, and paper would put the rest of the building in grave danger (figure 7.3). In 1906, offices were graciously supplied with cross-ventilation. Above the hallway door was a glass transom that could be opened to let air in from the hall. If, perchance, the transom were left open, the fire could easily be sucked into the hall by the elevator shaft, which was usually protected by an elegant, open grill within which the lift cars rode up and down. The fire found the oxygen and updraft it needed here, using the elevator shaft or the nearby stairwell like a chimney. The Call Building lit up like a smoking candle as the fire claimed the top floors.

If the temperature becomes high enough, as it did in the World Trade Center towers in September 2001, steel buildings can collapse. The standard method of fireproofing columns in 1906 was to leave an air space of a few inches around them and then encase them in a protective shell of metal lath and plaster, or of terra-cotta. A more expensive and more massive solution was to encase the steel column in concrete. All three methods were employed in buildings that went through the earthquake. In several tall buildings, like the Rialto Building, the fireproofing around the steel columns failed, the steel melted, and whole sections of the structures collapsed.

Class C brick buildings were far easier for the fire to destroy, not only because they were easier to breach, but because as the fire got hotter, even the structural timbers would begin to burn. When they did, the supports for the interior floors of the building often collapsed. Floor joists were often embedded in the exterior brick walls. As the joists fell in, they acted like levers and pushed exterior bricks out, wrecking the building, fueling the fire, and endangering anyone on the street below.

Fire-Resistant Redwood

The National Board of Fire Underwriters' Committee of Twenty had visited San Francisco and warned the city that it would not survive if a fire started in any of its wooden neighborhoods:

> Not only is the hazard extreme with the congested-value district, but it is augmented by the presence of a compact surrounding, great-height, large-area, frame-residence district, itself unmanageable from a fire-fighting standpoint by reason of adverse conditions introduced by topography. In fact, San Francisco has violated all underwriting traditions and precedents

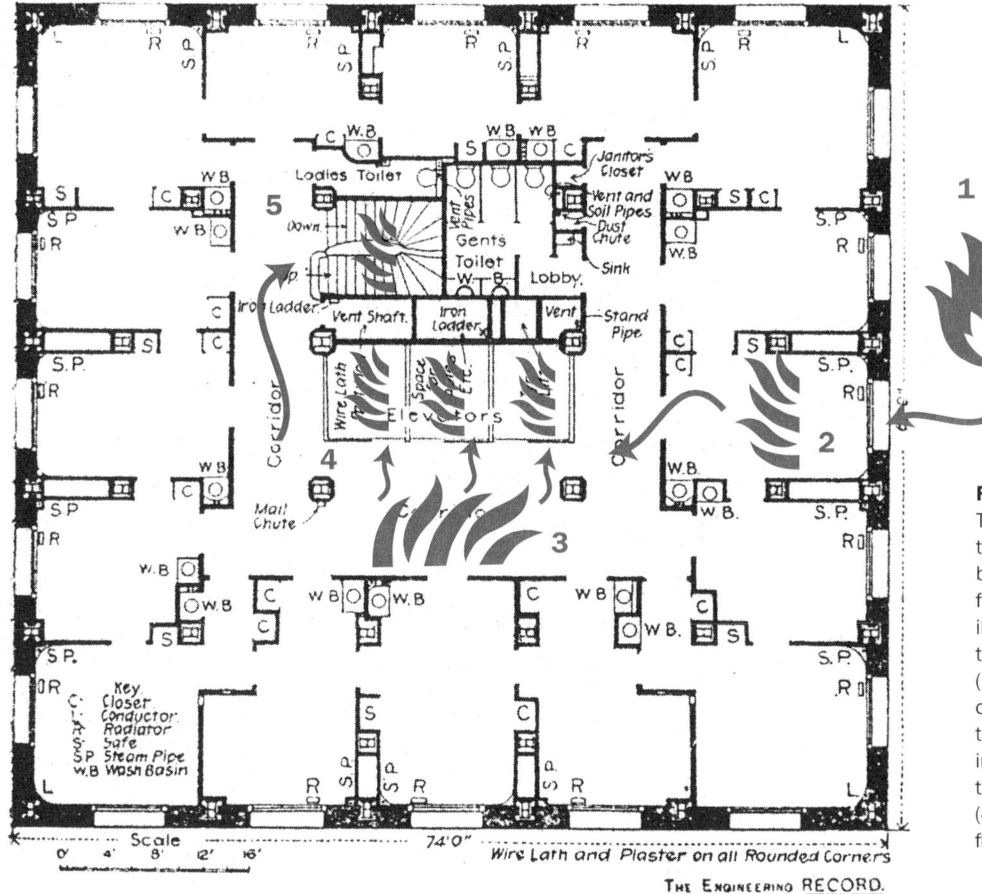

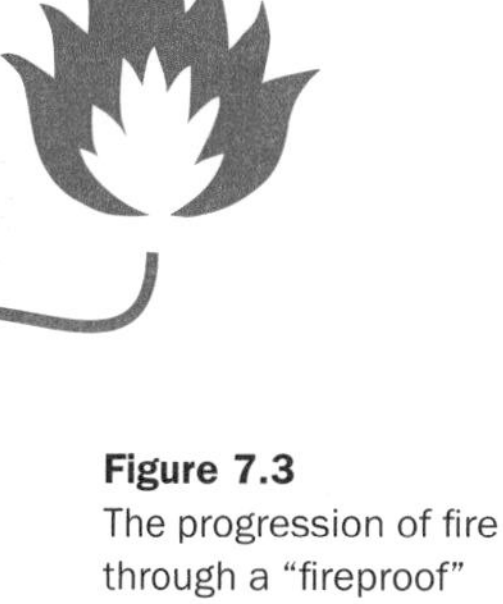

Figure 7.3
The progression of fire through a "fireproof" building: the plan of a typical floor of the Call Building illustrates how a fire enters through a broken window (1), burns the furniture in an office (2), is sucked through the glass transom window into the corridor (3), and then uses the elevator shafts (4) and the stairway (5) as flues.

> by not burning up: that it has not done so is largely due to the vigilance of the fire department, which cannot be relied upon indefinitely to stave off the inevitable.[50]

San Francisco's newly formed Board of Fire Commissioners pointed out that unlike Chicago, which had approximately seven brick buildings for every five frame structures, San Francisco's ratio was one to fifteen. They confirmed that no other large city in the country had such a relatively large number of frame buildings. More than 90 percent of the city's building stock was wood-frame construction, and much of it was redwood.

San Franciscans understood the vulnerability of wooden buildings, both outside and inside the fire district, but took some solace in the fact that they were built of redwood, which was thought to be fire-resistant. This is illustrated by a section of the city's 1903 building code that requires dividing partitions between buildings (common walls) to be "boarded with redwood" so as to effectually check all connections from one structure to another.[51] The presumption that any kind of wood might be used as a firebreak seems laughable today, and after the fire of 1906, San Franciscans' faith in the fire-resistant qualities of redwood was lampooned by many. However, even after the 1906 fire, San Francisco Fire Chief John Dougherty, who succeeded Sullivan, still believed that redwood was fire-resistant. And in explaining the favorable insurance rates given to owners of San Francisco's wooden buildings in the late nineteenth century, Englishman Robert K. Mackenzie wrote:

> The timber used for building purposes, redwood (*Sequoia sempervirens*), was of a peculiarly porous nature, absorbing moisture from the sea fogs which are common at night. "Redwood won't burn," was a phrase on the lips of every insurance expert in San Francisco. After all, there is something to be said for it, for, as a matter of fact, the record showed that fires in redwood buildings had been almost as successfully held in check as those in erections of brick, and redwood houses with blistered paint line the streets where the fire was checked.[52]

It is possible that the old-growth redwood San Franciscans were using had a far greater resistance to fire and rot than the second- and third-growth trees we presently use. These older trees have proven to be stronger than later growths because of their great age and girth when milled. Even today, heartwood in redwood trees is considered to be more fire-resistant than other woods, like pine.[53] But the wood used in frame structures, as opposed to hand-hewn buildings, is in smaller and more numerous pieces; it burned like kindling, not like slow-burning logs. When subjected to the terrific heat of a conflagration, redwood burns just fine.

Buildings Saved

In the midst of the conflagration in the South of Market, two federal buildings survived the earthquake and fire. Both were constructed to be earthquake- and fire-resistant. The U.S. Mint at Fifth and Mission Streets was buffered from other buildings by streets on all four sides. It was protected from the outside by firemen who used salt water from the Olympic Salt Water Company hydrant, and by employees who stayed inside the building and used water from an artesian well in the courtyard, putting out small fires with wet sacks.[54]

The new Post Office Building, built on the former swamp at Seventh and Mission, was badly shaken by the earthquake. Costly interior decorations crashed to the floor. Cracks, numerous but repairable, marred the building's internal and external walls. As at the Mint, employees stayed in the building and defended the roof and interior from the fire.[55]

On the other side of the burning city, the old brick Appraiser's Building, built to be earthquake-resistant, survived intact. It too was saved from the fire by the efforts of loyal employees inside, helped by U.S. Marines from the cutter *Bear* outside.[56] The mill construction California Electric Company, in the burned area to the south, was also saved by employees using a private water supply.[57] Unfortunately, the Palace Hotel, a Class C, earthquake- and fire-resistant structure with a huge reservoir, ample hoses, and plenty of personnel, did not survive. By the time the fire, already a conflagration, was burning fiercely near the hotel, the National Guard had used the water from the Palace's private reservoir in a futile attempt to protect other buildings.[58]

In a fire like this, to save even a "fireproof" building required a combination of additional factors: first, a structure built well enough to have withstood serious earthquake damage; second, enough available water from public or private sources; and third, people to fight the fire from the inside as well as outside. Sometimes luck played its part as well. Remember the small Burdette Building on the corner of Mission and Second Streets (see chapter 6) that was shielded from the heat and flames to the west by the Atlas Building and from the east by two streets. It survived the fire intact, amid gutted neighbors.

Desperate Measures

Brigadier General Frederick Funston began moving troops into the city immediately after the earthquake, when he saw fires beginning to burn. Whether Funston was justified in moving his troops, and whether he was, in fact, invited into the city by Mayor Eugene Schmitz, is still in question.[59] In any event, the troops entered the city mid-morning on April 18, evacuating and cordoning off the downtown to protect it from looters. This action, while perhaps wise from a law enforcement perspective, prevented employees and interested citizens from protecting more buildings from the inside.[60]

Market Street, with fires along both sides, was no longer a viable firebreak. It was clear that firemen had to create their own break or make a stand somewhere. When it was evident that little water would be available for fighting the fire, the army, presumably with the acquiescence of Mayor Schmitz, began blasting as-yet-unburned buildings in order to flatten out the area on the perimeter of the fire and thereby deprive it of heights from which to leap to other buildings.

Many procedural questions have been asked over the years about how the blasting was done, and whether it was all officially approved. Demolition was widespread. The fire department, private personnel, and the army were all engaged in blowing up buildings without any central controlling authority or guiding concept. They were roundly criticized after the fire for aggravating it rather than containing it.

The close blasting, and the difficulty of bringing buildings down completely, made blasting a losing proposition. In a desperate attempt to stop the fire, the army also used artillery to raze buildings, with no clear success. The fire reportedly spread to Chinatown when explosions from a botched demolition blew flaming bedding into the area.[61] Even if the blasts did not themselves spread the fire, they blew out windows that might have offered some small resistance to the flames.

The eastward movement of the fire was finally stopped on April 20 by a concentrated fire-fighting effort at Van Ness Avenue.[62] Van Ness, one of the widest streets of the city, was a natural place to establish a firebreak. Permission was given to blast undamaged buildings on the east side, the fire side of the avenue, to make the break a mile long and several streets wide. Water to soak the ruins was relayed with the help of citizen volunteers from nearby hydrants and from the bay, two thousand feet to the north. Fortunately, the wind changed, blowing the fire back on itself. But no sooner was the fire at Van Ness contained than the remaining unburned portion of the waterfront was threatened. Fire Chief Dougherty, exhausted, as were his men, led the fight to save the Pacific Mail Dock pier, driving the fire back on itself. At 7:15 a.m. on April 21, seventy-four hours after it all began, the last fire was finally extinguished. Although government troops had initially stopped private citizens from assisting in firefighting or from salvaging documents and possessions from the fire's path, they were finally included in the defense of Van Ness Avenue and Dolores Street. Without them, the last stand against the fire might not have been successful.[63]

Figure 7.4
Fire damage, looking east from the Fairmont Hotel down California Street (A) toward the Kohl Building (B), old St. Mary's (C), the Merchant's Exchange Building (D), Grace Church (E), and the Mills Building (F)

The fire destroyed 4.7 square miles, 508 blocks, and 28,188 buildings. Of those buildings, 24,671 were wood-frame; 3,168 were brick Class B and C; 42 were Class A steel-frame; 259 were unclassified brick and wood; 15 were stone; and 33 were corrugated iron and wood.[64] It must have been a terrible sight (figure 7.4). But the fire did something much more subtle and insidious that affected the city many years into the future. It obscured the destruction caused by the earthquake itself. It made assessment of earthquake and fire damage difficult to disentangle, and safe reconstruction practices more difficult to define, evaluate, and implement.

CHAPTER 8

Assessment of Damage in the 1906 Earthquake: A Centennial Perspective

How surprising to realize that this is the first comprehensive report on earthquake damage in San Francisco since 1909. Although there have been many books about the 1906 earthquake, a report like this has not been attempted for eighty-six years. Between 1906 and 1908, reports detailing different aspects of the damage, particularly in relationship to steel-frame buildings, were written by Richard Humphrey, John Stephen Sewell, Frank Soulé, G. K. Gilbert, John D. Galloway, Maurice C. Couchot, Christopher H. Snyder, Charles Derleth Jr., C. B. Wing, Edwin Duryea Jr., Franklin Riffle, and others. Building damage, rather than being a primary focus, was used as an indicator of shaking intensity, liquefaction, and settlement by the investigators from the California State Earthquake Commission, the American Geological Society, and a special committee sent by the Japanese government who were interested in the profile of the earthquake.[1] Engineering groups, including a subcommittee of the American Society of Civil Engineers (ASCE) and the newly founded Structural Association of San Francisco, were curious about the performance of steel frames in tall buildings and interested in the few reinforced concrete buildings that there were to study.[2] They were less concerned with brick construction and barely discussed frame buildings.

Fire insurance companies, which funded two reports, were interested in the effect of the fire on major buildings, few of which were brick and none of which were wood.[3] Private material suppliers, like the Roebling Construction Company, sent investigators, like A. L. A. Himmelwright, to assess the performance of their reinforced concrete floors, mostly in steel-frame buildings.[4] This was just one of many such reports generated by private material suppliers. Individuals, like engineering professor Charles Derleth Jr., wrote their own reports, but did not usually give the performance of brick and wood buildings much attention.[5] Had they been interested, they could have studied brick and wood buildings in the unburned portion of the city. Instead, based on cursory observations, they condemned the use of brick and approved of wood, which represented 90 percent of the city's building stock.

In the mid-1920s, engineers working for the clay products lobby did study the performance of brick buildings. They contested the prevailing negative assessment of brick and offered an optimistic study of its performance. These studies influenced John R. Freeman's 1932 *Earthquake Damage and Earthquake Insurance*, which established a new level of investigation into earthquake science in general and damage to structures in particular.[6] Freeman's treatment of the 1906 earthquake is exemplary, a brilliant text that stands today as the primary research document for the history of earthquake-resistant engineering. Yet even Freeman omits wood-frame buildings from his survey.

Since the 1930s, knowledge of earthquakes has increased as the ranks of scholars and practicing engineers have burgeoned, but new information on damage from the 1906 earthquake has been minimal. Scores of books have been published on the earthquake and fire, but their focus is "disaster"—not the specifics of damage to different building types.[7] Books by William Bronson, Eric Saul and Don Denevi, and Gladys Hansen illustrate the catastrophe vividly.[8]

The reconnaissance survey which follows, written in the style used by the Earthquake Engineering Research Institute (EERI), stands alone as a study of the earthquake damage from a centennial perspective and stands alongside earlier studies for comparison. There is no contemporary quantitative analysis of how much or what kind of damage San Francisco buildings sustained in the 1906 earthquake. While evidence at first seems plentiful, an inquiry into the damage to even a single major building quickly dispels that impression. The fire burned hot enough to melt and distort steel and to vaporize wooden beams and columns on the interiors of the brick earthquake-resistant structures, so even the most meticulous investigator would not have been able to write an accurate report detailing their performance. However, brief reports of the performance of steel-frame buildings immediately after the earthquake and fire rarely contradict one another.[9] The description of damage that follows is as accurate as I think is possible. It includes information from many early reports and new interpreted material. It correlates and integrates formerly disparate views to create a new understanding of the damage, particularly to brick structures.

A Reconnaissance Report of the San Francisco Earthquake of 1906

As is the custom, I will begin by discussing soil conditions:

Geologists studying ground motion and shaking intensity after the 1906 earthquake argue persuasively that filled land is the most hazardous soil for buildings in earthquakes.[10] Several filled areas in San Francisco can be used to prove their

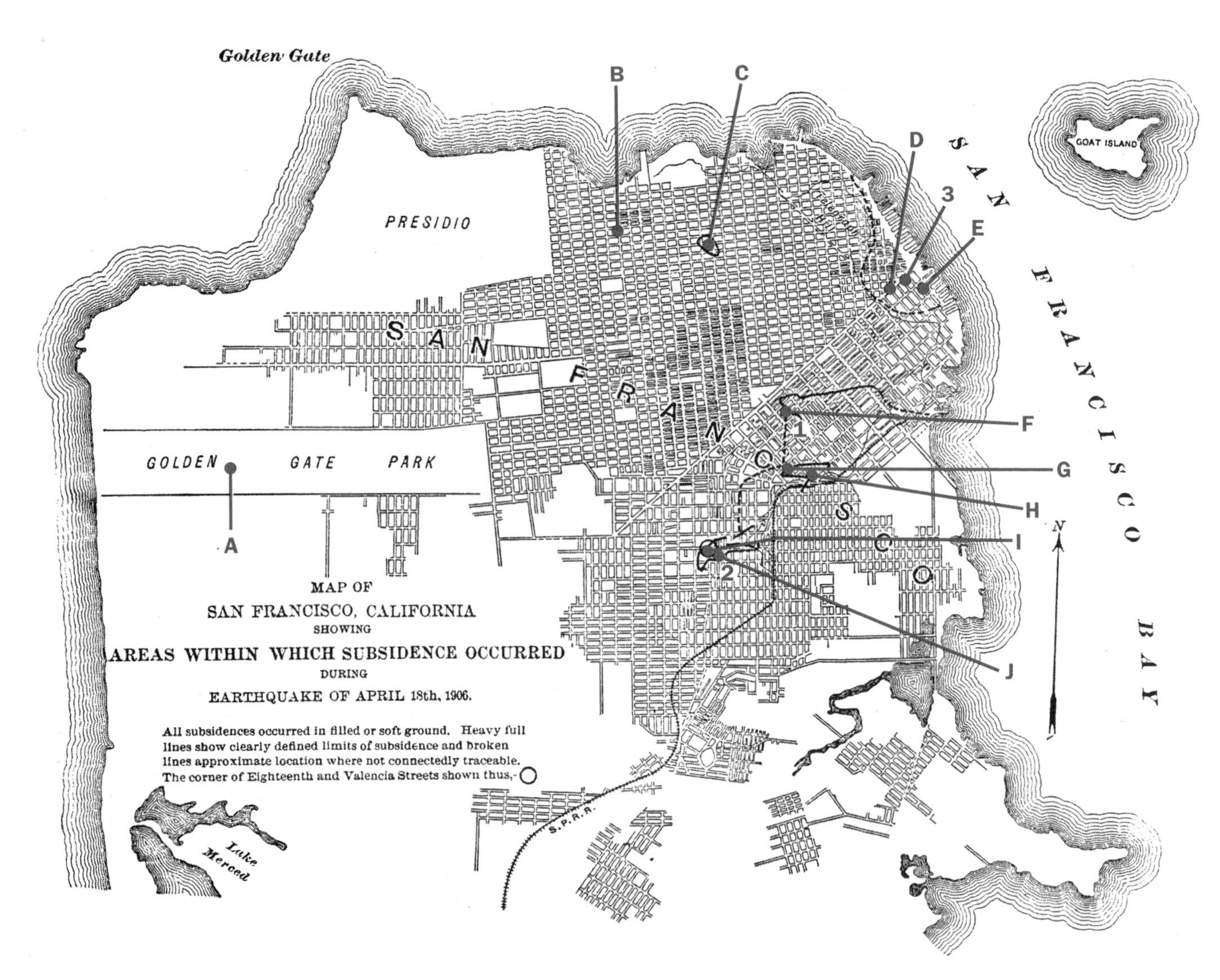

Figure 8.1

A map of earthquake damage due to subsidence in San Francisco in 1906, published by the American Society of Civil Engineers (1907). Areas of subsidence and ground failure are noted in color:

1. Marsh in the South of Market
2. Old Mission Lake and Mission Creek
3. Yerba Buena cove

A. Landslide on Strawberry Hill
B. Slip of fill at Union and Steiner Streets
C. Subsidence on Van Ness Avenue that broke water main
D. Subsidence on Market Street
E. East Street subsidence and surface rupture
F. Subsidence at Seventh and Mission Streets, at the corner of the Post Office
G. Subsidence on Dore Street
H. Subsidence on Brannan Street
I. Valencia Street
J. Howard (present South Van Ness) and Seventeenth and Eighteenth Streets

Figure 8.2
The Valencia St. Hotel, on Valencia Street, collapsed in the 1906 earthquake.

argument. Foundation failures caused by ground failure are easy to identify, as such buildings tend to settle and may collapse entirely. We find these foundation collapses in the marshes of South of Market ("Area 1" in the following discussion), the former river bed and lake associated with Mission Creek (Area 2), and the old cove of Yerba Buena at the foot of Market (Area 3, figure 8.1). Geologists analyzed not only building damage but breaks in the underground piping of the water system to map areas of liquefaction. Harry O. Wood's maps of earthquake intensity in the famous Carnegie Report on the 1906 earthquake pinpointed areas of liquefaction by collating building damage and soil types.[11]

The most dramatic failures occurred in the least affluent neighborhoods of the city. Poor workmanship in the damaged structures contributed to the loss of property and life. One of the most tragic failures occurred (in Area 2) on Valencia Street between Seventeenth and Eighteenth Streets when the wooden Valencia Hotel collapsed due to liquefaction and shaking (figure 8.2).[12] The hotel not only suffered foundation failure, but as its foundation dropped, it telescoped downward, collapsing on itself, and then fell outward. Tipped buildings on either side of the street indicate similar foundation failures. The situation worsened when soil movement ruptured the water mains under Valencia Street, pouring thousands of gallons of water into the already saturated and unstable soil, causing further pavement settlement and undermining already damaged buildings.

Similar foundation failures occurred in nearby streets where filled soil predominated. Also in Area 2, at Howard Street (now South Van Ness) between

Figure 8.3 (top)
Howard Street (now South Van Ness) after the 1906 earthquake

Figure 8.4 (bottom)
Detail of Howard Street (now South Van Ness) after the 1906 earthquake

Seventeenth and Eighteenth Streets, buildings seem to have dropped backward, some slipping to the side.[13] Concrete stair landings and masonry foundations close to the sidewalks didn't drop as much (figures 8.3 and 8.4). The buildings near the corner of Eighteenth fell to the north. Major damage occurred when buildings jumped off their foundations and smashed their cripple walls. The fact that few of these buildings collapsed, even when dropped sideways from their foundations, illustrates the strength of properly built wood-frame construction.

The earthquake left a record of its movement in Area 1, South of Market on Dore Street, in waves in the pavement. Modest houses with apparently very minimal foundations appear to have been tossed side to side, but they did not collapse (figure 8.5).[14] Sometimes inhabitants of the area were not so lucky. On Brannan Street not far from Dore, the shaking and the ground movement, which appears to have been a lateral spread, crumbled foundations, collapsing several wooden buildings and presumably killing their inhabitants.[15]

Although filled areas can be dangerous in earthquakes, with appropriate foundations, design, and detailing, buildings can survive. In Area 3, Alfred Mullett had built the reinforced brick Appraiser's Building in 1870 on a deep concrete mat foundation built to move as a unit. It was undamaged in 1906. In the same location fifty years earlier, architect Gridley Bryant had failed to tie the pile foundation and the structure of the U.S. Custom House together. This flaw probably contributed to the building's being severely damaged in the earthquakes of 1865 and 1868. Mullett knew of the Custom House, and he designed the Appraiser's Building and the more famous U.S. Mint to be earthquake-resistant. The nearby Montgomery Block, between Washington and Jackson Streets, constructed on a foundation of joined redwood timbers, also did well in 1906. The U.S. Court of Appeals and Post Office, built on a concrete foundation, was damaged but remained level despite drops of two to five feet in the land around it (figures 8.7a

Figure 8.5 (left)
Dore Street from the east, after the earthquake of 1906

Figure 8.6 (below)
East Street after the earthquake of 1906

and 8.7b). Tall, steel-frame buildings, even those constructed on fill, rode through the earthquake with little damage, indicating that although fill can be extremely dangerous, structures built either on piles or mat foundations can resist earthquake damage.

The earthquake also caused landslides and settlement outside the filled tidewater areas. In Pacific Heights, Union Street bulged over an old filled area west of Steiner Street (figure 8.8). The steel cable, reinforced concrete Cyclorama, a bicycle track and colonnade offering a 360-degree view of the city from Strawberry Hill, above Stow Lake in Golden Gate Park, was destroyed in the catastrophe by a landslide that undermined its foundation.[16]

Figure 8.7a (above, left)
Seventh Street and Mission, with the Post Office and Court of Appeals Building behind

Figure 8.7b (above, right)
Looking east from Mission at Seventh Street

Figure 8.8 (right)
Union Street landslide

A Review of Building Performance by Construction Type

To review San Francisco's 1906 building classifications: steel-frame structures were designated as fireproof Class A. Structures with load-bearing masonry walls with internal steel skeletons or other supports appropriately fireproofed were Class B. Buildings with load-bearing walls of masonry with internal combustible wooden posts and girders were Class C. A special factory system using large wooden members and brick facades was called mill construction. The last designation, frame buildings, included all the wooden buildings in San Francisco.

Tall Iron- and Steel-Frame Buildings

Although they suffered varying degrees of damage, tall iron- and steel-frame buildings, whether with exterior walls hung from the frame or with exterior load-bearing masonry walls, performed well in the earthquake. Professor Frank Soulé, dean of the College of Civil Engineering, University of California, Berkeley, wrote:

> So far as the writer can learn, *the architects had believed that in establishing solid foundations for high steel buildings, with good anchorage and bracing, adequate to take care of extreme wind force,* they had sufficiently guarded against the effects of any earthquake vibration which might occur. As a matter of fact, the *provisions thus made seem to have been ample and safe* so far as any disturbance of the foundations or any lack of support of the superstructure has been detected. Notwithstanding the severe vibrations these tall buildings have been called upon to endure, *they have remained plumb and very slightly damaged by the earthquake.*"[17] [emphasis added]

According to the ASCE, "no foundation damage was observed in any buildings" of Class A or Class B.[18] All foundations, which generally consisted of friction piles capped with concrete and covered by a grillage of steel, performed well in the earthquake. Earthquake movement side to side and up and down causes steel structural frames to bend and distort. Connections between steel members, like rivets or bolts in 1906 or welded connections in the twenty-first century, can be stressed beyond their breaking point: bolts and rivets can break off or shear, and welds can crack or tear.[19]

The Ferry Building (Union Ferry depot) tower is a good example of the damage that could occur to a Class A building (figure 8.10).[20] The Ferry Building was built to serve as a ferry terminal: a shed with a tower, constructed on a foundation of concrete piers on wooden piles, partially on fill and partially in the water. The structural system for the tower was a steel braced frame with diagonal steel bars attached to the frame by gusset plates to provide resistance to lateral (side to side) forces (figure 8.9). Both the shed and the tower were clad in stone. Structurally, the tower should be viewed as completely separated from the shed. It was designed

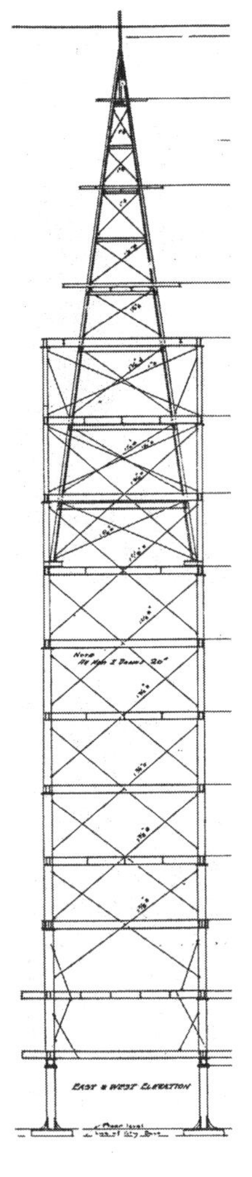

Figure 8.9
Ferry Building (old Union Ferry depot) tower structure, illustrating steel skeleton and diagonal bracing, after the earthquake of 1906

Figure 8.10
The Ferry Building tower after the earthquake of 1906

to resist 30 pounds per square foot (psf) wind pressure. Because the Ferry Building was severely shaken, did not collapse, and was outside the fire area, it is an excellent subject for study. The lateral forces on the tower generated by the shaking probably far exceeded the equivalent of 30 psf. The diagonal steel bars were pulled as the structure vibrated back and forth. They stretched and finally pulled so strongly that they sheared the rivets holding the gusset plates to the frame (figure 8.11). Why did this occur? Because the demand on the structure exceeded its capacity. That is, the forces acting upon the building were stronger than those it was built to resist. The building was clad in heavy stone and was built to be very stiff. The stiff tower didn't have the flexibility to dissipate lateral forces. The weakest link (the diagonal bars) broke, but miraculously, the tower did not collapse.

The sheared rivets of the first floor of the heavy Flood Building illustrate a different problem: a potential ground-floor soft-story mechanism.[21] The lateral forces of the earthquake pushed the column girder connections past their limits, breaking the rivets which held them together. The problem was probably created because the ground floor was designed with higher columns and fewer interior walls, in order to accommodate stores with larger open spaces.[22] As the whole building moved sideways, the movement was concentrated at the most flexible point in the structure, the first floor, rather than being distributed throughout the height. The steel frame attempted to resist this distortion at the column-girder connection and sheared the rivets joining them together, but luckily the ground floor did not collapse.

On the other hand, the Atlas Building illustrates how a building frame can bend back and forth over its entire height in an earthquake, again without collapsing. In 1906 the whole steel frame moved in a north-south direction. The movement was damped by the masonry walls, which cracked, releasing energy. A building must be able to dissipate energy through lateral movement, but not so far as to be unstable in any one story or to overturn.

Evidence of the Call Building's oscillation is preserved in the diagonal bars in the basement. As you remember, the steel skeleton of the nineteen-story Call Building was well braced. It had portal braces at each of its corners up to the fourth floor, all girders had gusset knee braces, and there were four sets of x-braced steel bars per floor from the basement to the fourteenth floor, and three above. In

1906 ASCE observers commented, "A few of the eye-bars were bent, as if the opposite ones had taken the strain," but there were no sheared rivets.[23] A careful examination of the steel bracing in the basement of the Call Building today shows some buckling toward the east (figures 8.12 and 8.13). Clearly, the steel brace, like this other bracing in the building, was tested by earthquake forces on April 18 and perhaps in the Loma Prieta earthquake.

Investigating engineers noticed X cracks in the exterior facades of many high buildings in San Francisco. Serious cracking was said to have occurred on the Atlas Building, the Flood Building, the Rialto Building, the Mills Building, the Merchant's Exchange, the Grant Building, and the Chronicle Annex, among others.[24] Not all the alleged damage was photographed, so it is difficult to comment on causes at a case-by-case level. Obviously, the Atlas X cracks were caused by drift (interstory movement); the cracks in the Rialto Building were caused by drift and pounding. Observers said damage was greatest at column splices; the Rialto Building seems to confirm this conclusion because it sustained significant damage on its second floor, where transitions in column size occurred.[25] In the Chronicle Annex, which incorporated deep spandrel girders intended to stiffen the building, X cracks appeared nonetheless, indicating drift in the columns. Most engineers realized the cracks indicated excessive movement of the steel frame but could not decide among themselves what the best bracing strategies ought to be (as we shall see in chapters 11 and 14). They were alarmed by the X-cracking and did not realize what we know today, that it is extremely difficult and expensive to design buildings that suffer no damage in large earthquakes. Structures must absorb the energy of shaking by cracking stiff walls or bending steel elements, both of which will cause damage, releasing energy while stopping excessive movement in the steel frame itself.

Engineers of the time were impressed by the excellent performance of the steel frame of the Whittell Building (figure 6.15), on Geary Street near Union Square.[26] But from our standpoint the frame should have performed well because it only had to support itself—it was not yet loaded, it had no floors or walls. Nevertheless, engineers were taken with the incomplete Whittell Building as a symbol for proper steel construction because it looked impressive among the ruins and was completely intact after the earthquake and fire.

The fire melted connections and buckled or collapsed columns, making it more difficult to pinpoint earthquake damage. A typical example of fire damage can be seen in the melted and buckled

Figure 8.11 (top)
Ferry Building tower, damage to steel frame in the 1906 earthquake. The diagonal brace has pulled its shoe and gusset plate so violently that all the rivets anchoring the gusset plate to the frame have sheared. Light shines through their empty sockets.

Figures 8.12 and 8.13 (middle and bottom)
Call Building, diagonal bracing in the east-west wall of the basement (illustrated in figure 5.4)

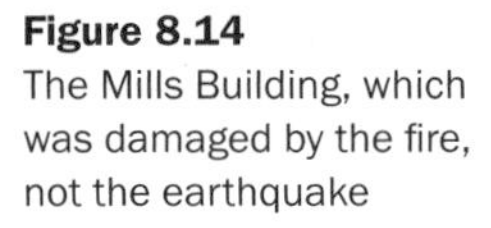

Figure 8.14
The Mills Building, which was damaged by the fire, not the earthquake

steel columns of the Fairmont Hotel.[27] But the cause of most damage is not so easy to discern. It is impossible to understand what happened to the interior of the Emporium Building, which may have suffered partial collapse and then burned.[28] The fire complicated matters further by causing explosions, which wrenched buildings again, like a large aftershock. Another cause of confusion was the attempted demolition with dynamite of many buildings.

Although most of the structural damage was immediately apparent, some lay hidden behind walls and column protection. The report by a group of Pacific Gas and Electric Company workers examining the Shreve Building seems to indicate damage to column protection but not to the columns themselves: "They found the building seriously shaken, its plaster walls cracked, and the sheathing of its steel columns shattered in many places."[29] More subtle damage took time to emerge. In 1910, four years after the earthquake, architect Julia Morgan and engineer Walter Steilberg investigated a newly opened wall and discovered new signs of cracking in the steel connections of the Merchant's Exchange, which they believed to be a result of the 1906 earthquake.[30] This late discovery illustrates how hard it is to accurately survey building damage. The problem still exists today, when, in many buildings, the seismic system is covered with cladding and can't be easily seen after an earthquake.

Brick Infill Masonry in Class A Buildings and Load-Bearing Brick Walls in Class B Buildings

The city's Class A and B buildings had either brick infill walls (walls held up by the steel frame) or load-bearing exterior brick walls, and many had brick facades as well, which would seem likely to create dangerous brick falls in the street. Yet looking through the photographs of San Francisco streets after the earthquake, one can see that the walls of individual buildings are largely intact, and there is little evidence of bricks in the streets; most brick facades in Class A and B buildings survived the earthquake without collapsing. The Call Building, the tallest Class A building the city with infill masonry, did not lose a stone.[31] The masonry in the uppermost floors was damaged by the earthquake, however, and had to be repositioned during the repair. The Mills Building (figure 8.14), with load-bearing masonry walls and a steel skeleton, also lost no individual bricks, terra-cotta, or stone, although pieces of exterior masonry were reported to have been cracked.[32] There was disagreement as to whether the load-bearing facade of the Mills Building was pushed out of plumb by the earthquake. One other facade, that of the Merchant's

Exchange Building, was reported to have lost some bricks. According to Professor Charles Derleth, "large areas of face brick fell from the west wall."[33] I believe this failure was, if not unique, at least one of the very few in Class B buildings with steel frames.

With the prevailing belief in the engineering profession today that unreinforced masonry is unacceptable as a building material in earthquake country, it is challenging to confront the surprisingly good performance of brick buildings in San Francisco in 1906.[34] Unreinforced masonry buildings (called "URMs" among structural engineers) performed poorly in the California earthquakes of Santa Barbara in 1925, Long Beach in 1933, San Fernando in 1971, Coalinga in 1983, Loma Prieta in 1989, Northridge in 1994, and San Simeon in 2004. In the 1906 earthquake itself, unreinforced masonry buildings at Stanford University, thirty miles south of San Francisco, were wrecked.[35] Not far from Stanford University in nearby San Jose, the Agnews Insane Asylum, constructed of unreinforced brick and stone, collapsed and killeds scores of patients.[36]

However, Class B and C load-bearing brick buildings with combustible interior supports performed better than modern engineers would have expected. The pictorial evidence suggests that a majority of brick buildings in the city did not lose cornices and fire walls, and did not collapse. In fact, complete collapse of multistoried brick dwellings or offices in earthquakes in the United States is a rarity because internal wooden partitions often keep the roof up even when the exterior brick walls fail.[37] These partitions take the load off the roof or upper floors even if they weren't designed for that. The more like a rabbit warren the interior is—the more walls—the more capacity. Sometimes the poorest rooming houses, those with the smallest rooms, survive partial collapse more easily because of their numerous partitions. In 1906 one industrial building illustrated this phenomenon: Rheingold's Warehouse (figure 8.15), with a heavy wood frame and brick facade, suffered extensive damage because it had no anchors joining its brickwork to its interior frame. The warehouse's heavily loaded floors knocked against its brick facade until it weakened and fell outward, exposing the interior. No diagonal bracing can be seen on the interior wooden structure, but the multiple interior partitions (unusual for a warehouse) take the load of the roof.

We have seen similar phenomena in hotels,

Figure 8.15
Rheingold's Warehouse, illustrating wall-diaphragm tie failure (see figure 4. 5) and out-of-plane failure. Diaphragms of floor and roof have kicked out bricks in the upper two stories. The bricks have fallen on the street and onto the building to the right, crushing it.

Figure 8.16
The California Hotel before the earthquake, with adjoining fire station occupied by Fire Chief Dennis Sullivan. Overlay depicts the trajectory of the chimney that fell on the firehouse on April 18, 1906.

rooming houses, and restaurants in San Francisco in which a facade detaches from the structure, leaving the lateral bearing walls and floor intact. The ASCE report explains that "a favorite method of constructing such buildings, when used as stores, was to build the front wall upon a girder, placed at the second-floor level, and supported on columns. As the interior form was generally independent of this girder and the wall, the result was the wall fell outward."[38] Examples of this kind of failure include the Delmonico Restaurant, on O'Farrell between Stockton and Powell; the Colonial Club, on Powell Street between Ellis and O'Farrell; A. Paladini Inc., at 520 Merchant Street; and several buildings in Chinatown.[39]

The problem with bearing-wall brick buildings without frames is that even if they do not collapse, they can drop their parapets, cornices, and fire walls. If only a relatively small percentage of all brick buildings lost a fire wall or a cornice, the result would be devastating for neighboring structures or pedestrians. What follow are documented cases of outward falling walls, parapets, or cornices. The witnesses are mostly from downtown San Francisco. Several incorrectly state that the cupola of the California Hotel collapsed into Chief Sullivan's quarters on the top floor of Chemical Company Number Three below it.[40] This is probably not true, because it is unlikely that the round turret would fail. Instead, a chimney from the California Hotel attached to the cupola fell sixty feet onto the firehouse beside it, smashing a hole all the way to the basement (figure 8.16). Other similar incidents include debris from the steeple of St. Dominic's church crushing a neighboring house (figure 8.17); a residence on Geary Street near Grant being crushed by a wall from the building next door;[41] and the firehouse on the southwest corner of Drumm and Commercial Streets being crushed by walls from the Ames and Harris Bay Factory.[42] Horace Dunn recounted how bricks and parapets from his brick rooming house on Montgomery Street fell, killing people and horses on Merchant Street (John Galloway mentioned this same event).[43] Officer Meredith described a wall from the Columbia Theater falling on a neighboring lodging house, crushing its roof and top two floors.[44] Nearby, a brick wall from a taller structure fell onto a Union Square lodging house called the Geary, destroying it.[45] Officer Meredith tried to rescue the inhabitants from the wrecked interior: "The stairs and hall were especially ticklish. The hall had been broken away from one wall, and was slanting at a heavy angle, threatening to drop at any moment." The rear wall of the Victoria Hotel, at the southeast corner of Geary and Williams, broke off in a similar fashion and crushed an adjacent two-story frame building "like a concertina,"[46] and the Girard House, a residence hotel on the corner of Howard and Seventh, was destroyed by bricks falling from a taller brick building, taking a toll of fifty to sixty lives.[47]

A few Class B and C brick buildings did collapse completely. Some pancaked, or telescoped from their roofs into their foundations. The buildings collapsed either because their brick walls fell away, eliminating support for floor structures, or they failed in plane, along the wall, also releasing joists. Some soft ground stories also played a significant role in wrecking buildings. Derleth described in-plane shear cracks in buildings he saw. Yet from surviving photographs, it seems that out-of-plane failure was the most common type. As the walls fell outward they would displace the wooden roof and floor joists. These joists then fell downward, knocking walls out as they dropped. This could occur in low, multistory brick rooming houses or commercial buildings, where one floor collapsed into the floor below it. Only a few cases of this type of collapse are documented, although we know more occurred.[48]

Figure 8.17
Rectory of St. Dominic's after the 1906 earthquake: detail of damage to wood structure caused by bricks falling from St. Dominic's

More frequent was the collapse of buildings with large open spaces, such as auditoriums, religious buildings, warehouses, or factories. These buildings were characterized by multistory brick walls that were more vulnerable to out-of-plane failure. They also had few interior vertical supports to hold up the roof when the walls failed. According to the investigators, these walls often appear to have been undersized, with the roof trusses improperly tied or not tied to the walls. Typical examples of this kind of failure were two buildings in construction on Geary near Fillmore Street, the temple of Congregation Beth Israel and Pike's Memorial Building next door (figure 8.18).[49] A case might be made that the mortar had not had time to set, thus causing the collapse. More likely, the roof trusses were not adequately attached to the walls and the walls themselves were weak. The diagonal cracks around corner apertures indicate the structure was racked by the earthquake and that the proportion of windows to walls was excessive. Even the remaining walls were beginning to fail in plane. The Beth Israel building burned in the late 1980s and I had an opportunity to examine the connections between walls and floors during the demolition. It lacked any iron reinforcement whatsoever—no tie bars, anchors, or tension rods.

Girls' High School (figure 8.19), erected in the 1870s according to the highest standards then available for earthquake-resistant structures, failed when roof trusses spanning an upper-floor auditorium pushed out the tops of its load-bearing brick gable walls.[50] The building did not collapse, but it was so damaged that it had to be condemned. Large areas of unsupported brick created by tall story heights or gables are very vulnerable to out-of-plane failure. In any case, if the walls are not

Figure 8.18 (top, left)
Congregation Beth Israel and Pike's Memorial, Geary Street near Fillmore, after the earthquake of 1906: collapse of two brick buildings. Neither building was reinforced with iron rods or bond iron.

Figure 8.19 (top, right)
Girls' High School after the earthquake of 1906, with gable failures indicated

Figure 8.20 (bottom)
The Hahnemann Hospital after the earthquake of 1906: failures due to poorly designed gables, poor brickwork despite bond iron, and poor configuration. As the ends of the U-shaped building wagged back and forth, they collapsed.

firmly tied to the trusses, they can also fall outward. This same defect caused several major buildings to collapse, including the Majestic Theater, the Tivoli Theater, and perhaps the Jackson Brewery.[51] A similar problem appears in the collapse of the end walls of the wings of the Hahnemann Hospital, at Maple and Washington Streets in the Western Addition (figure 8.20).[52] The irregular plan of the Hahnemann building contributed to its demise. The end gables collapsed, and despite the use of bond iron, one wing shed the corner of its brick wall. This is because of a problem of poor configuration, a concept not understood in 1906. The open, U-shaped plan cannot move as a unit in an earthquake. While the bottom of the U pins the wings at their base, they are free to wave back and forth and flap at the ends. The irregularity of the plan thus contributes to the wagging of the ends and the breakup of the brickwork at the corners. In addition, observers commented that below the open wall there were clean bricks, lying separately, not stuck together in clumps, indicating that the mortar and brickwork were of the poorest quality.

One of the most spectacular failures of the Western Addition was the brick church of St. Dominic on Pierce at Bush Streets (figure 8.21). Unlike the Hahnemann Hospital, the walls of the church were completely unreinforced. The towers collapsed, but other weaknesses can be seen as well. The incipient failure in the southern transept illustrates another example of a configuration problem. The back walls of the transept are rigidly connected to the main body of the church, while the end of the transept is free to wag back and forth. Like the Hahnemann, it began to collapse at the corner, where you can see a long diagonal crack. The gable of the transept nearly fell away from the church. Similar gable damage due to the same design failure can be seen on the south transept of the Mission Dolores Church north of the old Mission, and the assembly hall of St. Ignatius College at Van Ness and Grove Streets, where a collapsing gable broke off the central cornice and parapet.

Masonry Towers and Steeples

High towers and chimneys throughout the city cracked and collapsed, partially or totally. They were particularly vulnerable because of their height-to-width relationship. The brick exterior walls of such structures most often cannot remain integral during the large movement they experience during earthquakes. They crack and lose their stiffness, allowing greater movement. After only a few cycles of such movement, the brickwork loses its integrity and falls.

Many of the tall chimneys in San Francisco, like the Jessie Street power station chimney whose fall DeWitt Lipe witnessed, collapsed.[53] Most were made of brick and most served power stations, including those in the Mission District, South of Market, North Beach near the present site of Ghirardelli Square, and on Potrero Hill.[54] Many of the church and synagogue steeples in the city collapsed, including: St. Patrick's in the South of Market, St. Ignatius, the church which adjoined Mission Dolores, St. Luke's at Clay and Van Ness, the Unitarian church at Franklin and Geary, and St. Dominic's.

Along with church towers, tall domed towers were also vulnerable. As we shall see, the steel framework of the dome of San Francisco's City Hall, like the wooden framework of St. Dominic's, did not collapse, but shed its masonry cladding, destroying much of the roof and rooms below it. At Stanford University, to the south of San Francisco, the library's lantern, dome, and covering survived, held aloft by steel columns that, due to their sideways oscillation, destroyed the dome tower's walls and the masonry at its base.[55] However, well-built, low-domed buildings, like synagogue Sherath Israel, survived the earthquake with little damage.

In Favor of Brick

Although there were numerous brick failures, the majority of brick buildings did not collapse, that is "pancake" (dropping one floor onto another) or lose cornices or walls. Looking at photographs of main thoroughfares like Market Street, Powell Street, and Montgomery Street, it is clear that the brick buildings are externally and overall intact. The brick buildings in the area around Jackson Square, between Clay, Washington, Sansome, and Montgomery Streets, as well as the block bounded by Taylor, Broadway, Jones, and Green Streets, escaped the fire. These buildings were largely undamaged by the earthquake, though they were quite ordinary two-story brick buildings.[56] They remained standing, with few parapet and fire-wall failures. In 1906 the investigators didn't rigorously examine failures and successes in brick buildings one structure at a time, so they have left us only general observations, mostly negative. Freeman, in *Earthquake Damage and Earthquake Insurance,*

Figure 8.21
St. Dominic's, Pierce and Bush Streets, after the 1906 earthquake: vulnerability of a standard brick church without metal reinforcement. Gable failures, parapet failures, and tower failures wrecked this building.

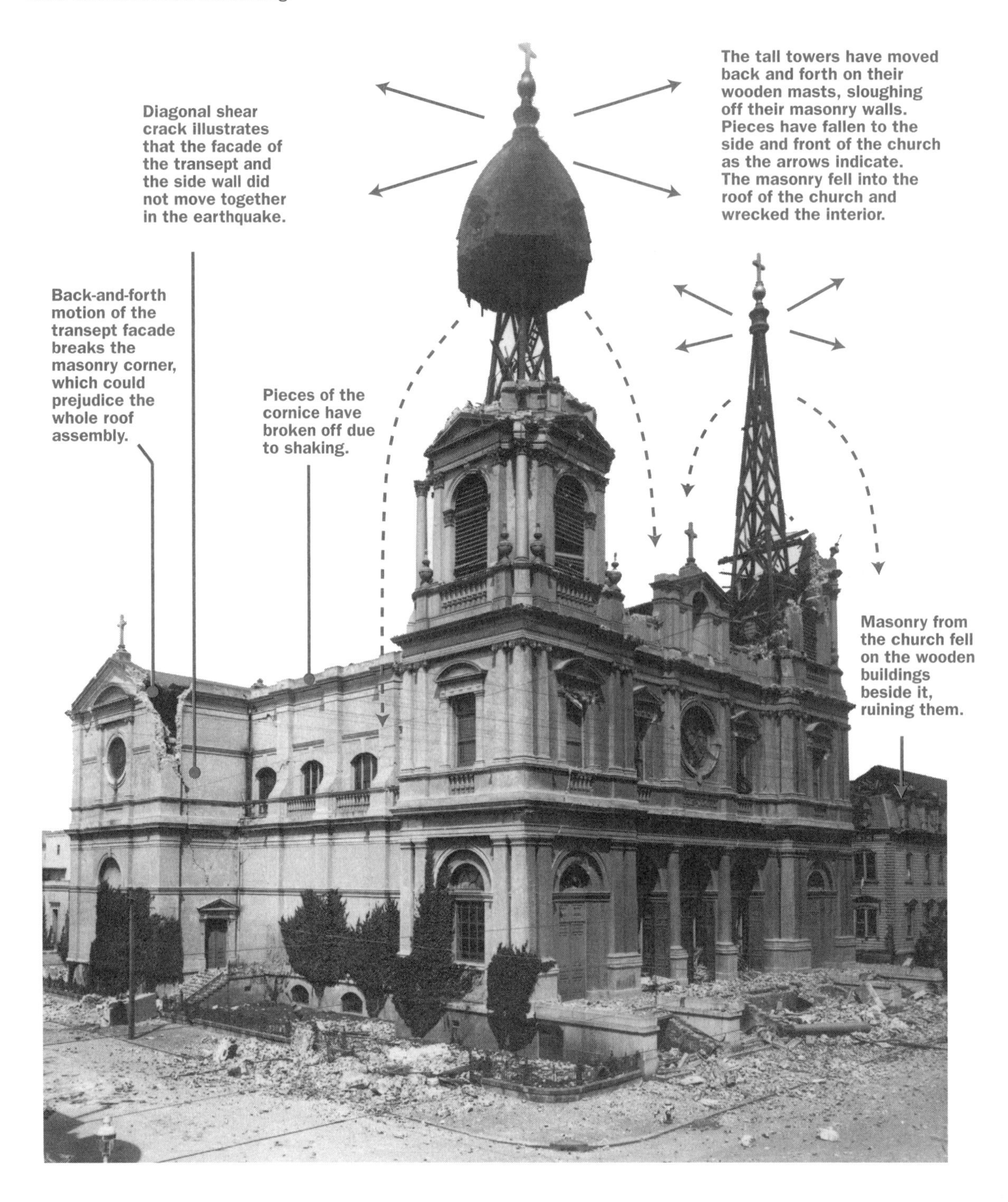

contradicts this negative evaluation by pointing out that many brick buildings survived the earthquake.[57] He lists as survivors undamaged by the earthquake the following load-bearing brick buildings: the high National Ice Company building (at Alameda and Rhode Island Streets), the brick Dunham, Carrigan and Hayden Co. Building, on fill (Kansas Street between Alameda and Fifteenth Street), a tremendous brick warehouse at the corner of Alameda and Rhode Island Streets, St. Mary's Cathedral (at the corner of Van Ness Avenue and O'Farrell Streets), the California School of Mechanical Arts (Utah and Sixteenth Street), and the Montgomery Block (Montgomery and Washington). Freeman points out that the Folger Building (at Howard and Spear Streets), with very thick brick walls and a partial steel frame, survived the earthquake with parapet and chimney damage but otherwise intact, although it stood on fill. He makes similar arguments for the Young Building, on filled ground at Spear and Market Streets, and the Lowry Building (at Mission Street between Steuart Street and the Embarcadero). He counters the examples of falling gables in churches with the gables of St. Francis's church and the Unitarian Fellowship, which did not fail. He argues that the brick chimneys of the U.S. Mint, the Clarendon Heights pumping station (on Seventeenth Street), and the Black Point pumping station (at Van Ness and Beach Street) did not collapse, and that the towers of the Grace and St. Francis churches are intact. To this list several score additional brick buildings could be added, including such survivors as the California Electric Company.

Brick Mill Construction

Investigators in 1906 were impressed by the California Electrical Company Building on Folsom Street between Second and Third Streets, which survived the earthquake but with some damage.[58] It was the only building in San Francisco that officially conformed to the Fire Underwriters' standards for slow-burning, or mill construction, now called "heavy timber," and fire protection. In reality, the timber frame, built massively so that it would char rather than burn, resembled the seismically resistant frame of the Grand Hotel. The California Electric Company, built on sand, lost a section of fire wall or parapet on the uppermost story (figure 8.22). This typical parapet failure was very dangerous to neighboring structures and pedestrians. In addition, its smokestack cracked at the base and the brick supporting its rooftop tank was damaged. The design of the building was flawed by the inclusion of a heavy elevated water tank, unwise in earthquake country; fire danger was more of a concern here than earthquake damage. The building's interior remained intact and successfully resisted the fire.

Figure 8.22
The California Electric Company Building, Folsom Street, considered a success after surviving both the earthquake and fire of 1906. Its interior was unharmed despite cracks in the water tank supports, the smoke stack, and collapsed cornices.

"Earthquake-Proof" Brick Buildings

None of the 1906 investigators knew that many buildings in San Francisco were constructed with features intended to improve their seismic resistance. In his letter to G. K. Gilbert recounting his experiences, Fred Plummer laments,

> During the past years I have several times courted the fate of one who persists in calling attention to the fact that the Pacific coast is an earthquake country and that engineers and architects should build accordingly. Frankly, the people do not want to know the facts. It hurts real estate. Yet I would as soon live in San Francisco as elsewhere, so far as this risk is concerned, provided the full lesson of the temblor had been learned, which it has not.[59]

Plummer may have owed his life to the earthquake-resistant retrofit of the Occidental Hotel (1869–1870) where he slept and was awakened by the earthquake and proceeded without incident on April 18, 1906.[60]

Researchers praised the Palace Hotel for its earthquake resistance without understanding that it was one of many buildings in San Francisco designed to be seismically resistant:

> The well-known Palace Hotel was built about thirty years ago, a few years after the earthquake of 1868, and before the introduction of steel-frame structures and concrete steel. It was intended to be earthquake-proof as well as fireproof, and was built with very heavy walls of brick, most of them being 2 feet or more in thickness, laid in cement mortar, and strongly braced by many cross and partition walls. In the brickwork, at every 3 or 4 feet in height, were laid bands of iron, riveted together at their ends and crossings. This building, although of the old type, successfully endured the great earthquake, its walls being practically uninjured."[61]

The same writer discusses the U.S. Mint and the Appraiser's Building without realizing they were in the same family of earthquake-resistant construction. "These buildings...entirely uninjured or only slightly injured by the earthquake...were well designed and constructed with the best materials and workmanship, upon foundations...found strong and satisfactory."[62]

The Palace Hotel, Grand Hotel, Appraiser's Building, U.S. Mint, Subtreasury Building, San Francisco Hall of Records, and Occidental Hotel all included special seismic design features and are documented to have survived the 1906 earthquake with more or less minor damage. This is an impressive list of "reinforced" brick buildings that seems to prove the strengthening methods pioneered in 1868 were valid. Of course it is important to remember that each earthquake is different and that soil conditions and accelerations in different areas in the same earthquake can skew data, but the success of these buildings indicates that while late-nineteenth-century understanding of seismic resistance was imperfect, we cannot dismiss their attempts.

Survival depended not on one "magic" element, such as bond iron, but on sturdy construction and an appreciation of how to design a structure to behave as a unit when moved by earthquakes. Bond iron, anchors, tie rods, a strong continuous foundation, and good design all played a role. That being said, even today we do not completely understand how such reinforcement works in brick structures, mostly because brick is not used, and therefore not studied or tested.

Not all supposedly earthquake-resistant buildings survived the earthquake without serious damage. The Lick Hotel, on Montgomery Street, had a wooden framework with brick walls that its owner claimed was "earthquake-proof." The building lost its cornice, "solid masonry, about 50 feet long, one and one-half feet in width and thickness," but did not collapse.[63] On the other side of the scale is the Mint, which survived the earthquake with its towers intact and no damage to interior or exterior walls.[64] Although the structure is veneered in granite and sandstone, the stone is backed by bricks tied with bond iron and held together by numerous horizontal and vertical iron rods. The deep concrete foundation, designed to survive an earthquake, sank slightly but remained intact, and no walls ruptured. Considering that the Mint was built on alluvial land next to the Mission marsh, it performed superbly.

To understand just how difficult it is to evaluate the performance of these older buildings constructed with special seismic features, let us return to David Farquharson's South Hall at the University of California, Berkeley. When it was slated for retrofit in the 1980s because it was classed as an unreinforced masonry building, the engineering firm of Rutherford & Chekene decided, because of certain design flaws in the building, that they could not depend on Farquharson's solutions. There were notable weaknesses. For example, the horizontal planes in the building—floors and roofs—were intended to act as diaphragms, distributing loads to the exterior walls. But the roof structure was poorly conceived and badly built.[65] The entire roof assembly needed to be rebuilt and many of its members replaced. The engineers also felt the many windows and fire flues in the facades weakened the wall planes to such an extent that they might fail in shear, that is "in plane." These weaknesses illustrate that even a superb example of 1870 seismic-resistant practice won't necessarily measure up to our standards today.

Most twenty-first-century San Francisco engineers would concur with the report that Sewell, Soulé, and Galloway made to the American Society of Civil Engineers in 1906: "It may be stated, as one of the most obvious lessons of the earthquake, that brick walls, or walls of brick faced with stone, when without an interior frame of steel, are hopelessly inadequate."[66]

This statement was hotly debated within the ASCE by other engineers who had witnessed scores, perhaps hundreds of successful brick buildings. One was engineer Franklin Riffle:

> During the first day of the fire the writer spent several hours in the district in which brick buildings were the prevailing type, and made note of the fact that most of these were in good condition before the flames gutted

> their wooden interiors and reduced their walls to ruins. If the committee had included in its investigations the brick buildings in the Potrero and Mission districts of San Francisco, which escaped the fire, it is doubtful if it would have condemned this type of construction in its otherwise excellent report. Almost without exception, these buildings either were not damaged at all or were damaged so slightly that the cost of repairs was trivial...It is the opinion of the writer that the one great lesson taught by the earthquake is not the inadequacy of any one of the well-known types of building construction, but the necessity of adhering more rigidly, in all types, to the fundamental principles of good design, good materials, and good workmanship.[67]

Engineer Edwin Duryea also challenged the conclusions of the committee, citing his own observations after the earthquake but before the fire, and providing a list of undamaged brick buildings.

In response to these criticisms, Galloway reiterated that he too was in the earthquake, and saw the destruction and loss of life from brick structures. In his rebuttal he dismissed quantitative evidence and based his argument on theory or engineering sensibility:

> Brick buildings may do in earthquake country. Some of them stood up. Outside of the fire districts, most of them were injured more or less, and many were destroyed. Their inadequacy, however, is not a question of counting a majority either way. It is purely an engineering question. It is a proper consideration of the stress involved and the resisting material. As such, the writer must insist that the non-flexible masonry, with *every joint a source of weakness* in the presence of a shearing mass, and with its dead, inert mass, is a "hopelessly inadequate" form of building to resist earthquakes. In the face of...[failures in San Francisco, at Stanford, and in Santa Rosa] it is strange that engineers will still champion the material which, analytically and by evidence, proves itself to be "hopelessly inadequate."[68]

Galloway's theoretical view of brick construction in earthquake country may be the correct one, but it is important to acknowledge that brick structures in San Francisco on the whole performed better than we might have expected in the earthquake of 1906. It is possible that the lessons of earlier earthquakes influenced the architects, engineers, and builders of even "average" structures to include seismic features in their brick buildings. Unfortunately, we cannot know today. But one building from the 1870s survives and might provide answers: the U.S. Mint. As of 2006 it is closed because it is judged to be a seismic hazard. Its walls should be investigated for the seismic features which, according to their builders, they include. The building should be evaluated considering the contribution or irrelevance of those features.

Wood-Frame Structures

Although wood-frame structures were only cursorily studied, researchers agreed that they performed well in the earthquake. The joint report of Gilbert, Humphrey, Sewell, and Soulé concluded that "well-constructed wooden buildings generally withstood the shock."[69] The committee of the American Society of Civil Engineers concurred: "Speaking generally, there was little damage to wooden frame buildings."[70] John Freeman summarized: "In general, wood-frame buildings resting on good foundations withstood the earthquake shock wonderfully well, even in places where the ground was moderately soft and mobile."[71] Photographic evidence corroborates these judgments. A photograph of the ruins of St. Dominic's church, in the heart of the Western Addition, shows the church surrounded by hundreds of undamaged wood-frame houses (figure 8.23). A view of the Hayes Valley fire encroaching on Grove Street from City Hall (figure 8.24) illustrates frame houses intact with the exception of fallen chimneys. Only an apparently brick facade on the north side of Grove Street has collapsed, cascading into the street. Scores of other panoramas tell a similar story.

Although wood-frame houses generally performed well, they were not immune from earthquake damage. They suffered because of five distinct design weaknesses: (1) poor foundations; (2) insufficient external and internal solid-sheathed walls; (3) poor vertical continuity; (4) unreinforced "cripple walls"; and (5) untied, poorly constructed chimneys. We have already seen how houses can be destroyed as their foundations settle. We have also seen how shaking can pull and push a wooden structure completely off its foundations, causing partial or complete collapse. Let's look at these five problems in more detail.

When a foundation moves, settles, and cracks, cross walls and exterior walls must remain integral, tied to one another. They must not disintegrate as they are pushed, pulled, and distorted. They must be able to resist shear, so they won't collapse in plane, and they must be able to channel forces through the floor and roof diaphragms. If they cannot, the building collapses, like the rooming house on

gure 8.23
ıe Western Addition as seen from Alta Plaza Park after the earthquake and e. Note outdoor kitchens. The ruin of St. Dominic's is on the left. Aside from llen chimneys, the wood-frame buildings look intact.

Figure 8.24
Looking east up Golden Gate Avenue toward the old City Hall after the 1906 earthquake. All the wood-frame buildings are intact, although one brick facade and several brick chimneys have fallen.

Golden Gate Avenue in figure 8.25. This building settled, but as it did, the interior walls collapsed in on themselves, pulling the facade toward the center of the block.

A common failure that plagued wood-frame buildings was the collapse of basement, or cripple stories: shorter than normal stories that usually serve as a kind of podium for the ground floor of houses in San Francisco. Cripple walls were often poorly reinforced, lacking any shear panels or diagonals. In the 1906 earthquake, scores of cripple walls probably failed, as in the three buildings in figure 8.27. Figure 8.26 illustrates a rooming house that was probably ruined by settlement and unbraced cripple walls. The wooden houses on Howard Street tipped at rakish angles, shown earlier in this chapter, have settled backward and fallen off their collapsing cripple walls (figures 8.3 and 8.4).

Infrequent, but visible in areas of the city, were wooden buildings that collapsed because of insufficient solid walls in the occupied stories. The building in figure 8.26 collapsed and smashed down to the right (the pediment over the door should line up with the stairs). The exterior bracing panels between the windows of the first floor were too slim to save the house. Apparently, there were few solid walls inside the house either. The walls between the windows could not resist the shear forces by themselves and ripped apart as the house fell sideways.

Figure 8.26 illustrates how a lack of vertical continuity in the load path of seismic forces can destroy a single story and cause the global collapse of a two- or three-story house. Two different kinds of framing were used in San Francisco's wooden buildings: the balloon frame, characterized by continuous studs running the full height of the building, and the platform frame, in which studs stopped at each floor level. One 1906 investigator blamed platform-frame construction for such failures and claimed balloon-frame construction was superior.[72] Even when distorted, the balloon frame shows its strength: in contrast to the Valencia Hotel, in which each floor collapsed, the walls of these buildings seem to have remained integral. Even the wall of the collapsed building in figure 8.28 bends between floors but does not come apart. The same can be said for the facade of the Golden Gate Avenue flats (figure 8.25), which fell backward. With balloon framing, even poor wood construction was very resilient and resisted total failure, as figure 8.28 demonstrates. This building is held together by its frame, even though its twisted shape indicates that the impact of the shock may have been amplified through the pilings of the wharf on which it stands. While these examples seem to argue for balloon-frame construction, platform frames were more efficient, cheaper, and more practical, given the growing scarcity of long lengths of high-quality wood, and could be constructed to perform well if properly nailed and vertically connected by metal ties.

The last and greatest danger confronting wood-frame buildings was the problem of brick chimneys. The most pervasive damage to wooden structures came from massive chimneys crashing through roofs.[73] Chimney ties that should have prevented this were specified in the 1904 code, but they do not appear to have been installed. Unreinforced brick chimneys are generally very vulnerable to earthquake damage, due to their brittle behavior and shape. The cross-sectional area of most chimneys is small enough that earthquake shaking will not only

Figure 8.25 (left)
Golden Gate Avenue after the earthquake: the collapse of a wood-frame rooming house

Figure 8.26 (below, left)
Collapse of wood-frame buildings in the 1906 earthquake due to cripple-wall failures

Figure 8.27 (above, right)
Collapse of wood-frame buildings in the 1906 earthquake due cripple-wall and shear failure

Figure 8.28 (left)
This wooden balloon-frame building on a pier was badly distorted when its foundation and diaphragms failed, but it did not collapse.

cause cracking and loss of integrity but will also cause upper pieces to completely overturn or slide off the lower portion, falling through the wooden roof or to the ground below. The bracing action of wood-framed structures is ineffective because of the different stiffness of the structure and the chimney. The dynamic response to earthquake shaking is much different for the structure and the chimney, making it difficult to adequately tie them together. If the ties fail, the entire chimney may become severely cracked and overturn or collapse. If the ties don't fail, the chimney often suffers damage throughout, but is very likely to fail at the roofline, with the top portion falling over. Photographs of the city suggest that ties were not very prevalent. One homeowner who did protect himself, Clemens Max Richter, rode out the earthquake in his wood house on Geary between Mason and Leavenworth. Built in 1887 for $12,500, Richter's house included a chimney that was tied to his roof.

> I was awakened by the terrific shaking of my wooden home. I remained in bed, but the force of the quake was so intense, the trembling of the building so violent, that a sudden collapse of the structure seemed imminent every second...the pel mel in the rooms was complete. The stairs were intact and the front door free of exit. The brick walls of some houses had partly fallen to the ground and the interior of the rooms came into view. My building on Geary had hardly any damage. The chimneys had been fastened with iron bands when built, as I had ordered, and escaped tumbling—the fate of almost every chimney in the city.[74]

Unfortunately, investigators did not compile a numerical survey of how many wood-frame buildings were intact and how many were damaged. It would have been useful to have such damage figures for cripple-wall failures, anchorage failures, or complete collapses due to lack of vertical continuity or solid walls. For example, how are we to interpret John Sewell's comment that "a considerable number of frame buildings had practically collapsed under the earthquake; some of them were thrown bodily from their foundations"?[75] What does he mean by "considerable"? Are most of these cripple-wall failures or some other type? The only reliable damage estimate is that most chimneys in the city, perhaps above 90 percent, broke apart.[76] Even this percentage is an estimate. The only detailed survey was for Peninsula and East Bay counties. Lawson collected data on 1,000 houses in San Mateo County for this study of earthquake effects and found that 88 percent of all chimneys collapsed.[77] Seventy-three of the 387 wood-frame houses he examined "moved," but he is unclear as to how they moved and whether any collapsed, because he was interested not in architectural response but in ground motion.

While there is a lack of reliable information about wood-frame failures, their relative success is everywhere to be seen in the available photographs. While some buildings may have partially failed, only a very few entirely collapsed. Most buildings survived the earthquake, and their biggest common problem seems to have been falling chimneys.

Case Study: The 1871 San Francisco City Hall

None of the engineering reports of the 1906 earthquake thus far have attempted to place the collapse of a building within its historical context. What historical circumstances of design and construction might contribute to a building's failure in the 1906 earthquake? The city's most spectacular ruin was also its most embarrassing: the new and extremely costly City Hall was seen as a symbol of the graft and corruption of old San Francisco (figure 8.29). Rotten to the core, it had to fail. Once again, using symbolism to explain building failure yields nothing but misconceptions. After all, the Hall of Records, where the misdeeds of the grafters were archived, survived the earthquake well. The City Hall failed because of a series of bad decisions that had more to do with its long construction history and the aspirations of the growing city than with corruption. Its configuration was poor, its conception flawed by frequent changes in program and design, its original earthquake-resistant features nullified, its exterior decorative features too heavy and untied, its dome too tall and either poorly detailed or carelessly constructed. While workmanship may have been poor in places, investigators did not feel that the building was badly built. It was doomed by its overall changing conception and by the early stage that earthquake engineering was in when it was built.

Officer E. J. Plume was inside the City Hall police station when the quake hit:

> As the earthquake progressed, the noise from the outside became deafening. I could hear the massive pillars that uphold the cornices and cupola of the city hall go cracking with reports like cannon, then falling like thunder. Huge stones and lumps of masonry came crashing down outside our doors; the large chandelier swung to and fro, then fell from the ceiling with a bang. In an instant the room was full of dust as well as soot and smoke from the fireplace. It seemed to be reeling like the cabin of a ship in a gale. Feeling sure that the building could never survive such shocks, and expecting every moment to be buried under a mass of ruins, I shouted to Officer Dwyer to get out.[78]

Figure 8.29
City Hall, San Francisco: wall failures, 1906, as seen from Larkin and Market Streets

A witness who was outside saw the dome drum lose its cladding and the huge exterior column drums fall (figure 8.30). The ruins of the shattered column drums, cornices, and walls fell away from the building on Larkin Street, creating a mountain of debris from one side of the street to the other.

The ruins themselves, which can be studied in several photographs, indicate that ground motion started the dome tower oscillating. As the oscillation increased, the tower's iron frame sloughed off the masonry from its sides, sending a rain of stone down on the rest of the building. About three-fourths of the dome tower's masonry fell on the City Hall's main body, mostly in a westerly direction, bombarding the roof and the already weak exterior facades. Newly installed iron trusses in the roof surrounding the dome did their own destructive work, pushing out the cornices they were attached to, which in turn pushed the tremendous freestanding iron columns away from the building.

The superstructure of the dome tower did not collapse: a careful examination of the rotunda after the earthquake by City Architect Newton J. Tharp showed that the dome had shifted to the northeast a distance of eight and one-quarter inches out of plumb.[79] The dome's roof and the lantern and statue on top of it were all sound, but immense damage was evident from the bottom of the dome proper to the springing point of an interior glass dome above the rotunda entrance. Most of the masonry walls had fallen away from the eight steel columns that supported the dome at this level, and the remaining masonry was extremely hazardous. About halfway up these columns, nearly all the rivets in the splice plates holding the columns together were sheared off, leaving only the horizontal lattice girders to hold them in position. These eight columns rested on four great girders, which in turn rested upon a complicated structure of masonry, steel, and iron. The four girders were not tied together or braced, and according to Tharp, the brick surrounding them had not been properly reinforced. He concluded that "without reliance on the inertia of the rotunda walls the whole upper structure bearing upon the main interior columns would topple in a summer wind."[80]

Figure 8.30
City Hall, San Francisco: condition of the dome after the earthquake of 1906

Of the dome tower's design, Tharp said, "On the whole... [the work] is well conceived and the workmanship of the best, but... it was certainly a great oversight not to have introduced diagonal bracing." He continues:

> For some reason the diagonal bracing in the vertical panels between the eight columns on four stages, thirty-two in all, was left out, although the pin-plates are in position on the columns. Directly above the four girders, the eight columns stand without effective bracing for a height of more than thirty feet.[81]

This omission is puzzling indeed, particularly since attachments for diagonal bracing seem to have been provided. The discrepancies between Tharp's statements about the lack of bond iron and court testimony that it had been used raise questions as to how closely the plans match what was actually in the building's frame. No legal inquiry into the unused attachments for diagonal bracing ever took place, and since no documents survive, we cannot know whether engineers, contractors, or workmen omitted it or why. We do know that even with diagonal bracing, the girder foundations would have had major problems, because the rotunda wasn't originally designed to support a circular drum.

Without the failure of the dome masonry and the fire, the core of the building would not have been a total wreck. The eastern part fared better than the rest, and its basement was reinhabited and used for the temporary city hall.[82] The exterior walls on all sides of the building, irrespective of date of construction, lost their parapets. The colossal engaged pilasters and freestanding columns decorating its exterior on the McAllister, Larkin, and Market Street facades all failed, causing enormous damage to the upper floors. Most of the small towers survived, but one on McAllister Street collapsed into the facade. The Hall of Records, reinforced with iron and isolated from the rest of the building, was not badly damaged. A group of citizens wanted to preserve the Hall of Records as a memorial to the earthquake and fire, but that initiative failed. The 1871 City Hall was demolished in 1909. Bricks from the ruins were ground into aggregate for the concrete of the new, 1916, City Hall.

Estimates of Earthquake Damage Versus Fire Damage

In closing, let us consider the allegation that the actual damage of the earthquake was far greater than authorities acknowledged. Neither my own investigation of earthquake damage nor the reports of the engineers in 1906 confirm this allegation. In fact, earthquake damage seems to be a small percentage of fire damage. Let us examine the loss estimates. As we do, we remember that the frame accounted for about 27 percent of a steel-frame building's total cost;[83] exterior decoration, interior finish, plumbing, electrical, and gas lines made up the rest. Now consider that these buildings were filled with valuable furniture, vaults, papers, and merchandise. After an earthquake, any building still standing would yield thousands of dollars of salvageable material. After a major fire, the same structure would be worth perhaps 40 percent of its construction cost.

Estimates of earthquake-versus-fire damage in 1906 varied considerably, with the earliest investigators agreeing on a figure below 5 percent for earthquake damage. Fire investigator Reed reported:

> The actual damage, though appalling to those who experienced the shock, was not as a general rule structurally serious as far as appearance went. Apart from buildings having ponderous architectural attachments, particularly the City Hall, where the damage was great and spectacular, the

apparent structural injury was mainly to tall chimneys, church towers, unbraced brick gables, copings and projections. Actual collapses were mainly confined to flimsy, frame structures...it was the exception that a building was rendered uninhabitable.[84]

The San Francisco Chamber of Commerce estimated between $300 million to $1 billion in property damage from the earthquake and fire combined,[85] while insurance companies estimated the same damage at about $350 million. The insurance figure became the standard, although it did not cover losses outside the burned area. That aside, we can use the $350 million official estimate to figure out how much destruction the various experts attributed to each catastrophe.

Properly constructed buildings in San Francisco were not damaged by the earthquake, and [I estimate] that the total of quake damage will not exceed $10,000,000 out of a total of $300,000,000 [or 3 percent].[86]

—Colonel Francis W. Fitzpatrick, International Society of Building Contractors, 1906

From the damages sustained in other portions of the city not destroyed by the fire I judge that the damage done by the earthquake of April 18th, 1906, would approximate about $15,000,000 [about 4 percent]. This large amount of damage I think was the result mainly from two causes, i.e. the heavy top wall and poor mortar which was shown by the clear appearance of the brick when thrown down.[87]

—Horace D. Dunn [whose eyewitness account can be found in chapter 6], engineer, 1906

The damage inflicted upon San Francisco from the direct and immediate effect of the earthquake was relatively small, being estimated at from 3 to 10 percent only of the total loss [$10,500,000 to $35,000,000].[88]

Architect and Engineer, 1907

I was located in San Francisco at the time of the earthquake and fire...and during the first couple of weeks immediately following the fire I was engaged in problems connected with the restoration of gas and electric service in the city...Immediately following this two weeks' period I was retained, together with my associate, by certain of the insurance companies to make an investigation and report to them as to the relative amounts of damage caused by the earthquake and the fire. This investigation was quite thorough, and while it was impossible to work out any exact figures, I came to the conclusion, and so reported, that the damage primarily due to earthquake was less than 5 percent of the total damage [or $17,000,000].[89]

—A. M. Hunt, insurance adjuster, 1925

Years later, Harry Wood alleged that Hunt was ultra-conservative and biased. But in assessing Wood's estimate forty-one years after the event, we must leaven our judgment with the reality of geological advocacy for increased study of earthquakes. Enthusiasm for waking people up to seismic danger created misstatements and misconceptions like Davidson's comments about the 1868 earthquake, which were often repeated. Speaking of Hunt's estimate, Wood wrote:

> I happen to know from correspondence and conversation with Freeman [that his estimate] is based on the estimate of two engineers of his acquaintance, and I know also that one of them may fairly be described as a California "booster." While *I have no competent estimate of my own* (that is made by myself) I was in on the multitudinous discussions of this matter at the Faculty Club during the months which followed the earthquake and I repeatedly discussed it with G. K. Gilbert. I do not believe that the amount fell below $50,000,000.00 and I suspect that $60,000,000.00 is nearer the true figure [which would mean 20 percent]. I would characterize $24,000,000.00 as ultra-ultra-conservative, and I hope that it will not prevail as an acceptable estimate.[90]

The last estimate of earthquake-versus-fire damage I will quote was made by Professor Karl Steinbrugge in 1982. Steinbrugge was a professor in the architecture department at the University of California and a Bay Area expert in earthquake engineering.

> A restudy of all available records...indicated that the earthquake losses amounted to perhaps as much as 20 percent of the total earthquake and fire loss in San Francisco [around $60,000,000].[91]

From my own observations, I think the 20 percent estimate is much too high and suspect the figure is closer to the initial 5 percent. But even if the 20 percent figure is used, there is still no doubt that an overwhelming portion of the damage was caused by fire. We tend to forget how frequent and destructive fires were in turn-of-the-century cities. In the early 1900s, fires claimed eight thousand lives a year in the United States.[92] In an average week they destroyed three theaters, four public halls, twelve churches, ten schools, two hospitals, two asylums, two colleges, six apartment houses, three department stores, two jails, twenty hotels, one hundred and forty flats, and sixteen hundred houses across the country. In a normal year, about $200 million worth of property burned, and in a single city, New York, as many as eighty-seven hundred fires occurred in one year. According to Francis W. Fitzpatrick, a leading fire expert of the time, buildings in the United States were mostly firetraps. He estimated that there were only three thousand properly fire-proofed steel structures in the country. "As regards homes," he added, "where we should spend the most time, and where we house those most cherished by us...I doubt if there are 300 properly built in the entire country, and there was not one in San Francisco."[93]

Fire was the greatest threat to property—not earthquakes—but the investigators made a valiant effort to study both earthquake and fire. They did an excellent job and I only wish they had had the time, inclination, and modern engineering skills to do more in-depth investigations on the one hand, and more quantitative surveys on the other. It is clear to me that the prestigious and independent committees and individuals who looked at earthquake and fire damage objectively certainly did not minimize earthquake damage.

Stephen Tobriner
Professor of Architectural History, University of California, Berkeley
EERI Reconnaissance Reconnaissance Report, virtual 1906 and actual 2006

CHAPTER 9

The Fire Did It: Recovery, Reconstruction, and Insurance, 1906–1910

Earthquake and fire had reduced San Francisco to a village of thousands of homes with no urban center. The city's major federal buildings—the Appraiser's Building, the U.S. Mint, and the Post Office—had survived earthquake and fire. Flames had laid waste to the financial district, the retail district, the wholesale districts, the manufacturing district, the produce district, city administrative buildings, and hundreds of apartment houses and other dwellings. The earthquake did its own damage, particularly to the city's infrastructure, destroying power plants, transportation facilities, the entire water, sewer, and gas systems, and even the city's new symbol of civic pride, the 1871 City Hall. The future of the city was compromised by the loss of major monuments and buildings, but its existence was threatened most gravely by the loss of commerce. Luckily, the wharfs were saved and the railroad terminal and many warehouses nearby were untouched. San Francisco was still the only sizeable port on the West Coast; its closest rivals, Seattle and Los Angeles, were not developed enough to capitalize on its temporary plight. Wealthy citizens, as well as local commercial and financial institutions, remained loyal to the city. But would the average family stay without housing or their former jobs?

This chapter follows the physical reemergence of the city, beginning in the emergency period with the modest, ephemeral community of shacks that can be seen as the starting point for the huge reconstruction effort to follow. The new San Francisco was shaped by political, social, and economic forces, and especially by payments from insurance companies, and their requirements for certain safety standards.

"Make no little plans," said Daniel Burnham, whose visionary plan for redesigning the city had been on the table when the earthquake and fire struck, providing a clean swath of land of the type that city planners have come to love. Was this the opportunity to beautify and reorganize San Francisco? Perhaps, but first its citizens had to be saved from exposure, hunger, and disease. It would take real shelter, not visionary plans, to revitalize them.

After major disasters, it is common for disparate populations to feel a sense of community; people are brought together by a shared experience of tragedy and disorientation for a short time during the emergency and recovery period. In San Francisco this community could be found in temporary shacks and sidewalk kitchens and, on the political front, in Mayor Schmitz's Committee of Fifty.[1] Schmitz and his ally, political boss Abe Ruef, had been on the verge of indictment for graft before the earthquake, but Schmitz won the city's conditional trust by appointing many of his enemies to this committee, whose mission it was to guide the city through the emergency period. James D. Phelan, perhaps Schmitz's worst enemy, was put in charge of the relief effort, staving off any allegations of graft in the organizing and dispensing of private and governmental funds.[2] Governor Pardee declared a month-and-a-half bank holiday so that he could call a special session of the state legislature. During this time no debts could be collected and banks had time to reorganize.[3] The federal government offered relief funds and guaranteed bank certificates. Faith in San Francisco's institutions and the financial stability of the city was bolstered by the survival of the U.S. Mint and the Post Office, both of which remained functional after the earthquake and fire. Two hundred million dollars in gold bullion were safely stored in the Mint. Postal service was able to resume almost immediately after the fire stopped. But local government was still in turmoil. The secret campaign to indict Schmitz and Ruef for graft, spearheaded by muckraker and *Bulletin* editor Fremont Older and joined by graft prosecutor Francis J. Heney and former mayor James D. Phelan, and financed by Rudolph Spreckels, had begun in January 1906. Schmitz and Ruef rode a wave of popularity following the earthquake, but an increase in urban crime in the fall of 1906 gave the group the opportunity to try for an indictment. The grand jury investigations began on October 20, 1906, wreaking havoc with civic government during the recovery and reconstruction of the city.[4]

Sidewalk kitchens were the first wooden and brick "buildings" constructed in the unburned parts of the city. They were erected as windbreaks after the mayor banned indoor cooking until all the house chimneys could be checked.[5] Crude fireplaces were made from grates and fallen bricks, and inverted sinks were sometimes used as stovetops, with the stovepipe rising from the trap. Cooking shacks and tent cities were adorned with humorous signs that bespoke a sense of egalitarian fellowship and poked fun at the plight of their inhabitants: "Un-Fairmont Hotel," "Open all night," and "Will exchange for country property." Cooking hovels bore signs like "The Inside Out," "The Step Inn," and "Goodfellows' Grotto," with mottoes painted on the exterior boards like: "Eat, drink, and be merry, for tomorrow we may have to go to Oakland."[6]

Makeshift shelters of blankets and rugs protected more than two hundred thousand homeless San Franciscans from the rain that fell four days after the earthquake, on April 22. As in the Vacaville-Winters earthquake of 1892, the United States Army soon provided tents for temporary shelter in the Presidio, Golden Gate Park, and other parks. General Funston had taken it upon himself to march federal troops into the city on the morning of April 18. With the consent of the mayor, the Army remained in the city, setting up tents, requisitioning and distributing food, and keeping order until the summer of 1906. Food flowed into San Francisco through donations, largely sponsored by the Red Cross, and was distributed free to citizens. While the number of people living in temporary camps after the earthquake was high, by June it had fallen to fifty thousand, and by July to twenty-five thousand. Bread lines were discontinued on August 1, 1906, and by the fall of 1906, the camps housed not more than seventeen thousand people.[7]

Many people who had been burned out did not take advantage of these military-operated tent cities, preferring to stay with relatives outside of San Francisco. Southern Pacific, regarded by many as a hostile monopoly, used the crisis to improve its image by helping the city in different ways, one of which was to offer free passage to a total of three hundred thousand people who decided to leave the city between April 18 and April 26, lessening the need for temporary housing.[8] Southern Pacific also funded promotional material to help stimulate the recovery.

The Committee of Fifty had been established in haste on April 18.[9] After issuing a few decisive initiatives, it lost its effectiveness, apparently through attrition. Within less than two weeks, it was replaced by the Committee of Forty. Ruef, who had not been a member of the Committee of Fifty, was installed on the Committee of Forty, which some saw as a return to business as usual. There were scandals over building safety. Temporary theaters, far below standard, were somehow approved for use by the Committee on Reconstruction.[10] The grand jury investigation that reopened in October would cripple city government for almost a year.

Provisional Shelter and Federal Government Relief

General Adolphus W. Greely, appointed to command the U.S. Army's Pacific Division headquartered at the Presidio, had arrived in the city in March 1906.[11] He had been away from the city during the earthquake but returned to find himself in charge of key aspects of the relief effort, along with Edward T. Devine, president of the American Red Cross, and the city's committees on Finance and on Housing the Homeless, both originally subcommittees of the Committee of Fifty. In the fall of 1906, the Department of Lands and Buildings of the newly formed San Francisco Relief Corporation, the San Francisco park superintendent, and the Army arranged for the construction of six thousand small, wooden houses in the areas formerly occupied by tents.[12] These one-story, two- and three-room houses built by San Francisco union carpenters have come to be called "earthquake shacks." They were built to shelter some seventeen thousand people still in the camps, mainly poor citizens from South of Market. By the fall of 1907, the camps had been disbanded. The earthquake cottages were sold for a nominal fee to the individuals who had occupied them and moved to privately owned lots. It is said that over five thousand of these small cabins were hauled away from the camps to be used as starter homes. Twenty-one survive in San Francisco today.

Notwithstanding these efforts, thousands of people remained homeless several years after the earthquake. Innovative, semi-permanent construction schemes were suggested by philanthropists and speculative builders. Cheap, mass-produced dwellings were advertised in the *Chronicle*, but these did not become popular.[13] Other schemes, such as the use of old trolley cars as temporary shelters, really did work.[14] The Housing and Shelter Committee, under the direction of Father D. O. Crowley, built fourteen hundred houses for people who could match the small sum spent on construction. Like the earthquake cottages, these were earthquake-resistant by default because of their one-story height, light weight, and small size.

Boomtown Recovery

With money in the banks, stable federal and state governments, a crippled but functioning local government, and insurance settlements rolling into the city, local entrepreneurs tried to reestablish their businesses as quickly as possible. The rush to rebuild resulted in much substandard housing and many marginal business buildings, laying the groundwork for later problems in some neighborhoods, once the businesses moved downtown again. The only undamaged streetcar line connecting the unburned Western Addition and the Mission District was on Fillmore Street, which became the obvious location for a temporary civic center.[15] The Fillmore neighborhood boomed, complete with real estate shortages and questionable construction practices. Buildings were stuffed with tenants. And when a building could absorb no more shops and offices, it was expanded by jacking it up and adding a new story at ground level. Some property owners made fortunes, but in every way the street's long-term safety was compromised. On Van Ness Avenue, one side of which had been burned out by the fire, a similar boom was on (figure 9.1). Within a month of the earthquake and fire, the White House and

Figure 9.1 (below, panorama)
A view of Van Ness Avenue. In the foreground are new stores built on the former residential street, before 1912. On the right the ruins of City Hall can be seen.

the Emporium department stores had relocated on Van Ness in huge temporary wooden structures; other retailers of all types followed.[16]

Many businesses throughout the city did not even wait for streets and lots to be cleared. They put up wooden posts in their old locations to support platforms and built temporary, one-story structures above the debris.[17] This was problematic because removing debris was vital for circulation through the streets, for repairing utilities, for clearing land for new construction, and for combating disease.

Plague Threatens Recovery

Nothing could have stopped recovery in its tracks more effectively than an outbreak of bubonic plague. A year after the earthquake and fire, in May 1907, a man aboard a tugboat in the bay died of plague.[18] By September 1907, fifty-five cases had appeared, and the future of the city was threatened by the specter of a major epidemic.

This was not the first outbreak of plague in San Francisco. It had appeared in Chinatown in 1900 and, in the tradition of San Francisco politics, the response was not straightforward. Mayor Schmitz refused to approve the printing of health reports and vital statistics and attempted to remove from office four members of the Board of Health who persisted in the statement that plague existed in the city.[19] Luckily, both the federal government and the business community mobilized, and the plague was finally arrested in 1904, after causing 113 deaths, largely Chinese.

This new outbreak, declared an epidemic in September 1907, was met with concerted community action. The new mayor, Edward R. Taylor, appointed on July 16, 1907, by the Board of Supervisors during Schmitz's graft trial, appealed to President Theodore Roosevelt for help, and the president directed Surgeon General Walter Wyman to take charge of the situation.[20] Wyman in turn appointed Dr. Rupert Blue, aided by a group of other doctors, to assist in combating the disease.

The anti-plague campaign had a massive effect on reconstruction efforts.[21] In order to kill the rats carrying the fleas that caused the disease, piles of garbage and debris had to be burned, and buildings that had survived the earthquake and fire but were rat-infested were demolished or burned to the ground. This was no small number: more than seventeen hundred houses were destroyed![22] The camp buildings and boardwalks were raised eighteen inches off the ground so that cats could more easily catch the rats. These modifications of the environment and a strong eradication program halted the plague by March 1909.

Demolition and Disposal

In the days following the fire, while the Army guarded the burned district, gawking gentlemen who tried to sightsee among the ruins were put to work clearing bricks from the streets.[23] Soon the Army Corps of Engineers, under the direction of the Lands and Building Committee, began to demolish the walls of ruined buildings on Market between the Ferry Building and Seventh Street.[24] They were a vanguard for the thousands of civilian workers who would be clearing the ruins in the months to come.[25]

Beginning immediately after the fire and through June 1906, the main workforce was composed of day laborers paid by the city who worked in the ruins, picking through debris and helping to load the trains running in and out of the city toward the bay (figure 9.2).[26] San Francisco had a large supply of laborers to draw upon, because many people were temporarily out of work. Reusable bricks were stacked and piled, metal was collected and sorted, and unsalvageable refuse was removed from lots.[27] The rest of the debris was thrown into railroad hopper cars to be hauled away. As early as April 27, six days after the fire, Mayor Schmitz had authorized the local rail companies to "lay track to facilitate the quick removal

Figure 9.2
The removal of debris by rail

of the debris."[28] Among the companies that laid track were Southern Pacific, the Ocean Shore Railroad, Santa Fe Railroad, and the United Railroad, which had owned a trolley franchise before the fire. The companies' new track connected to the state-owned Belt Railroad, which served the wharfs and remained intact after the fire. The terminus of the debris was the waterfront. On at least one occasion, it was loaded on a barge and dumped at sea. But most of the rubble was dumped at sites that were landfills before the earthquake: Mission Bay, North Beach, and Islais Creek Basin. Some was also used for "low areas" in various parts of the city. At least one load went to North Beach near Fort Mason. This is the closest to the Marina that rubble was reported to have been dumped.

As might be expected with so much major demolition, accidents were common. On November 31, 1906, a fourteen-year-old boy was killed when a wall of the Palace Hotel fell on Market Street.[29] The city announced that property owners would be held liable for injuries that occurred because of dangerous walls on their property. The grand jury, which was governing the city while the mayor was being prosecuted for graft, asked Louis Levy, secretary of the Board of Public Works, to order all owners of buildings that were still partly standing to remove dangerous walls and chimneys at once. If they should fail to obey, warrants would be issued for their arrest.[30]

However, the greatest danger was not to passersby but to demolition workers. Falling walls at the California Wine Association building at 189 Townsend Street killed one workman and injured several others.[31] Two workers who were eating lunch by the Hall of Justice laughed when their foreman advised them that the wall they were sitting by was marked as dangerous. Moments later the wall fell, killing them both.[32] A brick foundation collapsed on workers on Mission Street.[33] In the period from January to April 1908, nineteen workers died in construction—actually a very low number, given that approximately fifty thousand people were working in construction in San Francisco at the time.[34]

Two Failed Plans for the New San Francisco

Urban devastation of the magnitude suffered by San Francisco in 1906 often brings with it the wish to start over in a new way.[35] After the 1693 earthquake that destroyed forty Sicilian cities, several were completely replanned as Baroque centers with new streets and buildings, and a few were even moved to new locations. In 1755 Lisbon was transformed into an Enlightenment city and rebuilt on a new street plan with an earthquake-resistant building system. Without the spur of catastrophe, Paris was torn apart and reassembled as a modern, nineteenth-century city by Georges-Eugène Haussmann, who deeply influenced Chicago city planner and architect Daniel Burnham, whose firm had built the Mills and Chronicle Buildings, and who would now attempt to play a major role in post-disaster San Francisco.[36]

In San Francisco, former mayor James D. Phelan and the business community instigated two plans to make major changes, one architectural and one social.

The first, Daniel Burnham's plan, was being printed for general distribution in early April 1906, two weeks before the earthquake struck.[37] The plan's supporters sought to turn San Francisco from a provincial boomtown into a modern and elegant city.[38]

The Burnham Plan called for major streets to radiate from a series of nodes in the city (figure 9.3). These streets would carry traffic through the intervening grid to concentric rings of streets, the last of which would be on the peninsula's periphery. In the days of horse and buggy, Burnham accurately foresaw the need for such traffic arteries, and many of his ideas are reflected in later additions to San Francisco's highway system. A subway system was run under the diagonal avenues, linking the city together. The plan also included an extensive park system, which San Francisco desperately needed. Burnham envisioned a city whose hills would be crowned by parks and public monuments. Tree-lined boulevards and a centrally located civic center would symbolically tie the city together. The outer boulevard provided views of the bay and along it were public port and recreational facilities.

Burnham's plan epitomized the best and worst of turn-of-the-century urban planning. It was a grand and beautiful artistic conception. But it failed to address the needs of housing, commerce, and industry except in the most general terms. Other than suggesting that a board be formed to supervise public artworks, and conjecture that it would take fifty years to implement the plan, Burnham made no concrete proposals for implementation. Little consideration was given to the appearance or placement of individual buildings in the context of San Francisco's unique blocks, except to make them appear more unified. Although Burnham generally sketched a new proposal for increasing port facilities, he did not suggest specific placement of commercial and industrial infrastructure or their integration with port and rail lines. The plan's great hypothetical strength was that it honored the topography of the city rather than O'Farrell's grid.

Fire danger was considered in the Burnham Plan, but not earthquakes. A chain of parks was to extend from Buena Vista Park, near the Golden Gate Park Panhandle, to Twin Peaks Park, and finally to Bernal Heights and Potrero Parks to the south. In Italy, similar open spaces had been created to mitigate earthquake damage since 1693, but Burnham spoke of the parks only as a firebreak.

Earlier in San Francisco's history, it had been noted that the prevailing westerly winds could quickly bring a fire from the Western Addition over the hills into the downtown area. Financial District fires could move in the opposite direction. A barrier between these two areas would seem to make sense. But in a report on the fire by City Engineer Marsden Manson in 1908, such barriers were dismissed as useless for firefighting.[39] The report noted that in the case of a large 1907 fire in Superior, Wisconsin, wide spaces had not been effective in halting the flames.

The Burnham Plan was not only expensive but required condemnation of vast tracts of valuable real estate. It had many points to recommend it, but its enormous cost would not immediately be offset by adding value to San Francisco's location. A bond issue to pay for a scaled-down version of the plan failed. This was not a time to challenge property rights. Even a proposal to widen certain streets was

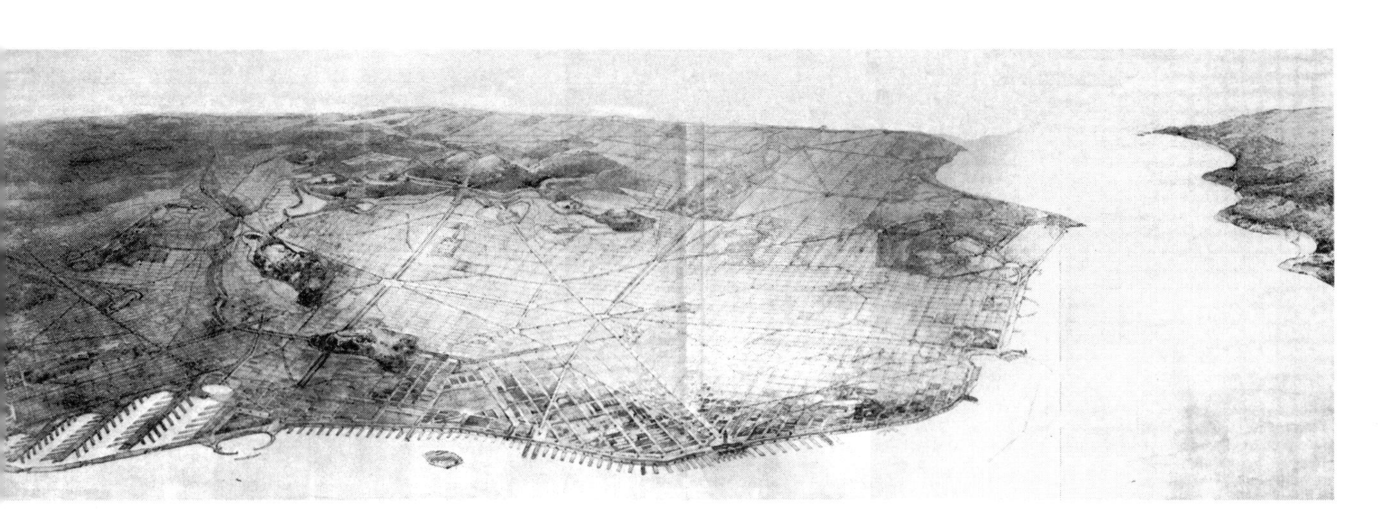

Figure 9.3
The Burnham plan for San Francisco, 1905

damned by association with Abe Ruef.[40] The Burnham Plan was like offering flowers to a patient that needed penicillin.

The second plan to renew San Francisco would have moved Chinatown to Hunter's Point. During the depression of 1873, Irish and German workers had begun to resent the Chinese because they were a source of cheap labor. In 1877 Dennis Kearney, an Irish-born San Franciscan, formed the Workingman's Party of California, which in the 1880s became the most powerful political force in the city and in 1901 supported the election of Eugene Schmitz. Kearney's group harassed Chinese immigrants working in the South of Market, and in the 1880s Kearney had been one of the political powers behind an attempt to destroy Chinatown.[41]

Discriminatory treatment had produced unsanitary conditions in Chinatown. Confinement to a small area meant overcrowding and the danger of diseases like the plague of 1900. Restrictions on the immigration of Chinese women encouraged prostitution. Not surprisingly, the move to banish the Chinese was largely based on race. Beginning immediately after the earthquake, the Chinese were constantly transferred from one temporary camp to another; the stated reason for these moves was that the white citizens of San Francisco were offended by the smell of Chinese cooking. The *Bulletin* had this to say on April 29, 1906:

> The Chinatown of San Francisco occupied one of the eligible quarters of the city, but it was a world within a world, a segment of the city of Canton...It was a city of dirt and color, of filth and squalor, of superstition and debauchery, of soft lights percolating from paper lanterns...it is certain that the Chinese quarter has been wiped out of existence. For this, at least, the people of San Francisco, when they rebuild the city, will give thanks.[42]

The attempt to move Chinatown ultimately failed because of opposition from other business interests. High population density meant high rents. Chinatown was a significant tourist attraction. The largely white owners of Chinatown property, calling themselves the Dupont Street Improvement Club, wrote a resolution to the Committee of Forty stating:

> What we want is a Chinese quarter constructed on a thoroughly sanitary plan...We do not want to be discriminated against and intend, if possible, to prevent the changing of the location of the Oriental quarter.[43]

But probably the most important reason for not moving the Chinese was that the Chinese Embassy lodged a protest with California Governor Pardee, threatening San Francisco with a trade embargo.

So Chinatown was rebuilt where it had been before, but with fire-resistant Class C brick buildings and carefully crafted facades with Chinese decorative features designed by white architects.[44] One of the Chinese merchants who may have influenced the appearance of the new Chinatown, Look Tin Eli, spearheaded reconstruction among the Chinese property owners. On October 15, 1906, Chinese and white owners and the San Francisco Real Estate Board met to discuss the reconstruction of Chinatown and agreed to rebuild using Chinese motifs to attract tourists and improve commerce. Despite the racism of the period, capitalism united the two groups in the common enterprise of rebuilding the city in an imaginative way which benefitted them both while creating a practical strategy for economic recovery.

Remaking the City

If the appeal of San Francisco was about location, location, location, the reconstruction of the city was driven by money, money, money. Choice of construction material was no exception. San Francisco, come what may, was going to be a wooden city again, because wood was still the cheapest material, the closest at hand, the easiest to work, and the safest in earthquakes. Time was of the essence; wood could be delivered within California by ship or rail. F. W. Fitzpatrick, who wrote extensively about the destruction of San Francisco, lamented this rebirth:

> I am afraid that the prophecy that San Francisco will be rebuilt a wooden city will be all too true. Apart from the timber being brought into that city exclusive[ly] for the temporary buildings, it is reported that two billion feet of lumber had been ordered, or is to be ordered, for the "permanent" reconstruction of the city. Forty million dollars' worth of kindling with which to make another $300 million fire.[45]

Additional building materials could be found in the wreckage of non-wooden buildings. Scrap iron was refashioned, bricks were cleaned, and many steel-frame buildings were reconstructed out of their own debris. One such structure was the Emporium department store, which lay in ruins after being shaken, burned, and dynamited. A crude foundry was built on the site and damaged steel columns were reheated, rehammered, and rerolled.[46] The recycled steel members were subjected to simple bending and shattering tests conducted on the spot. Architects and engineers, acknowledging that this rough-and-ready method of fabrication might not yield columns of uniform quality, doubled the amount of steel in buildings where these refabricated columns were used. In the Emporium, for example, reused steel is specified for ten feet on center as opposed to the average twenty feet.

Decisions to retrofit or not were made building by building and feature by feature. The Palace Hotel's brick walls were still plumb, but its floors, plumbing, wiring, and interior flues were wrecked. Its downfall, though, was that its facade, which featured bay windows, was no longer in style and could not compete with nearby neoclassical buildings. The Palace was quite difficult to demolish, a testimony to its fine masonry and iron bonding, both earthquake-resistant features.[47] The cistern under the hotel's courtyard, a fireproofing measure, was retained under the floor of the new garden court, and the old foundation was also incorporated into the new structure.[48]

Figure 9.4
Hale Brothers department store, 979–989 Market Street. Gutted by the fire, the building was rebuilt behind the surviving facade.

Some gutted buildings, with facades that barely stood by themselves, were resurrected from the inside out. The Hale Brothers department store, at 979–989 Market Street, suffered almost total internal damage (figure 9.4), but the exterior remained intact.[49] Using guy wires to hold the facade vertical while the remaining skeleton behind it was demolished, the contractor rebuilt the interior with new steel and then connected it to the old facade.

If remains of foundations, walls, or interior support systems were serviceable and appropriate for the rebuilding project, they were used. In scores of plans for buildings in 1906, notations explain that "old walls" or "former foundations" are to be used. Many of the existing brick buildings in the downtown area were rebuilt on the stubs of walls and the remains of foundations that survived the fire.[50]

Architect Willis Polk suggested that the city save money by buying the burned-out Fairmont Hotel to use as the new City Hall.[51] We can see from the facade today that it would have been an ingenious solution, the Fairmont crowning the hilltop like some modern Parthenon on San Francisco's Nob Hill Acropolis! That solution was not accepted, but Polk was given the job of retrofitting the sandstone walls of the Flood mansion across the street, and he was able to totally erase any evidence of the earthquake by replacing damaged stones with new stones from the same Connecticut quarry.[52]

Similar cosmetic work on numerous surviving buildings gives the modern observer false notions about earthquake and fire damage. Because of the intensity of the fire, the interior steelwork of many "survivors" had suffered extensively. In fact, almost no building remained unscathed; many pre-earthquake buildings survive today because they were retrofitted, sometimes extensively. Sometimes, as in the 1877 U.S. Subtreasury Building, at 608 Commercial Street, the damaged upper floors were removed and the first floor was reused (figure 9.5).[53]

Figure 9.5
Subtreasury Building, 608 Commercial Street, 1877. In order to save and reuse the building, it was reduced by three stories after 1906.

Some buildings survived the 1906 earthquake and fire only to fall to the wrecker's ball, but many pre-earthquake buildings still stand today.[54] They include not only the more famous large buildings on Market Street, like the Mills Building, the Flood Building, and the Ferry Building, but also the 1905 Pacific Telephone and Telegraph Building, at 445 Bush Street; the 1905 Shreve Building, at 201 Grant Avenue; the 1901 Kohl Building, at 400 Montgomery; the 1906 Howard Building, at 207 Powell Street; the Columbus Building, at 906 Kearny, in construction in 1906; the 1903 Aronson Building, at 700 Mission; the 1906 Grant Building, at 1095–97 Market; the 1905 Atlas Building, at 602–606 Mission; the small, 1905 Burdette Building at 90–96 Second Street; the 1851 Bank of Canton Building, at 515 Montgomery; Saint Boniface Church, on Golden Gate Avenue; the Church of Saint Francis of Assisi, on Vallejo Street; and scores more. Each of them was rebuilt to some extent, generally using materials from the ruins.

For a more detailed examination of how one structure was built, tested, and rebuilt, let us look closely at the Rialto Building, on New Montgomery Street near Mission Street (figure 9.6).[55] Commissioned by the Law brothers and designed by Frederick Meyer, the 1903 building had been a gamble on the future of the South of Market as a business district. The Rialto was a Class A steel-frame, ten-story structure with a U-shaped ground plan featuring brick, stone, and terra-cotta cladding and concrete floors. Meyer's conception was based on similar U-shaped buildings then being constructed in Chicago, with windows in rows around each floor and unpartitioned interiors that could be arranged to fit each tenant's needs. Throughout his life, Meyer constructed simple, efficient structures; he explained that he believed that a strong structure should be constructed "through science," with every possible safeguard against storms, earthquakes, and fires. He felt that a safe building always began with the frame, which "does the work," and the Rialto's steel skeleton was protected by hollow terra-cotta tiles in the basement and plaster and wire lath in the upper stories.[56] The basement foundation was reinforced concrete, as were the floors. Exterior windows were sheathed with tin to make them fire-resistant.

Figure 9.6
The Rialto Building: eastern facade, on New Montgomery Street, 1905

Despite all the earthquake- and fire-resistant features of this building, it was severely damaged by internal gas explosions shortly after the earthquake (figure 9.7).[57] The boilers probably sparked explosions in ruptured mains near the basement walls. Investigators stated that the Rialto's steel columns failed as a direct result of insufficient protection, but that conclusion is complicated by the explosions, which would have wrecked the fire-protective cladding and strained the building in ways similar to those of an earthquake.

In 1909 the owners commissioned the architectural firm of Bliss and Faville to reconstruct their building. Drawings of the damage were superimposed onto Meyer's original drawings to illustrate the extensive reconstruction program (figure 9.8). We can see that major column movement, creating both X cracks and infill damage, had occurred at the fourth and fifth floor (figure 9.9).[58] Pervasive breaks in the terra-cotta cladding indicate that virtually all the joints in the building between columns and beams must have been stressed. All exterior cracks and displacements of the infill were repaired. Interior columns and steel beams were also repaired where they had buckled or fallen. Concrete was added to the wire lath fireproofing, which was still extensively used in the building. For all intents and purposes, the Rialto Building that exists today is very similar to the one built in 1903.

Figure 9.7
The Rialto Building: internal collapse due to fire and explosion (photograph taken after April 18, 1906)

Figure 9.8 (above)
The Rialto Building: blueprint of the eastern facade. This is a rare example of specified post-1906 repairs. Bliss and Faville, the architects, noted damage (white areas) on a Frederick Meyer drawing. As the steel columns flexed, they broke the more brittle stone, tile, and brick moldings adjacent to them. This kind of superficial damage is unavoidable in a major earthquake and doesn't necessarily indicate major structural damage, but it warns engineers that damage may have occurred.

Figure 9.9 (right)
The Rialto Building: northern Mission Street facade, with reconstruction under way. Note damage between fourth and fifth floors, which may have resulted because of pounding between the Rialto Building and its neighbor to the west.

The Fate of the 1871 City Hall

The new City Hall had been a symbol of San Francisco in its splendor and in its devastation, but it could not be a symbol of its resurrection. It was the grandest ruin in San Francisco, and citizens flocked to be photographed in front of it. The steel frame of the dome still stood; on occasion it was strung with lights, which at night gave the illusion that it was undamaged. The east wing of the structure was temporarily rehabilitated to help house city government, but time was running out for the structure.[59] When the graft prosecutions began, the ruin came to be seen as a symbol of Schmitz's corrupt administration. Despite the fact that it would cost more to build a new City Hall than to renovate the old one, the citizens voted to demolish the building and eventually to build its replacement a block west, along Van Ness Avenue, as part of the new civic center inspired by Daniel Burnham's plan. Bricks from the walls of the old City Hall were crushed into aggregate and used in the foundations for the new building in 1912.

Insurance Companies Pay the Bill

If the eastern newspapers were to be believed, San Francisco would never have been able to rebuild itself. Early fears of financial failure can be inferred from articles in the San Francisco papers, *Sunset* magazine, and real estate advertisements that contained steady outpourings of "upbuilder" articles extolling the strength of the people and the city of San Francisco.[60] However, no major investors from the East poured money into San Francisco real estate after the fire. To save the city economically, insurance companies had to honor their commitment to reimburse the city for fire damage, despite the earthquake exemption. They obliged, reimbursing their policyholders in San Francisco and allowing the phoenix city to recover and rise again. Without the infusion of money from insurance companies, the city could not have rebuilt itself.

The San Francisco conflagration was the most destructive fire in America's history, but luckily, most of the loss was insured (at least $235 million of the $350 million in damages). While insurance companies had tried to warn the city of imminent danger from fire, they had also written policies as fast and furiously as they could.[61] When news spread that three thousand acres, including twenty-five thousand buildings in the central business district, had burned, the insurance industry shuddered.[62] Fires had forced insurance companies out of business in Chicago in 1871 and Boston in 1872 because those companies lacked the assets to pay claims, and things looked equally bad for many in San Francisco.[63] No company, even those supported by fellow insurers that had reinsured their policies, could have easily produced the money owed. More than ninety thousand claims were eventually handled by 233 insurance companies.[64] Since insurance settlements were the fuel for San Francisco's recovery and its reconstruction, it is important to understand the politics and economics behind the vast settlements the city received.

Insurance policies for a San Francisco building could be byzantine. A single building was often insured through as many as twelve separate policies, each with a different company. Some of these policies were also reinsured by a second company. When, a few days after the fire, insurance companies got word that about 80 percent of the fireproof safes in the city, theirs included, had failed, they were thrown into turmoil. Policies, records—essentially all proof of coverage—had burned in the fire. Fourteen of the 233 insurance companies had "earthquake clauses" that protected them from paying for buildings that collapsed before they burned, but they had no way to prove which buildings were still standing before the onset of the fire.[65]

The "Six-Bit" Strategy

Seeing how desperate the situation was, representatives of almost every insurance company met in Oakland on April 21, 1906, to establish a General Adjustment Bureau. This body would deal with the difficult if not impossible problem of disentangling earthquake and fire damage, recreating lost records, and setting uniform procedures for claim adjustments.

On the East Coast, the insurance companies' home offices were more concerned with their own survival. On May 31, 1906, as their adjusters worked in Oakland, company representatives met for a secret rate-setting session in New York. At that meeting, it was proposed that the amount of money to be paid to the stricken city be based on a simple, convenient premise: the earthquake had caused the fire, which had wiped out much evidence of earthquake damage. Further, it was because the earthquake had disabled the water system that the fire could not be promptly extinguished. Given that none of these conditions had been taken into account at the time insurance rates were set, the companies could not be liable for the full loss; they would share it with policyholders.

The suggestion was floated that 75 percent of the property value would be a fair settlement level. When a preliminary poll was taken, sixty-eight companies voted for the 75 percent "horizontal cut"; thirty-two voted against it. Then someone leaked this secret vote to the *Chronicle*.[66] The press soon dubbed these two groups the "six-bit" companies and the "dollar-for-dollar" companies, and how each had voted determined its moral character in the eyes of San Franciscans. Since the insurance industry was founded on the concept of trust (as well as good business), the thirty-two dollar-for-dollar companies moved quickly to separate themselves from the others. While still retaining their membership in the Oakland-based General Adjustment Bureau, they established their own "Committee of Five" on July 10, 1906, to adjust claims in San Francisco's Ferry Building.[67]

The most outspoken opponent of the six-bit companies was California Congressman Julius Kahn, who had this to say about "honest and dishonest insurance":

> I hope to place in the *Congressional Record* the name of every insurance company that refuses to meet its obligations in that city, in order that the people of the United States, the people who pay their premiums in the hope of recovering their losses in case of fire, may know the names of those companies that are unreliable and dishonest and that will not pay their obligations when the time comes for them to do so.[68]

The insurance companies responded to San Francisco claimants and denunciations like Kahn's in three ways. First, they issued a warning that no one could use an uninspected chimney in San Francisco. They threatened that all policies would be voided if a chimney fire started.[69] This very direct action prompted Mayor Schmitz to make the proclamation that sent San Franciscans into the streets to do their cooking. Second, the General Adjustment Bureau, in Oakland, and the Committee of Five, in San Francisco, went to work as carefully as they could to settle claims quickly and fairly. Last, companies brought in specialists to vouch for their honesty in adjusting claims. Some cases did end up in court when insurance companies claimed exclusion based on earthquake damage or, in others, because of dynamiting, which clearly was not covered.

For its part, San Francisco's chamber of commerce brought in Professor Charles Whitney of the University of California to assess the insurance process and publish his findings.[70] His study was clearly meant to remind insurance companies they were being watched. Whatever their initial reaction, the insurance companies themselves applauded Whitney's report when it finally came out in 1907.

Why They Paid: Policies, Politics, Fraud, and Good Business

All branches of government—legislative, executive, and judicial—converged on the insurance problem, and the insurance companies generally lost. California's Governor Pardee followed Kahn's lead in trying to influence the companies to honor their commitments as he saw them. He wrote a letter, published in all San Francisco newspapers, asking that the deadline for filing claims be extended by sixty days (doubling the filing period).[71] In case a company refused to give this extension, Pardee granted the insurance commissioner a new power: he could ask them to furnish complete records of all their policies in force. If they could not come up with this documentation, which was virtually impossible to obtain, because of the fire, their company charter would be revoked in California. Adding insult to injury, Mayor Schmitz and Governor Pardee sent a joint telegram to the head offices of all the companies:

> We appeal in our misfortune to your manhood, business integrity, and sense of justice to interpose your veto on the disreputable tactics of certain agents, who are irritating our people to the point of exasperation.[72]

California legislators were so antagonistic toward perceived insurance "welchers" that they began discussing the possibility of laws that would regulate the standard premium forms. Special attention focused on clauses, such as the "earthquake clause" and "falling building clause," that were central to a number of suits by and against insurance companies.[73] Generally speaking, the courts were about as sympathetic as Governor Pardee and the state legislature. A typical earthquake clause read:

> This company shall not be liable for loss caused directly or indirectly by invasion, insurrection, riot, civil war...or loss occasioned by or through volcano, earthquake, or hurricane, or other eruption...[74]

Only fourteen companies had earthquake clauses, and of those, eight used them to justify reduced payments. Three of the companies that attempted to use the clause to deny any liability were sued and lost.[75] One ruling by the U.S. Circuit Court in California held that the insurer had to prove that the earthquake caused the fire that burned a specific insured structure. Another ruling, by the U.S. Circuit Court in New York, held that fire spreading from a building destroyed by the earthquake was fire damage, and did not release liability.[76] The insurance companies were at a great disadvantage, in any case, as the burden of proof fell on them.

The falling building clause, unlike the earthquake clause, was standard in San Francisco policies. It read:

> If a building or any part thereof fall, except as a result of fire, all insurance by this policy on such building or its contents shall immediately cease.

The ambiguity in this clause arose from the question of what constituted a fallen structure. If a structure lost its parapets or fire walls in the earthquake but was otherwise intact, was it still insured? The San Francisco Superior Court found in the case of *Clayburgh vs. Agricultural Insurance Company of Watertown, N.Y.* that a building so damaged was still covered by fire insurance. The insurance company appealed this case to the California Supreme Court in 1909 and lost.[77]

Immediately after the 1991 Oakland–Berkeley Hills fire, a number of recalcitrant insurance companies were pilloried in the local press. Why was the process taking so long? Why were the adjusters so unreasonable and combative? I listened to the laments of many of my friends. Yet all in all, the settlements seem to have been generous. Several architects have told me in confidence that their clients used inflated claims to recover overly generous settlements. People who have suffered loss to their homes and possessions often feel entitled to inflate their claims because of the terrible trauma they have suffered. On the other side are the adjusters, sometimes overzealous in restricting reimbursement. Arguing with insurance adjusters after a disaster is part of the recovery process. Sometimes it is difficult to find the facts behind the settlements and even more difficult to assess whether the outcome was just.

Each claim submitted to the adjusters in 1906 had to be reviewed to determine whether the insured structure was standing after the earthquake and whether it had contained what the insured claimed. The fraud and deception in the adjustment process were phenomenal, and insurance adjusters were antagonized by the press, local government, and their policyholders.[78] With so little proof of any claim, sworn statements became vital to proving one's case. British insurance agent Robert Mackenzie observed that there were people in every bar who would sign affidavits; even the notaries who wrote these documents called themselves "affidavit factories."

Photographs were even better than sworn testimony. There was a brisk trade in both real and faked pictures taken after the earthquake but before the fire. George Brooks, an adjuster for the California Insurance Company, told the story of a $16,000 loss on a policy covering a stock of dry and fancy goods located in a building on Market Street.[79] The policyholder was assured that he would be paid 60 percent of the claim when he came to the office, and the balance in sixty days. Because $13,000 of the risk was ceded to other companies through reinsurance, Brooks filed claims with them. One of the other adjusters found a pre-fire picture of the supposedly undamaged building, showing the entire front of its second story thrown down into the street. Photograph in hand, Brooks asked again whether the building had been damaged. The insured said it had not, adding, "I'll sign it and swear to it. Not a brick in the whole building was disturbed!" When Brooks showed him the photograph, he said it was fake but did not contest the settlement.

Hundreds of questionable settlements must have been paid, considering that insurance companies often gave the claimants benefit of the doubt if they couldn't uncover conflicting evidence. Companies strapped for money, like Fireman's Fund, had to assess stockholders and give out payment in their own stock to stay afloat, but they weathered the earthquake, gained the community's respect, and some survived to become giants in the western insurance industry.[80]

Throughout this process, the insurance companies continued to be an influence for fire safety. They raised their rates to a far higher level than before the fire—partly to cover their risk in San Francisco, with its debilitated fire department and numerous temporary wooden structures, but also as a way of influencing community action.[81] Building owners who installed certain safety features received huge discounts, and the entire city was offered a sizable rate reduction if it improved its fire-fighting and water systems.[82] Indeed, insurance companies sent their own engineers to help evaluate the water problem and speed up solutions. Further, they sent another committee to the city in 1910 to monitor reconstruction progress.[83]

For their part, San Franciscans received the insurance money they felt they justly deserved. Were the settlements fair? San Franciscans had suffered psychological trauma, physical loss, and financial damage from the fire. Whatever their thoughts about the destructiveness of earthquakes, everyday citizens had to support their representatives who had stressed fire damage over earthquake damage, because without champions they would have lost their insurance settlements. There is no doubt that the fire was far more destructive than the earthquake, but it is also true that for the reconstruction of the city, citizens and their leaders had to be sure to emphasize the gravity of fire damage. Thus a dichotomy was established, making it harder for those concerned with earthquake danger to have a decisive public following. Still, engineers and architects could not ignore what they had seen. Many remained committed to building structures that could resist both earthquake and fire. They tried to implement changes in building practice. Unnamed citizens, like my grandfather, remained concerned as well. But casting a pall over any overt earthquake-resistant campaign must have been the perception among San Franciscans that they had, in a way, colluded against the insurance companies and that they mustn't disturb the basis of their settlement. From the distance of one hundred years, it is easy to disparage the upbuilders and the boosters who praised the city and talked about the fire in the days of recovery and reconstruction. But we must remember that while they did not aid earthquake awareness, they helped save the city from economic ruin, which would have put an end to new buildings and made seismic reinforcement a dead issue.

CHAPTER 10

Fire Codes, 1906–1915

Building codes, lists of legal construction requirements covering one detail after another, are difficult to read. However, the building code of San Francisco is as packed with cultural meaning as a cuneiform clay tablet in ancient Sumer. Each code is a snapshot of what the city thought was important about building standards at the time. Codes are records of the struggle between the rich and the poor, labor and management, materials providers and contractors, architects and engineers, property owners and the city. A code is a measure of how seriously a society considers a particular risk. In this chapter and the next, we see whether or not San Francisco used the law to mitigate the risk of future fires and earthquakes.

It is important to separate codes from architectural and engineering practice. Codes are the minimum standards to which buildings must be constructed. "Built to code" does not ensure a good building. Architecture and engineering practice, personal ability, and individual values differ from one engineer or architect to the next. The code and licensing movements of the nineteenth and early twentieth centuries attempted to establish standards to ensure safe buildings. They were designed to curb abuses in the building professions as well as to constrain dangerous cost-cutting measures on the part of property owners.

Architects and engineers, while bound legally by the code, are free to exceed it. The architects and engineers working after the earthquake of 1868 exceeded code requirements to build structures that they felt would be safer in earthquakes. Architects and engineers working on steel buildings in San Francisco in the 1890s also far exceeded the building codes, attempting to ensure added safety in earthquakes and fires. In our minds, we must decouple architectural and engineering practice from the code itself. In other words, our opinion of the requirements of the code should not affect our assessment of the individual buildings of the time, which are each very different from one another.

In examining the codes, bear in mind that several present-day authors, generally decrying the poor quality of post-earthquake buildings in San Francisco, have contended that the 1906 codes were weaker than those before the earthquake, and became weaker still over time. One writer blames this on the callous disregard of the capitalist oligarchy for the future safety of the people of San Francisco. Another writer states that the terrific speed of reconstruction precluded better codes. This chapter attempts simply to uncover the story of what happened, apart from the question of blame. Why, for example, weren't the fire limits enlarged? Why was reinforced concrete, hitherto forbidden, now accepted as a building material? Why weren't the heights of tall buildings limited when, as some people believed, that would make them more earthquake-resistant? Some people see the post-earthquake city as a firetrap. Why didn't citizens of the time make sure the city would be fire-resistant? Only an investigation of the codes can help to answer these questions.

Post-earthquake San Francisco had an urgent need for new codes. Until authorities established new standards, only buildings already under construction could be completed, and only Class A steel structures could be repaired without a permit. Not even wood-frame buildings could be started. Temporary structures were needed so that businesses in the burned section could continue operating and people could be housed quickly. New codes had to be written with all possible speed, and they had to meet the challenge of shaping a safer city without alienating investors and cash-strapped citizens.

Before the earthquake, the building codes of 1903 had already come under severe criticism for their limitations. A disorganized and poorly written patchwork, they addressed neither the contemporary problems of steel construction nor the use of reinforced concrete, thought by many to be a promising new material.

After the earthquake, the beleaguered mayor of San Francisco, Eugene Schmitz, pressured by insurance companies and by concern for citizens' safety, took action quickly. Only a few hours after the earthquake, he established the Citizens Committee of Fifty to direct the city's reconstruction. The bipartisan committee, while representative, was seen by some as unwieldy. As we have seen, within days, perhaps as a means to consolidate his power, the mayor disbanded the Committee of Fifty and reconstituted it as the Committee of Forty on the Reconstruction of San Francisco. Among the committee's twenty-five subcommittees was the one on Building Laws and General Architectural and Engineering Plans, charged with recommending new and improved codes.

John D. Galloway, the respected and experienced civil engineer who had designed the steelwork for the St. Francis Hotel, was the subcommittee's chairman. The members included Jeremiah Deneen, a builder; William Curlett, a prominent architect who had worked with Augustus Laver on the old City Hall; R. B. Berkeley, a respected engineer; William H. Leahy, a successful entrepreneur and owner of the Tivoli Opera House; Frank Shea, the architect of the Whittell

Building and the City Hall dome; and J. J. Mahoney, a powerful contractor.[1] Political boss Abe Ruef appointed himself to the subcommittee in order to influence laws affecting real estate, his major business interest. Seven city supervisors were also members of the committee.

In less than three weeks, between May 3 and May 22, 1906, the Subcommittee on Building Laws met with the Subcommittee for the Burnham Plan and with the Board of Supervisors' Committee of Seven.[2] It heard testimony from labor and business leaders, insurance companies, materials manufacturers, and the fire department. It reviewed building laws of Baltimore, New York, and Chicago, as well as model codes written by the National Board of Fire Underwriters. Public debate over the subcommittee's recommendations lasted until July 5, 1906, when the new code was rewritten and voted into law by the city supervisors. Five basic issues emerged in this period: enlargement of the fire district; use of reinforced concrete; reclassification of structures; limitation of building heights; and requirements for fire-resistant construction. This chapter addresses these issues as they relate to fire safety. Those relating to earthquake safety are covered in chapter 11.

Fire District Controversy

The fiercest debate was over enlarging the fire district, in which buildings were required to have fire-resistant features. Usually only buildings with noncombustible exteriors can be erected in a fire district, but San Francisco's pre-1906 fire district included wooden buildings dating from 1852, all of which were destroyed in the fire. Now, officials faced the question of whether to expand the fire district and exclude wooden buildings from that part of the city. Wooden structures had weathered the earthquake well. Noncombustible buildings, constructed of brick and later of concrete, were more expensive than wood. The safety of brick in earthquakes was also in question. Wooden dwellings and industrial buildings had often stood side by side, particularly in the South of Market. Residents who testified before the committee were often in conflict with businesspeople in the same district.

Acting Fire Chief Shaughnessy spoke as an experienced fireman and the department's probable next chief.[3] He had read the warnings in the 1905 reports from the National Board of Fire Underwriters and had encountered the difficulty of controlling a full-scale conflagration in San Francisco. He wanted the city to extend its fire limits to the west and especially into the high-risk South of Market area. His position was understandable from the standpoint of fire safety, but unattractive to those citizens who had little capital for rebuilding. Committee member Frank Shea opposed the acting chief's view on these grounds, predicting that if the original small property owners could not rebuild, "their property would be forced into the hands of real estate speculators, and injury greater than that effected by the fire would be the result."[4]

Abe Ruef proposed a compromise. The city's hilly areas would remain residential and would be excluded from the fire limits, while the flatter central areas would be included in new limits extending west to Polk Street and into the South

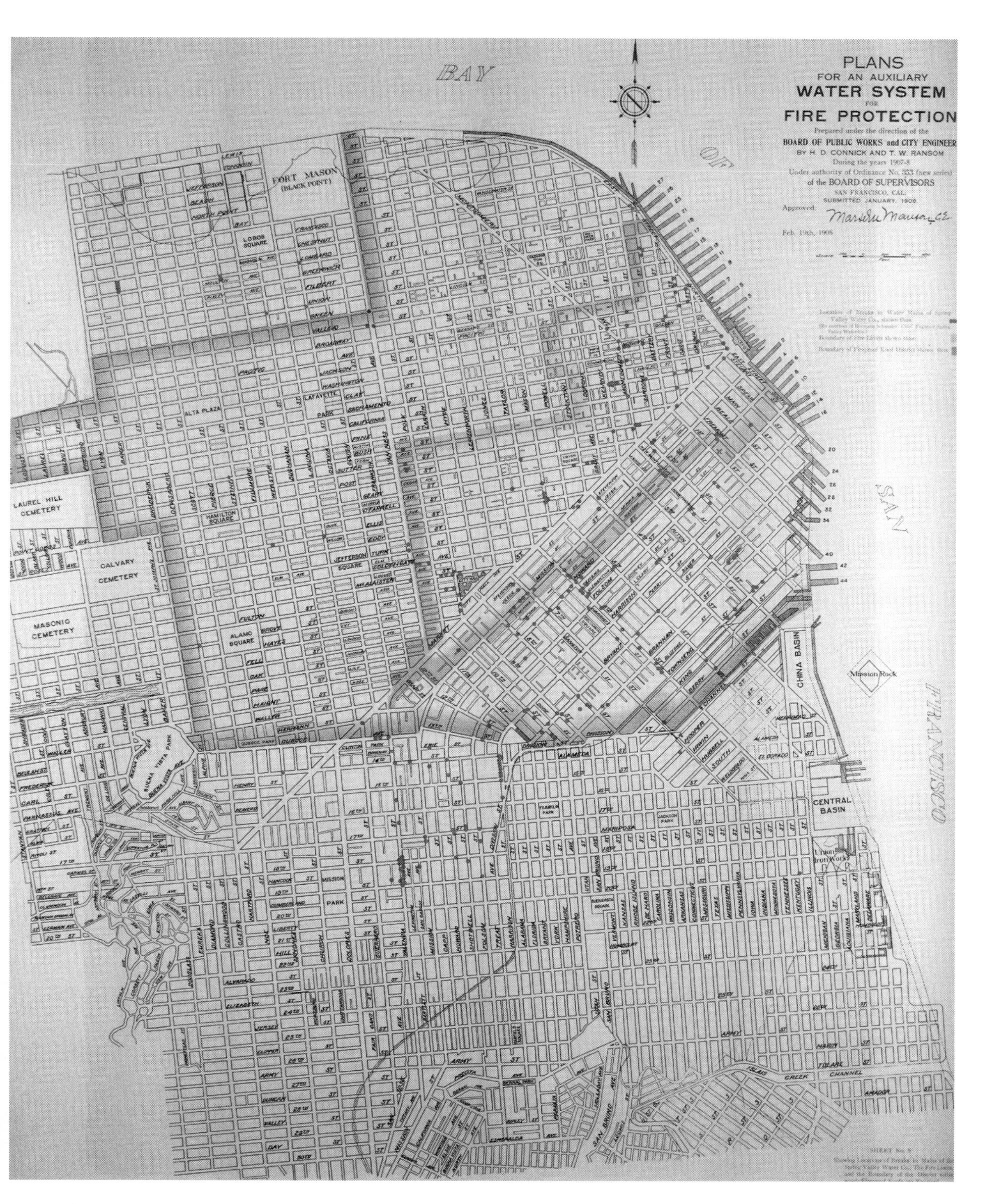

Figure 10.1
San Francisco's extended fire district and fireproof roof area, 1906

of Market between Third and Ninth extending to Bryant or Brannan.[5] This solution seemed reasonable to most middle class residents, who would not be affected because their property lay almost entirely outside the proposed fire district.[6] When some property owners in the Polk Street area east of Van Ness objected to their inclusion, the subcommittee pointed out that fireproof construction cost only 12 percent more than nonfireproof construction, and insurance rates for nonfireproof buildings were 100 percent higher. Mollified, the Polk Street owners withdrew their objections. In the end, the subcommittee carried Ruef's compromise to the board of supervisors over the objections of working class South of Market residents. As it turned out, this was only the beginning of the class struggle over the fire district.

By 1906, the South of Market was already in turmoil, due to the depression of 1873, the influx of transient working class men in the 1880s, and increasing industrial development. The area had been predominantly Irish and Catholic since the 1870s.[7] There were six churches and four parochial schools there, as well as an academy, two hospitals, two infant shelters, and Our Lady's Home for Old and Infirm Women, all operated by the church. Although all these institutions lay in ruins after the earthquake, the residents remained strongly attached to their neighborhood.

These citizens opposed the subcommittee's recommendations to extend the fire limits. They met in the shell of the ruined church of St. Joseph, at Tenth and Howard.[8] "Speeches in denunciation of what was termed a land-grabbing scheme on the part of the big capitalists were vigorous," according to the *Chronicle*:

> [Reverend John Rogers] deplored any laws that would compel men to sell their real estate on account of a lack of money to erect expensive structures, and attributed the proposal for the enlargement of the scope of the building ordinances to the spirit of greed of speculators...

Rogers continued the next day when the Board of Supervisors met, stating, "Back of it all, I see the agents of the insurance companies endeavoring to have a city of steel and stone built."[9]

A number of speakers at St. Joseph's pointed out that some brick buildings on the unstable land south of Market had collapsed, while wooden structures had fared far better. Reverend T. P. Mulligan summarized all the arguments:

> We should demand of the supervisors that the present fire limits be not increased. We are not opposed to a greater and more beautiful San Francisco, but we cannot forget that the calamity has left us poor. It is all right for the man from Chicago [Daniel H. Burnham] to camp on Twin Peaks and talk of cutting boulevards here and there, but who is to pay for them? We can hardly pay our taxes...I stand for the people before I stand for the beauty and grandeur of the city. *Better a city of shacks owned by the people than a city of sky-scrapers owned by Eastern capitalists* [emphasis added].

> The extension of the fire limits will mean our ruin. We cannot afford to put up Class A buildings and who wants to live in a...[brick] building? Frame buildings are our safeguard. The fire stopped at the wooden buildings at Van Ness Avenue and Twentieth Street [in the Mission].[10]

The Property Holders Protective Association was formed to meet with the subcommittee on the following day. At that meeting, representatives of both the association and another group, called the Associated Property Owners, argued against extending the fire limits. Garret McEnerney, representing Archbishop Riordan of San Francisco, spoke for the small property owners:

> While we may not ask you to diminish the limits, we can ask you not to extend the line and say to the small property owners, "Go on, and God be with you." What we need now is homes. Without homes you cannot have a greater San Francisco. Your small households are now either in tents or living in Oakland, Alameda, Sonoma, Marin, or other parts of the state. We want them to come back, we want a home city and not a warehouse city.[11]

It is interesting that, considering all of the controversy, no one is reported to have spoken in favor of extending the fire limits. Except for the extension west to Polk Street, the fire district remained the same. Why didn't the supervisors press more strongly for an enlarged fire district? Although extending the limits would have been a simple and effective fire-prevention strategy, they had other issues to worry about. Fire districts were hot political issues with enormous economic impact. There were similar battles in Chicago after its fires. The labor movement, with its large Irish component, was a strong force in San Francisco. The worker-supported Union Party had put Mayor Schmitz and his graft-ridden administration into power. It would have been politically imprudent for the supervisors to confront this group. The issue of recovery hovered over these discussions as well. Fast recovery would facilitate the return of the permanent population, while a recovery slowed by safety requirements might alienate those ready to return.

Powerless as they felt they were to create a larger fire district, the supervisors did at least designate a large portion of the burned area as a "special fireproof roof area," which meant that the same requirements that applied to the roofs of the fire district also applied to a wider area.[12] Many of San Francisco's roofs, roughly from Van Ness Avenue in the west above Green Street, Broderick to the west below Green Street, south to Duboce, to Bay Street in the north, to San Francisco Bay in the east and including some of the South of Market, were to be made of asphalt, tile, slate, asbestos, terra-cotta, or metal. Flammable roofs are a major danger in large fires. Firebrands blown skyward from burning structures often ignite fires on nearby roofs.[13] In a conflagration or a firestorm, firebrands can leapfrog from one building to another, even across wide streets.[14] The requirement for noncombustible roofs was a major step in preventing conflagrations like the one the city had just experienced.

Concrete Enters the Codes

During the late nineteenth century, pioneers like the French inventor François Hennebique had been experimenting with combining concrete and reinforcing steel, or rebar.[15] Roman engineers had used concrete combined with brick as a structural material, but since the fall of the Roman Empire, concrete had not been used as a building material in Europe. Roman concrete was strong in compression, like rock, but weak in tensile strength; alone, it would be unsuitable in beams. But by placing steel bars into the molds in which concrete was poured, Hennebique and others improved its tensile strength.

In pre-earthquake San Francisco, Ernest L. Ransome had been using concrete with rebar for years.[16] In 1884 he patented his most important innovation, the placement of cold-turned steel rebar in concrete for reinforcement. He went on to pioneer the use of reinforced concrete in sidewalks in the 1880s, and San Francisco's sidewalks were soon considered the best in the nation. In 1889, he built what appears to have been the first reinforced concrete bridge in the world, the Lake Alvord bridge in Golden Gate Park. Ransome and others used concrete before 1906 in the construction of a dry dock at Mare Island, San Francisco's sea wall, the foundations for the Ferry Building, the interior columns and floors of the Academy of Sciences, and the Cyclorama bicycle track in Golden Gate Park.

While concrete was accepted in East Coast codes as early as 1903, it was regarded with suspicion in San Francisco. The brick lobby portrayed it as rough and undependable. Prior to 1906, it had been allowed in low-rise buildings for columns, and it was generally accepted as excellent for floors in steel-frame buildings, but it was not allowed in high, load-bearing walls. The earthquake changed all that. Engineers had high expectations that concrete was fire-resistant, based on its performance on the East Coast, and they were convinced that reinforced concrete could perform well in earthquakes. A search for local exemplars yielded an incomplete one-story Bekins warehouse with a brick exterior and reinforced concrete columns and floors, and the Academy of Sciences, also with a brick exterior and concrete-filled iron columns and reinforced concrete floors.[17] Farther afield there was a successful Ransome building: the reinforced concrete Stanford Museum, which performed well in the earthquake while nearby substandard masonry buildings failed spectacularly.[18]

Opposition from labor unions and terra-cotta manufacturers, combined with public skepticism, made the admission of reinforced concrete into San Francisco's building codes a battle.[19] Newspaper coverage of the proceedings of the Subcommittee on Building Laws reveals the vehemence of the opposition.[20] Lawyer James C. Sims took a "commonsense" approach, claiming that "re-enforced [sic] concrete is yet in the experimental stage, and I believe it is not feasible to erect a building of that material over four or five stories high." The reply from William Hammond Hall, of the Structural Association, was scathing: "The gentleman announced that he was neither an architect nor an engineer. If that is the case, then his opinion on this subject is not worth anything."

Underlying the arguments of those who supported adding concrete to the codes was the hypothesis that concrete was better than brick. Speaking for the opposition, E. J. Brandon of the bricklayers' union demanded tests to determine which was superior. Several major professional organizations supported the use of concrete, including the Structural Association, the Architects of San Francisco, and the San Francisco Chapter of the American Institute of Architects.[21] Ultimately it was the urgent need to start rebuilding that defeated the hesitation, caution, and out-and-out aesthetic dislike voiced by concrete's detractors, and concrete was added to the code.

The Reclassification of Structures

At first, concrete was categorized as Class B (buildings with load-bearing walls of steel or masonry, floors supported by steel frame or wooden posts, and all surfaces protected by metal lath), but once its particular strengths and requirements became apparent, the whole classification system had to be rethought. By 1909 Class A buildings had steel frames and could for the first time have concrete walls that were either self-supporting or hung from the frame; the walls could also be brick or terra-cotta.[22] Class B buildings were defined as having a frame of reinforced concrete that carried all wall and floor loads. Class C was similar to the pre-1906 definition, comprised of "buildings defined as having exterior walls of brick, stone, or reinforced concrete and an interior frame of combustible material."[23] Thousands of concrete buildings rose in San Francisco between 1906 and 1933, carrying the hope of fire- and earthquake-resistant construction.

The Limitation of Building Heights

One of the most time-tested, tried, and true earthquake mitigation strategies is height limitation.[24] The argument is that if buildings are uniformly limited to one or two stories, they can be built to be more earthquake-resistant. Another benefit is that fires can be fought more easily in two stories than in ten; lower buildings can be entered more easily, and volumes of water can be directed at them more effectively, particularly if there is a water-pressure problem.

Height limitation was a national issue in the early twentieth century, debated in metropolitan centers like Boston and New York on the grounds that tall buildings cast shadows and otherwise negatively impacted their neighbors.[25] The 1907 Boston code states, "No building, structure, or part thereof shall be of a height exceeding two and one-half times the width of the widest street on which the building or structure stands."[26] New York chose not to limit building heights in relation to street widths. In San Francisco the height limitation was linked with the plan of Daniel Burnham, who had taken for his model the uniform facades of Paris boulevards, an aesthetic that was antithetical to the anarchic skylines of America's burgeoning cities.

The trouble with linking safety with building height was that the tall, steel-frame buildings of San Francisco had performed admirably in the earthquake. For firefighting, standpipes and other devices could be used to bring water to the highest of the city's buildings. (A standpipe has connections at street level and hose connections at each floor that can be filled by private or city water supplies. In a serious fire, a fire engine can attach its hose to the street-level valve and pump into the standpipe, providing added pressure for fighting fires on multiple floors.) In the end, as we shall see in chapter 11, engineers and architects helped to defeat height limitations.

Fire-Resistant Construction

Building codes requiring noncombustible roofs and allowing the use of reinforced concrete in construction would help increase fire safety in San Francisco. How else did the codes address this issue? Remembering Chicago's 1903 Iroquois Theater fire, in which 602 people died from smoke inhalation and trampling while searching for unmarked exits, the Subcommittee on Building Laws included an extensive section on fire safety in theaters.[27] Exits must be plainly marked, doors must open outward, theater programs must include a diagram of the building showing the locations of exits, and fire-resistant materials must be used for structural members.

Fire escapes were required on all structures, *except* steel-frame buildings and steel or concrete buildings with load-bearing facades. If these buildings had two interior staircases, no fire escape was required.[28] However, the San Francisco codes still did not specify that stairways had to have automatic fire doors to isolate them if a fire broke out. There was a good deal of discussion about fire escapes, with some professionals discounting them because people would be afraid to use them if buildings were too high. The National Board of Fire Underwriters endorsed fire escapes, but faith in fireproof steel construction was strong: in New York in the 1890s, workers had remained at their desks in the World Building, confident of their well-being, as fire burned around them.[29] Luckily, the fire was extinguished without injury.

The new San Francisco codes provided for fire-fighting devices and fire barriers within steel, concrete, and brick buildings.[30] The most important addition was the requirement for standpipes in virtually all buildings exceeding fifty-five feet in height.

The fire after the 1906 earthquake had been so hot that steel members bent and buckled. Investigators noted that coverings of brick, terra-cotta, and metal lath and plaster around columns had cracked in the earthquake.[31] The best solution was concrete, which had cracked less and therefore provided better fire protection. But because other more stringent codes in New York and Chicago included brick, terra-cotta, and metal lath and plaster, San Francisco included them too. The specific San Francisco experience, in which earthquakes damaged column protection, was overshadowed by the appeal of national standards.

The worst omission in the new 1906 codes was in regard to wood-frame buildings. There was no mention of fire-resistant walls or firebreaks in contiguous buildings, which might have helped prevent the spread of fire from one frame building to the next. According to former San Francisco fire chief Emmet Condon:

> One of the greatest building defects [in the code] from the standpoint of fire hazard [was allowing] row housing to be built with wooden side walls abutting the adjacent building without barriers to arrest the spread of fire from one building to another. For example, in the outer residential districts in the western part of the city, a block usually consists of 24 houses built with wooden side walls touching one another. As far as fire protection is concerned this is effectively a single block-sized structure. Each of these blocks represents a single fire load of combustible construction exceeding 30,000 square feet. Under major fire conditions this fire load would tax the best efforts of an entire metropolitan fire department.[32]

Fire walls had appeared in the 1903 codes.[33] They stipulated that adjoining walls between houses be constructed of redwood, which was thought to be fire-resistant. But this stipulation is not included in the 1906 codes, nor are brick, terra-cotta, asbestos, metal sheathing, or other fire-resistant materials required.

Other fire codes, like those of New York, did require fire-resistant walls:

> Any exterior wall of frame construction, hereafter erected within three feet of a side or rear line of the lot or plot on which it is located, or hereafter erected as the side wall of any frame tenement house, shall have the space between the studding filled in solidly with brickwork or other approved incombustible material.[34]

This is similar to the model code prepared by insurance companies, which included the most sophisticated fire-resistant methods of the period. But in San Francisco, where over 90 percent of the dwellings were wood, the lack of any mention of fire walls or firebreaks is astounding. There were no required setbacks from adjoining wood-frame buildings, either. The only fire-related requirements for wood-frame construction were that exterior sheathing over studs should continue from roofs to foundations, and that the area between studs be blocked by bridging, to stop fire from burning up these vertical hollows.[35] Perhaps authorities felt that a requirement such as New York's for brick infill of wood-frame walls would be too radical a change in local building practices, especially when brick was seen as unstable in earthquakes. Or perhaps, having seen the ferocity of a firestorm, local builders simply felt that fire stops were useless.

The Evolution of the Building Codes

When the first version of the new building code became law on July 5, 1906, it was greeted with little enthusiasm. Bricklayers felt that concrete walls were allowed to be too thin; they wanted them the same size as brick walls.[36] Engineers were unhappy because concrete was still linked to brick, and not treated autonomously. Firefighters were unhappy with the restricted fire district, but satisfied with the new building heights, which were lower than architects and engineers would have wished. An anguished John Galloway, trying to balance good building practice with the need to rebuild as quickly as possible, fired off this retort to criticism by his colleagues in the American Society of Civil Engineers:

> The distant observer may ask why, with virgin ground before it, the city did not cut avenues, widen streets, and build nothing but incombustible buildings. Such a comment is most superficial. The city had suffered from the greatest fire in history. Most of her industries were wiped out of existence, all business buildings destroyed, goods burned, streets wrecked and filled with debris, sewers broken, the water supply badly crippled, and the transportation system destroyed. Comment on civic responsibility, in the face of such conditions, is mere froth. What San Francisco needs is the cheapest building possible in which business may be done, to insure the community enough to eat. The other subjects can wait.[37]

While the codes of 1906 were framed in the frenzy of reconstruction, the revisions that followed were not. Within a year, the height limitation on Class A buildings was rescinded, to the dismay of firefighters but much to the relief of engineers and architects, who rightly saw it as unjust. The first major revisions, made on December 22, 1909, involved making the codes more explicit on the wider acceptance of reinforced concrete, defining terms more clearly, and improving the organization of the sections on building classification.[38] There was little change in the sections dealing with fire safety, other than a requirement that all buildings, irrespective of class, have fire escapes.

In 1910 a commission chaired by Dr. Langley Porter noted that appalling conditions in tenements had developed as a result of loopholes in the 1906 codes concerning population density and air circulation.[39] A 1913 state law attempted to curb abuses. Code revisions in 1915 included more sections on patent fireplaces and gas and electrical lines and fixtures. The codes of 1923 were a great improvement. Finally, an index of all code subjects appeared. Many requirements were updated, although those for frame construction remained insignificant. New York codes were still more comprehensive in the specification of construction standards.

There were any number of reasons for the weaknesses in the San Francisco codes, but one of the most compelling was the drive for national standardization of building codes, which had the unintended effect of discouraging local variation.[40] In general, the code revisions of 1906, 1909, 1915, and 1923 succeeded in making San Francisco a safer city by adopting building practices and standards from more advanced codes, like those of Chicago and New York. But the worst of San Francisco's fire problems concerned its many wood-frame structures. Nationally, codes focused on business districts of steel, concrete, and brick; wood construction practices were the least professionalized and largely unregulated. They were commercially unimportant and, in most cities, outside the fire district. It was left to contractors and carpenters to determine the detailing for wood buildings. None of the frame buildings outside the fireproof-roof district had significant fire-resistant features.

The post-earthquake fire codes of San Francisco tell the story of a city that did not want to relinquish its traditional building material, wood. Wood construction was the proverbial elephant in the room that nobody wanted to address. Building professionals concentrated on making the important business and commercial center of the city as fire-resistant as possible, but they did not regulate wood construction except to stipulate fire-resistant roofs in a large designated area of the city. Few if any San Franciscans would have consented to give up their wooden houses. Many thought of redwood as a fire-resistant material. The long-held San Francisco belief that wood was safer than brick in earthquakes seemed to have been borne out on April 18, though brick buildings did better than expected. As Reverend Mulligan said, "frame buildings are our safeguard," and brick was expensive. Narrow San Francisco lots made firebreaks or setbacks an economic hardship and posed a definite problem for fire mitigation. Many of those lots west of Van Ness were occupied by wooden structures with side walls that touched each other. Except for requiring studs to be covered by sheathing, this problem was not addressed in the code. Let us imagine for a moment that the city did require owners to fireproof these walls. Ironically, the logical choice would have been asbestos—in the long run, more dangerous than fire danger itself.

The codes did little to change the look of San Francisco. They did not impose an unwanted aesthetic on downtown San Francisco by limiting heights, they did not greatly extend the fire district, which would have increased the prevalence of brick and concrete buildings. However, they did succeed in raising construction standards to the national level for class A, B, and C buildings while letting the city rebuild itself quickly. At first the codes might be dismissed as not stringent enough to save San Francisco from future fire. But while fire was on everyone's mind, earthquakes were more frightening, even if the external propaganda emphasized fire. If wooden buildings are important for earthquake mitigation, the building code essentially passed responsibility for protecting these buildings from fire to the fire department and the new auxiliary water system.

CHAPTER 11

New Earthquake Codes

If engineers and architects investigated the 1906 earthquake and understood the potential danger of future temblors, why didn't they press for earthquake-resistant requirements in the new building codes?[1] In California, no codes explicitly labeled as seismic appear until the municipal codes of Santa Barbara and Palo Alto in 1926, and the regional Uniform Building Code in 1927, all prompted by building failures in the Santa Barbara earthquake of 1925. The state of California issued an even more comprehensive code in 1933, in response to the Long Beach earthquake of that year.

Cities in Italy had developed earthquake-related codes in the eighteenth century. Palermo limited the size of stone balconies after the earthquake of 1726. After the earthquake of 1783, the Bourbon government established an earthquake-resistant building code that mandated insertion of timber frames in masonry structures. This concept of a flexible wooden frame embedded in masonry also appeared in nineteenth-century codes after an earthquake on the island of Ischia, in the Bay of Naples, and major earthquakes in Calabria. After the 1908 Messina-Calabria earthquake, the government of Italy instituted a new seismic code stipulating lateral resistance for all buildings as well as requirements for timber-frame masonry and ferroconcrete.

It has been said many times that San Francisco architects and engineers didn't see, didn't understand, and didn't act to codify what could be learned from earthquake damage.[2] A close look at the 1906 codes shows that earthquake resistance was taken into consideration, but the codes were not labeled as such. Perhaps this explains why the efforts of so many professionals have been obscured, overlooked, and forgotten. The real story includes ongoing discussions among these professionals about earthquake-resistant design which, while not codified, did influence construction practices.

We have reviewed the changes that the Subcommittee on Building Laws and General Architectural and Engineering Plans made in the fire codes. Now let's examine those changes that might be considered earthquake-related from two viewpoints, that of the most knowledgeable engineers of the time, and from our present-day understanding of earthquake-resistant construction.

For a view of the 1906 codes at the time they were written, there are two principal sources: lengthy discussions of the Structural Association of San Francisco, and a small book written by University of California engineering professor Charles Derleth Jr., who was secretary and a founding member of the Structural Association. Derleth later became dean of the School of Engineering at UC Berkeley, and the consulting engineer on numerous structures, including the Golden Gate Bridge. In *The Destructive Extent of the California Earthquake*, published in 1907, Derleth discussed earthquake damage and proposed thirty-six points that for him were important in improving structural performance.[3] Derleth had nothing to do with the framing of the San Francisco codes; he was an unbiased outsider. By using Derleth's observations, we can know what a reputable engineer of the time considered to be reasonable and effective earthquake-resistant solutions.

Cornices

The first earthquake-related change in the 1906 codes addressed cornice support. Cornice failure was so common in the 1906 earthquake that Derleth made it the first of his thirty-six points. He, among others, did not advocate abandoning cornices altogether but thought they should be more securely attached. He wrote:

> Cornices and top walls of buildings, especially of brick and stone structures, should not be heavy nor have great projection. In the future, cornices and top walls should be more securely anchored with metal, their masonry bond should be made with care, and the cementing materials should be of high quality.[4]

The 1906 code improved on a vague cornice regulation from 1903 that stated, "Cornices of stone, brick, or other masonry shall be properly supported on and well secured to the wall..."[5] The new, clearer code specifications of 1906 read, "All cornices should be of metal, stone, or terra-cotta, secured to steel or iron brackets, which shall be supported by and connected to the steel frame of the building."[6] The improved law's impact can be seen in the structural steel cornice supports on numerous structures built between 1906 and 1909. In 1909, the law changed again. Cornices no longer had to be attached to the building's frame, but more stringent tying was required: "Stone and terra-cotta cornices shall have every piece anchored to backing with heavy anchors, and where necessary supported on steel supports."[7]

Architects practicing around 1906 would never have consented to a ban on so basic a decorative element as the cornice without strong evidence against it. Many cornices had survived the earthquake, and the State Board of Architects assured that:

> Safety is not a question of style of architecture but quality of workmanship...Cornices and arches need not be excluded from the new city. Where they were properly anchored and built they withstood the shock and the fire both. It is the opinion of the board that the city beautiful need not be without its picturesque cornices and decorations.[8]

The cornice provision remained unchanged until the 1970s, when cornice projections were severely limited and special ties were required.

Parapets

The second earthquake-related change in the 1906 codes addressed numerous failures of parapet walls. These failures have plagued us recently as well, notably in the Whittier and Loma Prieta and other strong earthquakes. Derleth counseled:

> Projecting brickwork and fire walls of ordinary brick and stone structures have proven themselves an abomination, especially where lime mortar was used. With cement mortar the destruction was nearly as great, the only difference being that the material fell in large masses.[9]

While the 1903 code did not stipulate any reinforcement for parapets, iron brackets were required after 1906. The 1909 code further required a three-by-three-inch continuous steel angle built into the wall and connected to metal rods or other anchors extending from and attached to the roof.[10] Both codes call for the proper mixture of cement mortar as well.

Lateral Bracing

There was much discussion concerning lateral bracing of steel-frame buildings—where and how much of it should be used—although it did not enter the codes. Class A steel-frame buildings were judged to have survived the earthquake relatively unscathed—they were readily visible rising over the ashes after both earthquake and fire. Engineers were elated by their performance. Himmelwright, while underscoring the necessity of bracing, nevertheless noted that steel-frame buildings "generally sustained earthquake shock without damage."[11] William Hammond Hall wrote, "Steel frames...without exception, as far as is known, withstood the effect of this violent and really vicious earthquake practically uninjured."[12] Another engineer, Couchot, while noting a number of X cracks, nevertheless said of the city's steel-frame buildings, including the Flood Building, "all these magnificent buildings were practically intact after the earthquake."[13]

Although Class A buildings did not collapse, many showed signs of stress on their masonry exteriors. X cracks and dislocation of bricks around openings and at joints in the steel frames were common. In 1906 the reason for these cracks was not clear: Derleth attributed most of the cracking to improper bond or anchoring in the brickwork, or to lean mortar. He contended that internal walls of reinforced concrete would be far better than hollow clay tile or brick in an earthquake because they could be made lighter and, of course, they would be integral rather than made of separate blocks. He acknowledged that "high structures were subjected to a considerable racking [that is, interstory drift], as is evidenced by the lines of rupture in brick walls at the floor."[14] Derleth, as well as many of his colleagues, understood the necessity of flexibility:

> The amount of earthquake stress produced in a member of a structural steel frame is directly proportional to the resistance offered. The stiffer a structure, the greater will be the induced stress produced by earthquake vibrations. The more a structure is capable of yielding, like a willow tree to a storm, the less will be the tendency for earthquake rupture or collapse."[15]

In general, Derleth's observation here is correct. Today we would say that these cracks result from the inability of the masonry elements to bend as much as the more flexible steel frame, which separates from and crushes the nearby masonry damping the structure, dissipating energy, and controlling drift as the building vibrates from side to side.

The problem for engineers in 1906 and today is how to ensure flexibility while limiting drift. What kind of bracing is appropriate and where should it be positioned? Each engineer in 1906 had his own opinion. Christopher Snyder, the future engineer of San Francisco's new (1912) City Hall, urged the use of broad lattice girders to stiffen buildings.[16] John Galloway, the chairman of the Subcommittee on Building Laws, agreed, saying that the question of bracing was the most important one structural engineers faced. In a rare and candid statement, Galloway conceded, "I will own up to some cases where I have designed buildings in which the proper consideration of bracing was not taken."[17] Derleth advocated that "steel frames for Class A buildings should be provided with considerable knee-braced framing in the vertical planes between the main floor girders and columns to which they are riveted, and whenever possible, diagonal framing should be introduced similar to that provided to resist wind stresses."[18] However, he also cautioned his readers that "diagonal framing is not desirable for high buildings in earthquake countries" because it is too stiff and unyielding. Using the San Francisco Ferry Building tower as an example, he contended:

> For earthquake conditions, triangular framing for high buildings is not so desirable as a rectangular framing with stiff joints and continuous members. Unlike the triangle, the rectangle can change its form without changing the lengths of its sides... With spandrel girder and knee bracing,

therefore, the main columns, by their elasticity and continuity, can yield and vibrate to a considerable extent without endangering the integrity of the building's frame.[19]

Derleth approved of the latticed spandrel girders being used in the new, post-earthquake Humboldt Bank Building (see chapter 13). When he built Sather Tower at UC Berkeley in 1916, he used diagonal bracing and rectangular framing on alternate stories to ensure both flexibility and stiffness. The discussion of braced frame versus moment frame is still going on, except now we have much more experience, from quakes and from testing.

While many other engineers agreed in principle, the problem of where to place the proper bracing proved difficult to resolve. Engineers did not (obviously) have our sophisticated computer analyses, or expensive tests like those on the shake table, or today's knowledge derived from observations of many earthquakes around the world. However, they did what they could. Some, Galloway among them, championed maximum stiffness on the first floor: "If wind bracing is left out of the first story, it might as well be left out of the entire building. It is an attempt to sustain a weight by a chain, one of the links of which is removed." [20] After studying all of the cracks and displaced masonry, a number of engineers dissented. Many buildings seem to have cracked in only their upper stories. For example, we know that because of drift, the top of the Call Building was badly shaken, loosening courses of masonry, while the bottom did not sustain the same damage.[21] Himmelwright recommended, "The mid-height of tall buildings should be particularly well braced on account of the racking tendency in that zone."[22] Another engineer, H. C. Vensano, in a paper delivered before the Structural Association of San Francisco on July 12, 1906, agreed with Himmelwright but felt that in order to lessen the stress on the upper portions of the frame, the bottom floor should be made extremely flexible:

> I do not believe rigidity is entirely desirable, but it is absolutely essential in certain parts. It is essential that the part of our structure carrying our curtain walls be made as rigid as possible. I would like, then, to make this suggestion; that we make the part above the first floor (the part which carries our masonry) as rigid as possible by the introduction of deep girders, gussets, knees, or other means, and then attempt to obtain as much elasticity or flexibility as possible in the only part which remains, and that is in our basement story...Make columns long in the basement length; this will make them more flexible and lessen the initial pull on the building. Design them so that the total maximum deflection can occur below the first tier of beams and thus avoid shear of rivets at this line when columns move in opposite directions.[23]

There is clear danger in Vensano's approach. The bottom story can be made so flexible that it becomes a soft story and collapses. He and the others did not yet know how strong earthquake forces are. He is proposing base isolation, but of the wrong kind. When Christopher Snyder designed the steel frame for the new 1912 City Hall, completed in 1916, he wrote that he omitted bracing from the first floor through the third floor to absorb earthquake forces, sparing the upper parts of the structure. In the 1989 Loma Prieta earthquake, the flexible first story performed exactly as Snyder had expected. In the view of engineer Eric Elsesser, the soft story may have saved City Hall by dissipating some of the earthquake forces.[24]

The wide divergence of views on how a steel frame should be braced probably explains why more detailed directives did not appear in the code. The engineers couldn't agree. The 1906 code does, for the first time, include stringent standards for bolt stresses, column detailing, and floor loads, which were copied from the New York and Chicago codes and were already standard practice for most engineers.[25]

The wind-load requirement for steel-frame buildings was the most important addition to the 1906 building code. Most San Francisco steel-frame buildings constructed from the 1890s to 1906 had a wind-load capacity of 30 pounds per square foot (psf), even though it was not required by law. In 1906 this standard became part of the building code:

> In buildings over one hundred feet high, or where the height exceeds three times the least horizontal dimension, the following provisions of this section shall apply: The steel frame shall be designed to resist a wind force of 30 pounds per square foot acting in any direction upon the entire exposed surface. All exterior wall girders shall have knee bracing connecting to the columns. Provision shall be made for diagonal, portal, or knee bracing to resist wind stresses, and such bracing shall be continuous from top story to and including basement.[26]

We know from deliberations in the 1890s that engineers believed that wind pressure and earthquake vibrations caused the same kind of stress on Class A buildings. They used the term "wind pressure" to refer to both earthquake stresses and wind forces. Galloway reiterated what every engineer knew:

> There are no means of calculating the stresses in a building due to an earthquake, but, judging from the behavior of buildings such as the Shreve Building or the Claus Spreckels Building, I would say that if a building is properly designed for a wind pressure of 30 pounds per square foot on its superficial area, that it would be sufficient to withstand an earthquake of an intensity equal to that of April 18th.[27]

The addition of a wind-load paragraph to the 1906 building code was not labeled as a seismic provision, but it was clearly considered earthquake-related at the time. As the years went by, wind load was considered just that—wind load—but to the engineers working in the years immediately following the 1906 earthquake, this was a seismic building code requirement, even if it also appeared in New York and

Chicago codes. This is a very important point to emphasize: architects and engineers understood that excessive drift had to be controlled to avoid collapse. They had seen structures built to the 30 psf standard survive severe shaking. They were cognizant of the problems of demand and capacity, they knew that extremely stiff buildings could be dangerous in an earthquake.

In the buildings being constructed in San Francisco after 1906, obvious attempts were made to increase strength by using lattice girders and deep spandrels to maintain "wind-load" resistance while retaining flexibility. Due to a fortunate limitation in their knowledge in that day and age, engineers only included the frame in calculations for wind resistance. When the infill of the exterior walls and some interior walls and partitions is included in these calculations, we realize that these buildings are far stiffer and tougher than the 1906 engineers knew. The Empire State Building, constructed in 1931 to withstand wind loads of 20 psf, is 4.65 times stiffer than its builders had anticipated.[28] This unanticipated strength, or stiffness, that walls or infill provide will probably save the majority of post-1906 steel-frame structures in San Francisco from collapse in future earthquakes. However, in some cases the infill also produces soft stories and can even cause severe torsion or twisting.

Code changes regarding cornices, parapets, and wind load were, as we have seen, valuable. A fourth change, height restrictions on tall buildings, was not motivated by earthquake danger, but by concern with firefighting and urban aesthetics.[29]

Supervisor Jeremiah Deneen, a prominent member of the Subcommittee on Building Laws, wanted to impose height limitations in order to move property owners toward accepting elements of Daniel Burnham's Beaux Arts plan for the city. The fire underwriters had their own formulas, which would have limited buildings on Van Ness Avenue to 300 feet, on Market Street to 250 feet, and on Montgomery and most other Financial District streets to 150 feet. San Francisco's fire chief believed that fires could be fought to 255 feet. But the technical members of the committee were outvoted and the final formula was based on Burnham's Parisian model: Class A structures could not be taller than one and one-half times the width of the streets on which they faced. This limited structures to 180 feet on Van Ness, 150 feet on Market, and 90 feet on Montgomery.

Fire-fighting concerns and the negative impact of height on neighbors' light and air had led the city, in 1903, to limit steel-frame buildings to 221 feet. Now, engineers were outraged, believing that the development of steel-frame construction and safe elevators made height constraints completely unnecessary. Architect Frank Shea and engineer R. C. Berkeley expressed the general view of building professionals—the same debate was going on in cities throughout the United States—in a newspaper article: high buildings had resisted the earthquake as safely as low ones, and no height limitations were necessary. Under sustained attack from the technical community, all height restrictions for class A structures with loads carried by steel frames were deleted from the codes in spring 1907.

Discussed But Not in the Codes

Several seismic issues were discussed at length but not addressed by the new codes: the use of terra-cotta as cladding and column protection, brick as a structural material, and the design of wood-frame houses.

Years before 1906, the Chicago fires of 1871 and 1874 had demonstrated that metal columns melted at high temperatures. Because of the vulnerability of steel frames, they had to be protected.[30] Terra-cotta, along with wire lath and plaster, brick, and concrete, was considered an acceptable material to use around metal columns for fire protection. Lath-and-plaster fireproofing, used in the newly completed Fairmont Hotel, had behaved very poorly. Derleth, the voice of engineering at the time, reminded his readers that brittle materials were hazardous in earthquakes. He condemned terra-cotta for ornament and column protection, and also hollow clay walls.[31]

Why did no restrictions on the use of lath-and-plaster and terra-cotta fireproofing and claddings enter the code? Concrete, while effective for column protection, was heavy and expensive. It was as yet unproven as facade material, and at that time it was aesthetically unacceptable. Until plastic became available, no other material was as malleable as terra-cotta. An overwhelming number of San Francisco's facades are faced with terra-cotta, often cast into decorative shapes, sometimes masquerading as stone. When tied with metal straps to a steel frame, terra-cotta had done quite well in the 1906 earthquake. Hollow clay tiles and lath-and-plaster wall protection were both respected as fireproof materials in the United States and hard to discredit. In the absence of an ideal solution, the codes did not address cladding and fire protection issues—but engineers and architects were vitally concerned about these questions.

The second material widely discussed when the new codes were being developed was brick. Common brick construction with poor ties and no reinforcement is extremely hazardous in earthquake country, but some brick buildings with special construction provisions did well in the earthquake of 1906. Architects and engineers were puzzled by the material's idiosyncratic performance, and the brick survivors of 1906 fueled a debate within the professional community. But, focused as they were on the new concrete and steel-frame technologies, engineers had little interest in brick and wood buildings, and the performance of brick buildings was not systematically studied. Derleth's critique was typical. He focused on poor workmanship and low-quality lime mortar. He also condemned the lack of ties and anchors and likened brick buildings to "weak boxes" built with "no adequate provision for transverse stiffness, because the structures entirely lacked transverse brick walls or frames." And then there was the matter of flexibility:

> The prime requisite for a structure to withstand earthquake shock is elasticity; that is, the ability to return without serious damage to its original shape and position after being distorted. It should vibrate without offering great resistance to distortion; in other words it should yield

> readily...structures of brick and stone built of blocks, like brickwork and cut-stone masonry...do not answer the requirements of yielding and elasticity.[32]

The prestigious reconnaissance team of the American Society of Civil Engineers agreed:

> It may be stated, as one of the most obvious lessons of the earthquake, that brick walls or walls of brick faced with stone, when without an interior frame of steel, are hopelessly inadequate. As a method of building in earthquake countries, such types are completely discredited.[33]

Several brick buildings that had survived in 1906 were used as examples of brick's superior behavior in earthquakes: the Palace Hotel, the U.S. Mint, and the U.S. Appraiser's Building. The Appraiser's Building was doubly significant because it was built on fill. What most engineers didn't know in 1906 was that all three of these buildings were replete, even redundant, with earthquake-resistant features. Thousands of other brick buildings were not seriously damaged, which implies that they were probably reinforced in some manner.

Since the larger issue of whether or not to limit the use of brick in buildings could not be agreed upon, what entered the 1906 codes were specifications for better brickwork and cement mortar. More than 90 percent of the new construction after 1906 was either flammable wood or fire-resistant brick. Banning plain brick construction would have been impossible, even if the experts had agreed. It was only in 1933, after the poor performance of brick in the Santa Barbara and Long Beach earthquakes, and when new technologies had become available, that unreinforced brick and other masonry would be banned at the state level.

In evaluating brick buildings that survived the earthquake, Derleth uncovered the lost history of earthquake-resistant brickwork in San Francisco. He saw what his colleagues missed at the Palace Hotel, which had survived the earthquake but not the fire:

> The Palace Hotel was built immediately after the earthquake of 1868 and was intended to be earthquake-proof. Looking down upon the ruins of the structure one saw a honeycomb of brick walls giving lateral stiffness in all directions. Iron rods were embedded in the walls. The old Palace Hotel therefore should not be classed as a simple brick building, for it contained some attempt at reinforcement.[34]

Derleth understood what he had seen at the Palace but didn't understand how widespread the use of brick reinforcement had been after the earthquake of 1868. He did not discover the earthquake-resistant reinforcement and special foundations in the U.S. Mint or the Appraiser's Building. Not knowing about the reinforcement in the walls, neither Derleth nor his colleagues could evaluate the performance of bond iron, vertical and horizontal rods, or metal ties. The Structural

Association debated a section of the 1901 code that mandated the placement of bond iron under the joists of all brick buildings. T. H. Skinner, a member of the association, asked whether this kind of reinforcement worked:

> Now I want to ask a question of the Society in general. That is: What is the consensus of opinion as to the use of steel bands in brick walls called for by the building ordinances? I would like to know if there is an example which has been of any use whatever. In the Palace Hotel, I understand that there were bands every three feet in height in the center of the wall; but there was sufficient thickness to these walls to carry the bond. But of the average work throughout the city, I think you will find the greatest point of weakness just at these iron bands.[35]

His comments elicited this response from another engineer, F. C. Davis:

> Our great fire and earthquake have forcibly shown us the folly in omitting to apply the well-known principles of using sufficient amount of metal bands and anchors to securely bind our walls together...If we people of the Pacific Coast wish our walls of masonry to be rigid, we also must use plenty of metal bands and anchors to so bind our walls together that they will become one homogeneous mass.[36]

The initiative to change the law was lost in the absence of consensus. The 1903 bond-iron ordinance was reinserted into the 1906 codes with no further discussion or tests. In fact, because of its placement under joists, at a weak point in walls, bond iron (which was also called out in codes in nonseismic areas), may be ineffective in holding buildings together in earthquakes.[37] But this remnant of earthquake-resistant lore was incorporated into the codes all the same.

Wood-Frame Buildings

As we learned in chapter 10, directives for the design of the most common construction type, wood-frame, are minimal. The design of these buildings was not the province of engineers or architects but of builders. Balloon- and platform-frame construction were lightly regulated, and framing plans rarely appear in permit applications. Derleth, like most investigators, felt that ordinary frame construction could resist earthquakes:

> Properly constructed wooden buildings withstood the earthquake with the exception of their brick chimneys...Some frame buildings collapsed upon favorable ground due to improper underpinning [cripple walls]. There should be more continuity in the frames of wooden buildings, especially at the floor levels, and the underpinning should be more carefully attended to in the future.[38]

No specific requirements for continuity, proper underpinning, or attachment to foundations were prescribed, although brick or concrete foundations were required for the first time. But the codes of 1906 include a single sentence included in the pre-earthquake code which helps to explain why wood buildings performed well in the earthquake. The code states, "All outside walls and cross-partitions shall be thoroughly angle braced."[39] No such requirement appears in the New York codes of 1917, which state only that a building should be "braced in all angles or the sheathing is put on diagonal."[40] The difference is that in San Francisco both outside and interior walls are to be "thoroughly" braced, not just the corners, which would improve the lateral resistance of the whole wood frame. This sentence might explain why diagonal braces appear throughout some post-1906 San Francisco Bay Area buildings (see figures 13.33 and 13.36). This now forgotten sentence may have been inserted as an earthquake-resistant requirement before 1906. Even as earthquake engineering progressed in the twentieth century, the investigation of the behavior of wood-frame construction lagged behind that of every other material because it is not of great interest to most engineers.

There is another significant observation of Derleth's that never entered the codes: "Important buildings...should not be placed on filled ground or treacherous ground." And he urged engineers and architects to pay particular attention to soil in their foundation designs.[41] Again, it would take years for these observations to become a formal part of seismic design. Even as late as the 1971 San Fernando earthquake, many people were surprised by the overriding importance of soil conditions in earthquakes.

Examining the books and reports that appeared after the 1906 earthquake and reviewing the debates, it is clear that most building professionals were seriously trying to see and understand, not to deny or conspire. Although only cornices, parapets, and wind factors were addressed in the new codes, the broader discussion of construction practices stimulated architects and engineers in the community to consider other elements of earthquake-resistant construction. They could use this information, much of which was still too raw to become a part of the codes, to set their own standards for seismic safety. Responsible professionals usually surpass the codes. By the late 1920s, the post-1906 earthquake discussions and the reasons the code changes were made were largely forgotten, and the belief that there were not seismic provisions in the 1906 code became a tradition. This chapter asserts the opposite, that many engineers and architects saw and understood earthquake problems and tried to address them in the 1906 code, though not explicitly naming them. They addressed the problems nonetheless: parapets, cornices, cement, wind pressure, and new bond iron specifications were all added to the 1906 code.

CHAPTER 12

A Tale of Two Water Systems

Since our great losses on April 18–21, 1906, the price of insurance has been raised to such a figure that it is a severe handicap to all business and imposes upon this community a burden of over $4 million annually...As the protective [auxiliary water supply] system will benefit both the insurer and the insured, the cost should be borne by both...the owners should be guaranteed a very substantial reduction of the price of insurance upon the installation of the system herein proposed.[1]

—*Marsden Manson, Chief Engineer, City of San Francisco, 1908*

The proposition for an auxiliary water system for fire protection in San Francisco... is the one best suited to the present needs of the city; one [which will]...show itself to be a wise investment in the curtailment of loss by fire and in the substantial recognition such a curtailment must receive from the fire insurance companies doing business in San Francisco.[2]

—*W. C. Robinson, Chief Engineer, Underwriters Laboratories, 1908*

Manson and Robinson made their points. Special oversized fire hydrants with red and blue bonnets stamped "SFAWS" stand throughout the downtown area of San Francisco and beyond, bearing mute witness to the city's attempt to confront its history of fires and earthquakes and to appease the insurance companies (figure 12.1). These colorful hydrants are the surface manifestations of a huge water system comprising reservoirs, tanks, pipes, and special buildings, only to be used for catastrophic fires. The creation of the San Francisco Auxiliary Water Supply System, also called the High Pressure Auxiliary Water System, was, from

Figure 12.1
A San Francisco Auxiliary Water Supply System fire hydrant. These hydrants are larger than those of the domestic system and have distinctive bonnets. According to the design specifications, if the hydrant connection breaks under the street, it is automatically cut off so that water will not bleed from the system.

a safety standpoint, the most important engineering and architectural work ever to be undertaken in San Francisco. To this day it is the centerpiece of fire-fighting strategies for the city, yet its existence remains unknown to many San Franciscans.

The Spring Valley Water Company

Having long desired a municipally owned and operated water system, the city now claimed that the resources of the Spring Valley Water Company, which had supplied it with water at the time of the earthquake, were insufficient for present needs. The problem was not just fire protection; the limited supply of water would restrict the future numbers of people who could settle in San Francisco. The Subcommittee on Water Supply and Fire Protection, one of Mayor Schmitz's post-earthquake subcommittees, stated the fire protection argument in no uncertain terms:

> The protection against fires afforded by the system of the Spring Valley Water Company was inadequate, even as it existed before the "earthquake-fire"... The system is in a less efficient state now than before the fire, and as shown by the earthquake, the emergency requirements are much more severe than hitherto realized.[3]

The committee concluded that it was "imperative" for the city of San Francisco to own its own water supply. They cited the deteriorating relations between the city and the Spring Valley Water Company, which, astoundingly, had refused to tell the city engineer where the water valve was on the Calaveras intake pipe. The National Board of Fire Underwriters, which endorsed municipal water facilities in every city where they wrote insurance, had recently cautioned the city about its continued battles with its private water supplier.[4]

Thirty-two years earlier, during a particularly dry year, the city had begun the search for its own water supply. In 1874, the San Francisco Water Commission went to the Sierra Nevada in search of water and decided on the area around the Mokelumne River.[5] But the following year, the commission considered a cheaper proposal to acquire rights to Calaveras Creek, in northern Santa Clara and southern Alameda Counties.[6] The Spring Valley Water Company, which had been supplying San Francisco with water since the 1860s, bought the riparian rights to Alameda Creek, the main tributary of Calaveras Creek.[7] Outmaneuvered, San Franciscans either had to capitulate to the Spring Valley Water Company or buy it out. The buy-out proposal failed at the polls in 1876.

San Franciscans settled into their relationship with the Spring Valley Water Company with such misgivings that other water sources were investigated every decade, but no plan actually materialized.[8] Dry years during these decades had no doubt given rise to schemes to improve the city's water supply. In 1900 the new city charter directed city administrators to find new water sources.[9]

Figure 12.2
Hetch Hetchy Valley before it was dammed.

In 1900 and 1901, Mayor Phelan, City Engineer Manson, and the Board of Public Works had proposed damming the Tuolumne River in Hetch Hetchy Valley.[10] Hetch Hetchy was a narrow meadow in the Sierra Nevada surrounded by spectacular, nearly perpendicular granite walls, 170 miles east of San Francisco (figure 12.2). It closely resembled its world-famous neighbor, Yosemite Valley. When Yosemite was designated a national park in 1890, Hetch Hetchy was included within its boundaries.

Under federal law, a private citizen could apply for water rights on public land. Phelan, as an agent of San Francisco, secretly applied for the rights to use Hetch Hetchy water, but when it became known that his scheme included building a dam inside the park, the government turned him down.[11] The Department of the Interior ruled against the city not once, but twice. The city then appealed to Congress. A bill to allow San Francisco to build the Hetch Hetchy Dam was in committee when the earthquake struck. The earthquake gave new energy to the stalled initiative.

Days after the earthquake, the Spring Valley Water Company was challenged to explain the failure of its system and it did so with complete candor.[12] City engineers and officials, usually anxious to discredit the company, actually found themselves agreeing with its findings. The supplier lines from all three reservoirs—Pilarcitos, San Andreas, and Crystal Springs—had broken in the earthquake, but the reservoirs remained intact (figure 12.3). The thirty-inch Pilarcitos line tore apart violently and was telescoped (figure 12.4). According to engineers' reports, it

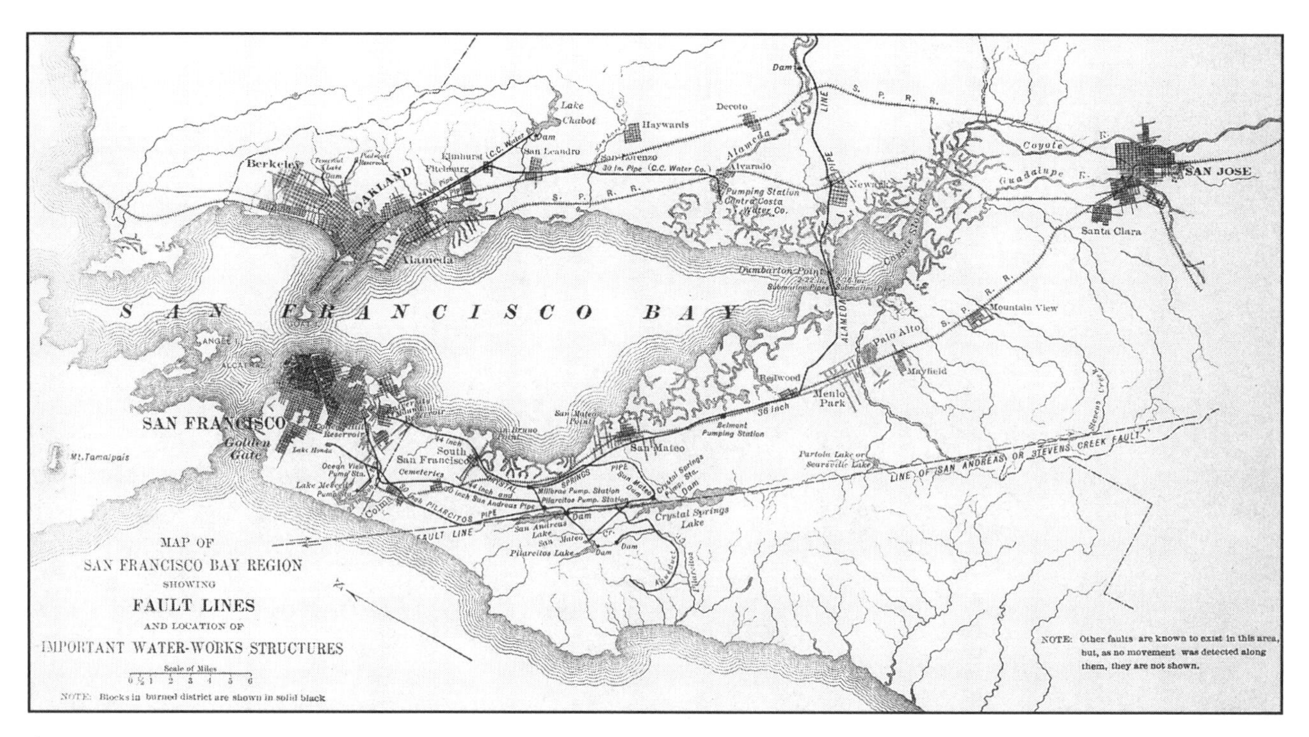

Figure 12.3 (top)
Water lines and reservoirs of the San Francisco Bay region. The Spring Valley Water Company supply lines within San Francisco are, from east to west (top to bottom on this map), the Crystal Springs pipe to University Mound Reservoir, the San Andreas pipe to College Hill Reservoir, and the Pilarcitos pipe to Lake Honda. The supply lakes and dams on the fault line remained intact after the earthquake.

Figure 12.4 (bottom)
The Pilarcitos pipe, emptied of water after the earthquake. The displacement along this pipeline was two or more feet.

is doubtful that any kind of pipe could have resisted the earthquake forces in the locations where it was laid.[13]

The Spring Valley Water Company's engineer summarized the effect of the earthquake on city pipes:

> As you know, the principal breaks caused by the earthquake in our City Distributing Pipe System occurred in those regions where the streets crossed old swamps and deep fills, while on the solid ground the main pipes suffered but very little. The city sewer system was also badly hurt in these identical sunken streets, particularly south of Market. But with the exception of the tearing off by the earthquake of the 22-inch and 16-inch pipes on Valencia Street (caused by Valencia Street sinking about five feet between Eighteenth and Nineteenth Streets), which, within a comparatively short time, emptied the College Hill Reservoir, the damage done by the breaks in the balance of our pipe system was not any more serious than, as the conflagration progressed, the gradual tearing off, by the burning of the falling buildings, of over 23,200 service connections, such as house services, factory supply pipes, standpipes for hose reels, large elevator pipes, automatic sprinkler pipes, etc. The immediate effect...was to take away the pressure in the burning region.[14]

Volume and Pressure

While some members of the subcommittee were scrutinizing the source and delivery system for San Francisco's domestic water, others were debating the most effective method of meeting the fire department's need for a guaranteed water supply. Two criteria had to be met: adequate volume and sufficient pressure. The insurance underwriters recommended that fire protection for the downtown, including the fire district (also called the "congested value district"), required a hydrant volume of twenty thousand gallons a minute.[15] But in the heart of San Francisco's downtown, at Market and Second Streets, only thirteen thousand gallons per minute were available. By redirecting water from another, higher, service point to this location, the volume could be augmented to nineteen thousand gallons per minute, but this procedure was not the preferred one. Adequate volume was supposed to be available in every single major branch, or service.

Inadequate pressure also posed a problem. Pressure in any water system varies, but the average acceptable pressure hovers around 60 to 150 pounds per square inch (psi). The pressure in the mains in the congested area of San Francisco (the downtown) before the earthquake averaged 52 psi. The National Board of Fire Underwriters stated:

> The average pressure of 52 pounds in the congested value district is too low for good fire protection, inasmuch as it will not properly serve automatic sprinklers, interior standpipes, and other invaluable protective devices, even in buildings of moderate height, without the aid of special fire pumps.[16]

They wanted a constant pressure of 75 to 100 psi for domestic service in the congested value district. But even this pressure was not, in the eyes of the insurance underwriters, enough for complete fire protection. In their last report on San Francisco, in 1905, they had endorsed the totally new auxiliary water system then under consideration by the city. It would be independent of the domestic water system and have a pressure of upwards of 200 psi and a volume of twenty thousand gallons per minute.[17]

Research into a dedicated auxiliary water system for firefighting had begun in 1903.[18] City Engineer C. E. Grunsky and Consulting Engineer J. C. H. Stut studied the auxiliary water systems of Cleveland (under construction), Milwaukee (1889), Detroit (1893), Buffalo (1897), Providence (1897), Boston (1898), Chicago (proposed), Philadelphia (1903), and New York (Coney Island 1905–1906, Brooklyn and Manhattan under construction).[19] San Francisco was not as far behind the East and Midwest as it might at first seem. It was only in 1903, after several large fires in New York and several others nationwide, that Manhattan envisioned a high-pressure auxiliary water system.[20] This system didn't become operational until July 6, 1908.[21] In four of the earliest systems—Milwaukee, Buffalo,

Detroit, and Boston—fireboats pumped water from rivers, lakes, or bays into a special series of pipes for use by the fire department.

To imagine the difference between firefighting with a dedicated water system and a domestic system, think of a central vacuum cleaning system. Instead of lugging a vacuum cleaner from room to room, all one is required to do with a central system is plug a hose into special receptacles in the walls. Just so with an auxiliary water system: the pressure was already in the pipes and did not need to be pumped out by individual steam engines. The engines could be used, however, to increase the already high pressure still more, so that water could be thrown even farther. This possibility gained importance as buildings became taller. And with an auxiliary water system, greater quantities of water could be concentrated upon a fire than ever before.

The economic arguments for an auxiliary system were convincing. In 1905 it was estimated that fires were costing the American public the equivalent of a $2.47 per capita tax.[22] Insurance premiums on new construction in San Francisco had risen, in two years, from $1,606,204 per year to $3,876,430: 141 percent. Insurance adjusters promised to lower the rates 25 to 33 percent if an auxiliary water system were installed.[23]

As early as one month after the earthquake, the concept of an auxiliary water system was endorsed, and its design was being debated. By the summer of 1907, city engineers Woodward, H. D. H. Connick, and T. W. Ransom, as well as W. C. Robinson, the representative of the National Board of Fire Underwriters, had agreed upon a plan.[24] However, the agreement collapsed in October when the city insisted that the auxiliary water system also be used for flushing sewers and sprinkling streets.[25] Stressing the importance of a fully independent system, the insurance underwriters stood their ground, maintaining that the system was to be used exclusively by the fire department. The integrity of the system rested on its exclusivity. The supervisors soon capitulated.

A joint report issued in 1908 by the city engineers explained the auxiliary water system in detail.[26] In a bond measure the same year, the citizens of San Francisco voted to spend the $5.4 million to build it.[27] This was a great victory for proponents of an earthquake-resistant fire-fighting system. Earthquake and fire needs would be combined. For the first time in America, a public water supply system incorporated the best in earthquake engineering, including such features as flexible pipe connections and built-in redundancies to allow for the inevitable failures in a severe earthquake.

The late fire chief Dennis Sullivan had supported building a reservoir on the city's highest mountain, Twin Peaks, at an elevation of 758 feet, as the centerpiece of an auxiliary water system. But instead of the saltwater system envisioned by the chief, the new system would use fresh water from Spring Valley Water Company. Salt water extinguishes fires better than fresh, but it is more destructive to merchandise and fire equipment.

Gravity flow from the reservoir on Twin Peaks would feed two distributing and pressure-regulating tanks, each holding five hundred thousand gallons: one at

Ashbury Heights (elevation 462 feet) and another at Clay and Jones Streets (388 feet). If the water supply to the Twin Peaks Reservoir failed, then pumps from the Ashbury tank could shuttle water back uphill for storage. If the whole freshwater system failed or more water was needed, two saltwater plants at the waterfront would pump water through the mains, creating sufficient volume and pressure. If these saltwater plants failed, or if still more water were needed, then fireboats could pump water into the system. If all else failed, a system of cisterns, distributed throughout the downtown area of San Francisco and kept fully supplied with water, would be available. Redundancy of supply would be repeated in every aspect of the auxiliary water system.

The Auxiliary Water System

The auxiliary water system was built almost precisely as delineated in the report of 1908, and with a few modifications, including extended service to new areas, it is exactly the same system that guards the city today. A map of the system (figure 12.5) illustrates the components that still exist. The letter "A" on the map is the Twin Peaks Reservoir, and "D" is the supply reservoir of the Spring Valley Water Company, which pumps into the Twin Peaks Reservoir. "B" is the Ashbury Heights tank, which services the upper, or higher, zone of the city (above 150 feet in elevation). "C" marks the Jones Street tank, which distributes water to the lower zone. "No. 1" is the saltwater pumping plant at Second and Townsend Streets, and "No. 2" is a similar plant at Fort Mason, or old Black Point. The reservoir and pumping plants supply a grid of pipes under the streets of San Francisco whose only manifestation in the cityscape is the aforementioned red and blue hydrants.

The structure and design of each component illustrate how the engineers thought they could defeat the twin plagues of earthquake and fire damage. The well-built brick tunnels of the Spring Valley Water Company, which actually crossed

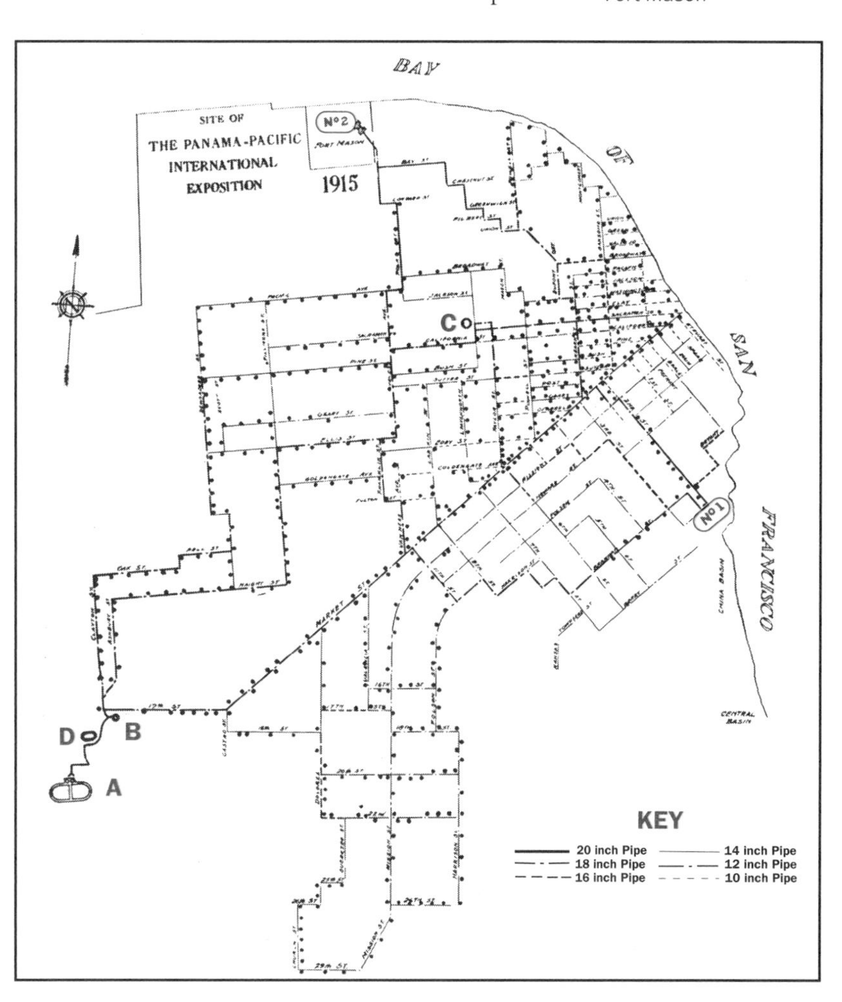

Figure 12.5
San Francisco Auxiliary Water Supply System, 1914: pipes, hydrants, reservoirs, and pumping stations
A. Twin Peaks Reservoir
B. Ashbury Heights Tank
C. Jones Street Tank
D. Spring Valley Reservoir
Pumping Station No. 1, Second and Townsend
Pumping Station No. 2, Fort Mason

Figure 12.6
Twin Peaks Reservoir shortly after construction. The earthquake-resistant construction of the reservoir and its designed redundancy are extremely impressive. Built of reinforced concrete, considered an earthquake-resistant material at the time, the flexible joints of the slabs are backed by secondary drains in case they leak. A sump pump under the reservoir collects seepage. The reservoir is divided in two, in case one side is damaged in an earthquake.

Figure 12.7
Auxiliary Water Supply System pipes from the Twin Peaks Reservoir. The pipe joints are flexible, yet the rods guarantee the connections will remain intact in an earthquake. Redundancy is present in the design: if one pipe breaks, the other can still be used.

the San Andreas Fault, had shown unusual strength. The bricks adhered as a unit, breaking in place. The whole tunnel almost behaved as a monolith, but ultimately it was torn apart by the tremendous earthquake forces. No brick walls were used in the AWSS: after 1906, concrete began to replace brick as a building material in earthquake engineering works. Extraordinary care is evident in the design of the Twin Peaks Reservoir (figure 12.6). Shaped as a section of an inverted elliptical cone, it is made of reinforced concrete slabs six inches thick. Twenty-foot-square slabs form the bottom, and slabs ten feet across at the top, tapering to the bottom, form the sides. The slabs are joined together with flexible joints that can shift in earthquakes without losing their seal. Each joint is backed by a drain, so that if a leak does occur, it will not undermine other slabs. A sump pump under the reservoir collects seepage. In case of serious rupture, the reservoir is divided into two autonomous parts by a concrete wall; half of the reservoir could function if the other half were damaged.

Below the reservoir stands the steel Ashbury tank, constructed on reinforced concrete foundations. Its gatehouse is built in the Mission Revival style, with reinforced concrete walls, fire-resistant iron decoration, and a tile roof. Below the Ashbury tank stands the buried Jones Street tank, constructed of reinforced concrete.

The pipes that connect these vital supply reservoirs and tanks to the system are made of cast iron, some with a diameter as great as twenty inches and weighing as much as forty-four hundred pounds per twelve-foot length. In addition to being caulked, sections of pipe are bolted together to provide extra strength and flexibility (figure 12.7).

Noting the breaks in the Spring Valley system that occurred on fill, or made ground (figure 12.8), a map was made that identified the worst sites in the city for laying pipes. Pipes that had to cross this infirm area were specially designed to move but not break. They had a double sleeve sheathed in steel and caulked with lead (figure 12.9). The large quantity of lead gave the joint great flexibility. Each

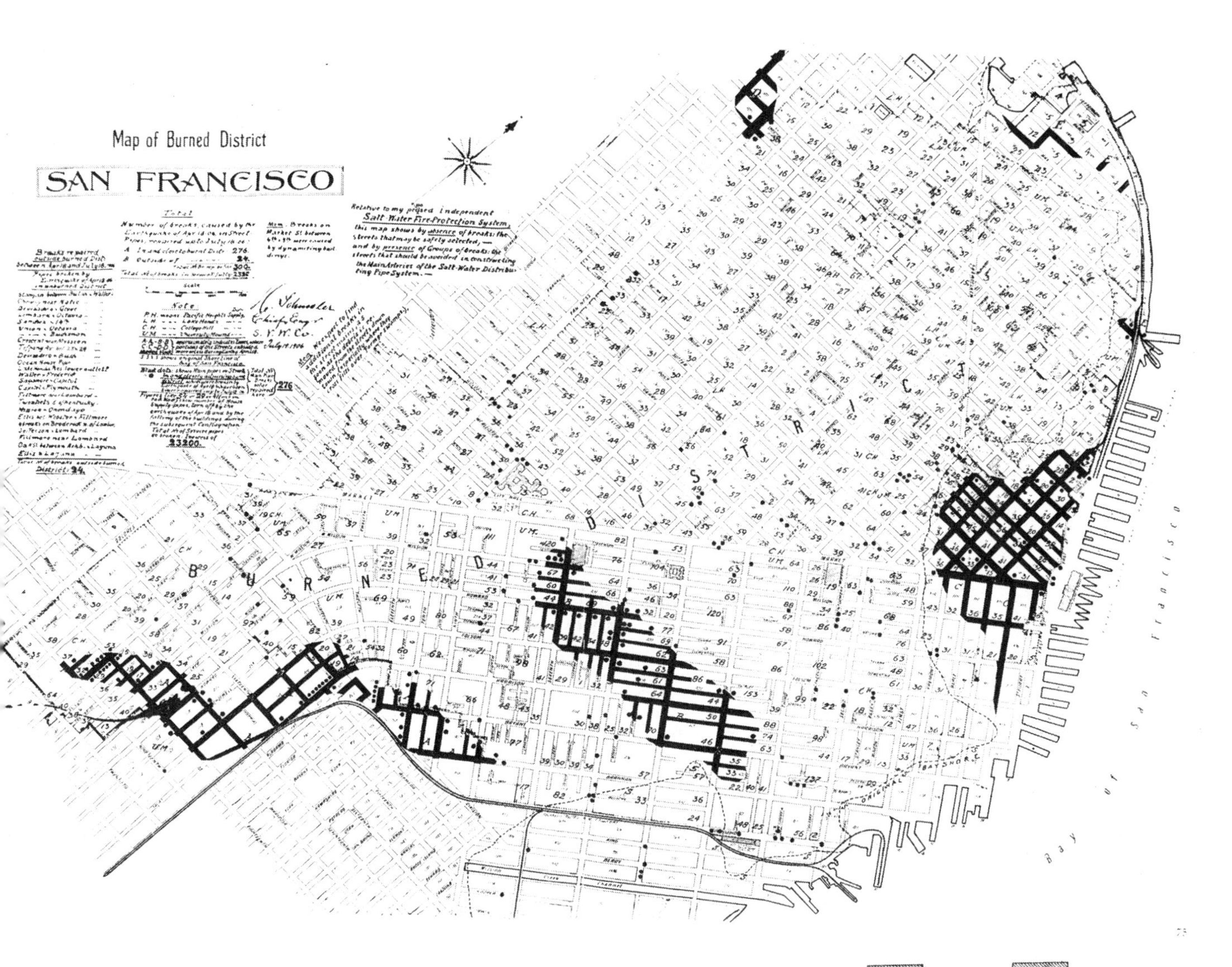

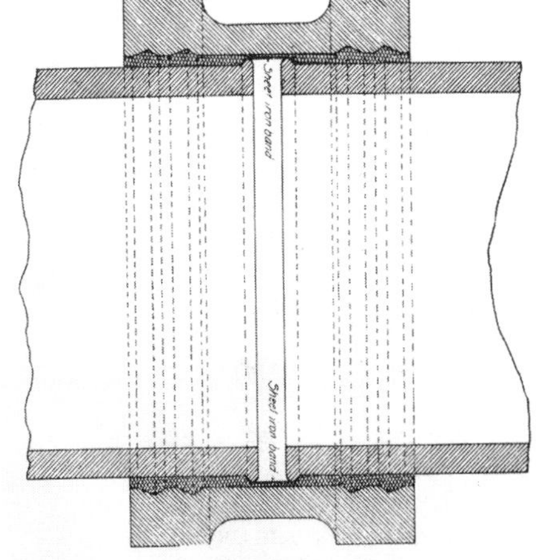

Figure 12.8 (above)
Major breaks in the water system repaired after the 1906 earthquake. Settlements and breaks in the pipes laid in fill are shown in black. This map provided the basis for identifying unstable ground in San Francisco. Spring Valley Water Company engineer Hermann Schussler indicates here how he would improve the water system.

Figure 12.9 (right)
A diagram of the flexible Auxiliary Water Supply System pipe joint to be used in the fill indicated in figure 12.8. Engineers learned of the danger of pipes breaking in made ground and tried to incorporate flexibility into the connections so that future earthquakes would not break them.

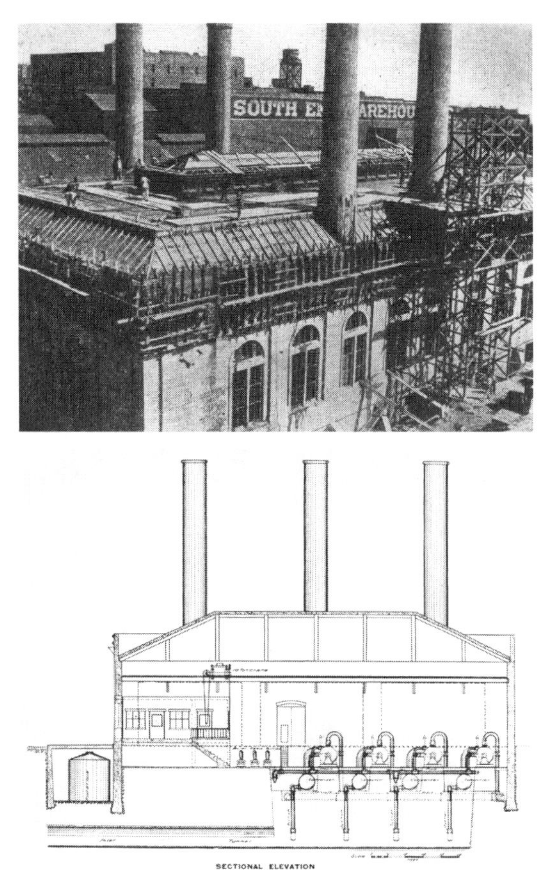

Figure 12.10 (top)
The exterior of Pumping Station No. 1 (Second and Townsend Streets), in construction in 1909. Engineers and architects chose concrete over a steel frame for the walls to ensure earthquake safety. The pumping station was built on rock. The roof was designed to withstand buildings that might fall on it, and the reinforced concrete chimneys were designed to resist earthquakes.

Figure 12.11 (bottom)
Section of Pumping Station No. 1, showing the intake tunnel from the bay, 1908. The pumping station was designed so that in an emergency it would suck salt water from the bay and pump it into the pipes. The boilers, since replaced by modern equipment, were kept hot around the clock.

joint was secured by twenty-four inch metal rods which were attached to the two lengths of pipe to restrain them while allowing flexibility. These special pipes also had strategic valves in the grid and could be quickly isolated if a break did occur. Earthquake-resistant features extended to the two saltwater pumping stations, built on rock foundations, that would begin to pump water into the system if the freshwater system failed. These stations still function around the clock today, waiting on standby in case of emergency. The largest of the two is a Secessionist-style reinforced concrete building at Second and Townsend Streets (figure 12.10). Not just a utilitarian pumping plant, this building was built to showcase the system beneath the streets. Until recently, when the fire department converted the building into an administration center, large wire-glass double doors allowed pedestrians to view the elegance of the interior, with its shining tiles and batteries of machinery. A balcony inside gave a closer view. Today the diesel pumps are underground and the main floor has become the headquarters of the fire department. An underground, reinforced concrete intake tunnel brings water into the structure (figure 12.11) and once carried it to four large centrifugal pumps built by Bryon Jackson Ironworks in Berkeley. Four huge Babcock and Wilcox boilers powered the Curtis steam-driven turbines that ran the pumps. They were magnificent to see but they all have been replaced by smaller, more modern equipment.

This pumping station was designed to withstand the impact of falling debris in an earthquake, even though it stood apart from other buildings. The steel frame, with its reinforced concrete walls, is massive. The building is protected from fire by its steel roof, reinforced concrete exterior, and metal sashes. The huge, reinforced concrete chimneys, since removed, were designed to resist ground motion.[28] The openings of the building were protected by wire-glass and steel shutters made by Kinnear that roll down over the windows when fire threatens. A similarly designed pumping station, in Mission Revival style, can be seen at Fort Mason on the west shore of Aquatic Park (figures 12.12, 12.13, 12.14). There the original machinery has been preserved and can still be viewed.

If, somehow, the pumping stations failed, two new city fireboats, the *Sullivan* and the *Scannell*, could plug their hoses into specially designed connections at the end of San Francisco's wharfs (figure 12.15). The fireboats could pump into the system, building enough pressure and water volume to maintain it at acceptable levels.

The cisterns, the last and most primitive resort, to be used in case the entire water system and all the machinery failed, were positioned under the intersections of San Francisco's streets. These circular, reinforced concrete tanks (figures

Figure 12.12 (top, left)
Pumping Station No. 2 (Fort Mason) in construction. Similar to Pumping Station No. 1, this building is built on rock with a steel frame and concrete walls. Everything in the system has a twin for redundancy in case of earthquake damage. Pumping Station No. 1 has been remodeled for a fire department administrative building, but this pumping station is relatively intact.

Figure 12.14 (bottom, left)
Fort Mason Pumping Station in 1986 during a test of its pumps. While the old equipment remains, the new turbines installed in the building do the work today. Note the removal of the chimney with the change to gas turbines.

Figure 12.13 (top, right)
Fort Mason Pumping Station in the 1920s

Figure 12.15 (bottom, right)
Hose connections for fireboats at the end of a pier at Fort Mason. If the pumping stations fail, or more pressure is needed, fireboats can pump salt water from the bay.

Figure 12.16 (top, left)
Cistern in construction, 1908. If an earthquake causes multiple failures in the system, the cisterns can be used for water supply.

Figure 12.17 (top, right)
Cistern in construction, 1908. The positions of these reinforced concrete cisterns are marked at street level by circles of bricks in the pavement.

Figure 12.18 (left)
A demonstration of the San Francisco Auxiliary Water Supply System at "Newspaper Corner" in the 1920s. On the right is the Call Building, and behind that is the Examin Building. With the new system, water pressure and volume could be increased so that even fir in tall buildings could be extinguished.

12.16 and 12.17) were an updated version of the square wooden cisterns that San Franciscans had constructed after the fires of the 1850s. Pumpers could siphon water from them in case water was not available in the mains, or use them as backup water sources.

By 1912 the first pumping station and a majority of the pipes for the system were completed. Mayor James Rolph requested that the insurance companies lower their rates. The National Board wanted proof that the system they had encouraged would work.[29] Tests were successful and, finally satisfied, insurers began to lower San Francisco's rates in 1913.

Hetch Hetchy

While the High Pressure Auxiliary Water System was being constructed, San Francisco lobbied furiously for independence from the Spring Valley Water Company, which served both the high-pressure system and the domestic water system. In the aftermath of the earthquake, the city government's bias against Spring Valley had made it easy to minimize the fact that the company's three large reservoirs had held enough water to successfully fight the fire—if it had been available. But the failure of the major water mains and the connections to individual houses within the city lowered water pressure and cut water supplies to 6 percent of what should have been available in the downtown area. Seeming water shortages in 1907 once again "proved" the need for better sources of water.[30] The fight for Hetch Hetchy continued. Opposed not only by Spring Valley but by the Sierra Club, the famed naturalist John Muir, and irrigation concerns that used the Tuolumne River, the city fought a pitched battle. Even Professor Charles Derleth Jr. condemned the long-distance pipeline that would be required to tap Hetch Hetchy, noting that it would be very risky in earthquake country.[31] In a 1909 pamphlet, an impassioned Muir wrote, "Dam Hetch-Hetchy! As well dam for water-tanks the people's cathedrals and churches, for no holier temple has ever been consecrated by the heart of man."[32] In answer to Muir's cry that San Francisco was destroying one of the wonders of nature to save money on water, ex-mayor Phelan responded passionately:

> It is not "cheapness" but abundance and purity and reliability that move the people of the Bay of San Francisco to provide themselves and their posterity in time; but if it were partly a question of reasonable cost, I think that the people of the country would consider it no small element in the petition of a city that has just been destroyed by fire, *due to a lack of water* [emphasis added], and which is now using her last available penny, under the limits of taxation, to re-create a municipal plant, and to worthily serve the nation as its western gate. She should not be falsely condemned for the lack of esthetic appreciation of the beauties of nature, nor reproached for thrift.[33]

Phelan's refutation again used the erroneous argument of water supply shortage to justify damming Hetch Hetchy. Conveniently overlooked was the fact that Spring Valley's allegedly inadequate Crystal Springs Reservoir was now supplying the AWSS. Hetch Hetchy was in fact not needed for firefighting.

Nevertheless, in 1913 San Francisco finally achieved its goal, acquired the Tuolumne River water rights, and was allowed to build a dam at Hetch Hetchy. Much later, in 1928, the city bought out the Spring Valley Water Company. San Francisco now had two separate municipal water systems: the old Spring Valley system for domestic water and everyday fires, and an auxiliary high-pressure system in the event of a catastrophe.

There was never any opposition to the auxiliary water system. Despite opposition to damming Hetch Hetchy from the environmentalists of the time, for San Francisco it was a win-win situation. The AWSS was for the most part hidden underground, waiting to save the city in the event of another earthquake and fire, and it satisfied the insurance underwriters, which gave reconstruction a boost when insurance rates were lowered significantly. Finally, there was a straightforward way to mitigate earthquake and fire damage: resilient and redundant water pipes, and sufficient water under sufficient pressure. The city embraced the AWSS with practical urgency and conscious foresight. They had voted to pay the price.

CHAPTER 13

San Francisco: The Phoenix Rising, 1906–1915

Between the earthquake and fire of 1906 and the beginning of construction for the Panama-Pacific International Exposition in 1914, San Franciscans erected 174 new steel-frame buildings, 194 reinforced concrete buildings, 2,699 brick buildings, and 25,440 wood-frame buildings to replace structures that had collapsed or burned, and to meet the needs of an expanding city.[1]

Every structure built in the new San Francisco was an expression of what its owner, architect, engineer, or builder considered to be beautiful, useful, and/or safe. Critics of the reconstruction have argued that the desire for quick profits and the urgent need to rebuild the city triumphed over safety concerns—that lessons that could have been learned from the earthquake and fire were disregarded.[2] While a survey of the more than twenty-eight thousand buildings that went up during that period is impossible, a sample of available architectural records suggests this blanket condemnation is incorrect. Of course there were poorly constructed buildings, even by 1906 standards, and more that now look poorly built by our standards, but many buildings demonstrate that architects and engineers did learn from the earthquake and fire. The insurance companies paid, and San Francisco was rebuilt. Fire-resistant and earthquake-resistant construction practices were implemented by an impressive number of individuals, even under pressure for a quick fix.

The phoenix city that rose from the ruins was of a piece, new and elegant as the boosters claimed, not a hodgepodge of disparate styles. Burnham's Parisian planning scheme had been dropped, but Beaux-Arts aesthetics were all the rage for individual building design. The French school's impact on American design produced what has been called the American Renaissance style, in which every conceivable version of Renaissance and Baroque architectural ornament was married to every building type prevalent in San Francisco. The Ecole des Beaux-Arts

Figure 13.1
Looking east down Market Street in 1915 to the new Humboldt Savings Bank Building (with "1915" on its tower) and the retrofitted Call Building, we see a mix of old and new buildings constructed to harmonize aesthetically with one another. On the right, closest to the picture plane, is the rebuilt Emporium with three earthquake-resistant buildings behind it: the Commercial Building, the Pacific Building, and the Humboldt Building. The Humboldt Building repeated more than the tower motif of the Call Building; it rivaled the Call Building in earthquake and fire safety as well.

(the most influential architecture school in the world) championed not only appropriate historical borrowing but also a sense of hierarchy in organization, symmetry in plan, appropriateness of style and proportion, and the sense of the *marche*, the experience of procession through the building. Architects were taken with this style: they designed buildings with Doric, Ionic, Corinthian, or Composite orders in faux stone terra-cotta over steel frames. Buildings were often intentionally related to the proportions and decoration of neighbors. Where the two grids came together on Market Street, each architect tried to introduce different variations on the Classical theme (figure 13.1).

These facades are the bread and butter of professional architecture, but exquisite though they may be, they are not going to tell us whether their architects and engineers considered fire or earthquake safety. Neither do the facades of the hundreds of brick and thousands of wood-frame buildings constructed during this period. There was little intervention from city government during 1906 and 1907 when the first buildings went up; architects, engineers, and their clients were on their own, even after the new code was approved, because of the lack of enforcement. In all, 42 steel-frame structures (Class A), 69 brick with load-bearing walls or reinforced concrete structures (Class B), 892 brick and wood structures (Class C), and 5,928 wood-frame structures were granted permits in the fiscal year ending June 1907.[3] On the quality of these new structures, the head of San Francisco's building inspection department simply stated, "Our work force is entirely inadequate to thoroughly inspect and enforce the building laws."[4] Plans weren't checked as thoroughly as they could have been, and accusations of incompetence and bribery clouded the permit process.[5]

In that first year, the city government could neither legislate safety nor enforce it. Not only was the terrific devastation from earthquake and fire a problem, but charges of graft and ensuing grand jury indictments tore apart the civic government. Fortunately, there were many architects and engineers who wanted to build to standards far above the codes.

Class A Steel-Frame Buildings

Of the more than 150 steel buildings constructed between 1906 and 1916, five in particular illustrate the prototypical solutions to problems of earthquake and fire. (Others are referred to tangentially to amplify particular points.)

The Humboldt Savings Bank Building, 1906–1908
783–785 Market Street
Frederick H. Meyer, architect; Christopher Snyder, engineer

In this period of frenetic building and administrative chaos, Frederick H. Meyer, the local architect who had designed the Rialto and several other prominent pre-earthquake buildings, began construction on the new earthquake- and fire-resistant Humboldt Savings Bank on Market Street (figures 13.2a and 13.2b). Meyer, a principal in the firm Meyer and O'Brien, was a native San Franciscan who had never received a degree in architecture, and in fact had no formal education after the age of fourteen. He learned about construction by working on buildings and about aesthetics and design through one of San Francisco's architectural clubs.

Meyer was no doubt aided in making structural decisions by the contracting engineer for the West Coast office of Milliken Brothers Steel Company, Christopher H. Snyder.[6] Snyder, a native of Illinois, designed buildings for Milliken Brothers between 1902 and 1912. Like most engineers working in San Francisco until around 1913, Snyder did not have an office of his own but worked with a steel manufacturer. Architects sent their drawings either to the manufacturer or to the local representative, who would do the calculations, design the steelwork, and then return the drawings. Snyder had attended the post-earthquake meetings of the Structural Association of San Francisco and coauthored the definitive earthquake report for the American Society of Civil Engineers. He wrote several articles endorsing strong lateral bracing for steel frames. After 1912 he established his own office in the city and was one of the most prolific and influential engineers in San Francisco.

Because the Humboldt Savings Bank Building was in the early stages of construction when the earthquake hit, it could have been erected without a new building permit. Here was a chance for speedy reconstruction with no intervention from city building officials. The original design of the facade remained unchanged (figure 13.2a), following Beaux-Arts principles. A beautifully articulated Classical entrance led into the lobby of the narrow building, with the Humboldt Savings Bank on the left and the entrance into the office building on the right. Above the ground floor, sixteen floors of offices with repetitive windows were tied together by banded

Figure 13.2a
Design for the facade of San Francisco's Humboldt Savings Bank Building, 783–785 Market Street, 1906

Figure 13.2b
The steelwork for the Humboldt Savings Bank Building in construction, 1906

pilasters, culminating in an elaborate Baroque dome. The Humboldt Savings Bank dome intentionally mimicked the larger Call Building dome several blocks east on Market Street, providing a second vertical marker to establish the primacy of the street.

Meyer had been very much affected by the poor performance of his earlier projects, the Rialto Building, the Monadnock Building, and the Hahnemann Hospital. He made changes in the Humboldt accordingly. He was surely the architect to whom the Japanese seismic expert Dr. T. Nakamura referred when giving his impressions of earthquake damage on June 16, 1906:

> Architects in this country have had little experience of earthquakes and, in designing their buildings, they have paid very little attention to that subject. If I remember correctly, one architect, after the earthquake, altered his specifications of a building which was in the course of construction at the time of the earthquake. That was a wise act.[7]

Figure 13.3
The Humboldt Savings Bank Building in construction, with spandrel girder and knee bracing for lateral resistance to earthquakes, here labeled "wind bracing."

The foundation for the Humboldt Savings Bank had been under construction on April 18, 1906, and the contractor, George Wagner, was ready to begin the steelwork.[8] Although the foundation was not damaged in the earthquake, the shock and subsequent fire altered Meyer's conception of the structure. The artistic design of the facade remained the same, but he drew large red X's over structural drawings and wrote "VOID" on them, changing many features of the original design to make the building safer. He and Snyder stiffened the building, altered its foundation, changed brick walls to concrete, and incorporated many new fire control techniques and devices. These were not casual changes, but related directly to failures in Meyer's own buildings.

The Humboldt Savings Bank Building was tall (244 feet) and slender (50 feet at its least width, on Market Street), requiring special reinforcement. The front portion, on the Market Street side, ended in an elaborate tower topped by a dome above its seventeenth floor, while the office block behind was thirteen floors. The narrow tower was designed to be carried on a steel frame (figure 13.2b), which was braced along its width by spandrel girders and knee braces (figure 13.3). This kind of lateral strengthening already existed in the frame of the Whittell Building, which did well in the earthquake and fire. To further strengthen the structure, the steel columns were spliced at alternate floors. These two-floor runs, it was hoped, would help the structure retain its integrity if pushed from side to side. There is no corresponding spandrel or knee bracing in the length of the building, presumably because it was felt that the reinforced concrete curtain wall, combined with the strength of the steel, could handle the load.

Originally, the narrow tower of the Humboldt Savings Bank was to be made of brick and stone, but seeing the poor performance of these two materials in earthquake and fire, Meyer changed to concrete on the entire exterior: plain on the sides and rear, and a concrete backing for the stone and terra-cotta veneer on the front. Colusa sandstone, which proved very vulnerable to spalling—breaking off in chips—in fire, stopped at the third floor, and the terra-cotta veneer continued up the rest of the building.

Meyer took great care in regard to fire safety, which he often promoted through talks and articles.[9] He used virtually every available fire-fighting precaution that could be adapted to San Francisco's situation (figure 13.4). The columns of the building are protected by concrete. The floors, originally specified to be hollow tile, are reinforced concrete. The reinforced concrete floors of Meyer's Marston, Rialto, and Monadnock Buildings had all done well, but the wire-and-plaster column protection on the Rialto had failed, while concrete column protection on the Marston Building had held.

Figure 13.4
The Humboldt Savings Bank Building in an advertisement for the "earthquake-proof" Kewanee water supply system.

The Call Building had burned from the top down because the heat of the fire had burned out its top windows. Metal trim would have offered a barrier to fires. In spite of the expense involved, Meyer used metal trim, which was about twice the cost of first-grade oak. In fact, he took care to design all exterior surfaces to be incombustible. The exterior windows in the light court were protected by wire glass. Wire glass is exceptionally thick and incorporates a "chicken wire" membrane that ensures that if the glass cracks, it will not shatter and loosen from the frame.

As we have seen, elevators are a hazard in fires because their shafts can act as flues. In successive drawings made between March 31 and December 22, 1906, Meyer progressively isolated his elevator shafts by cutting them off from the rest of the building with automatic "underwriters' doors." Two of these tin-clad doors with heat-sensitive releases adjoined the elevator core (figure 13.5). If fire broke out, they automatically closed, isolating the whole elevator core. The elevator core was of concrete with metal covering over all wooden surfaces. Even the draft up

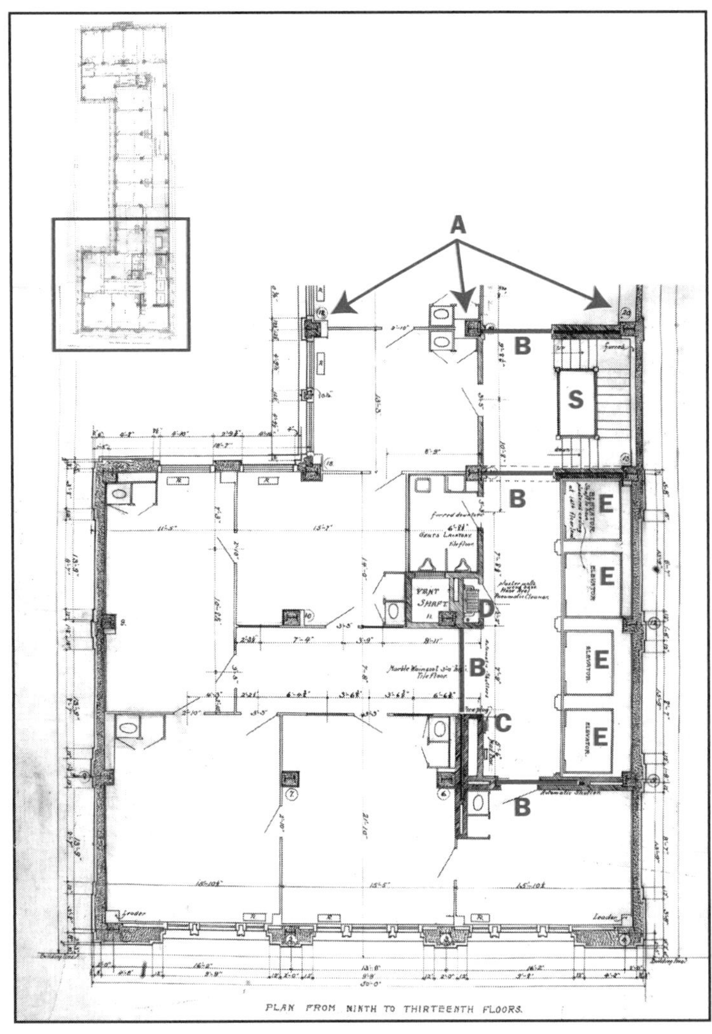

Figure 13.5
The floor plan of the Humboldt Savings Bank Building illustrates fire prevention methods: A. concrete around steel columns; B. automatic fire doors for enclosing stairwell (S) and elevator shafts (E); C. fire standpipe for fire department hose plug; D. internal fire-fighting hose.

the isolated stairwell was reduced: a note dated April 9, 1906, directing "all risers except those of main flight from first to mezzanine and those from 17th to 18th to be perforated" was rescinded. All the risers were solid in the final drawings, so air could not filter through them and provide draft for fires.[10]

The whole building had a system of standpipes and hoses on each floor, supplied not by gravity but by a pneumatic pump and tank in the basement. Meyer saw that standard roof tanks were dangerous in earthquakes because they sometimes fell from their supports onto the building. Even if this catastrophe were averted, they often shook loose from their connections. Meyer's solution was novel for San Francisco. He opened wells in the basement so that if the Spring Valley Water Company pipes failed, the building would have its own supply, which could be pumped directly from its wells. If the pneumatic pump failed, a special backup "underwriters' pump" could be pressed into action. These precautions were expensive, but in a sense they paid for themselves: the Humboldt Savings Bank was able to save 47 percent on its fire insurance because of its fireproofing and fire-fighting system.[11]

In both its seismic design and its fire-resistant qualities, the Humboldt Savings Bank Building was outstanding for its day. According to Charles Derleth Jr., writing in 1908, it was "one of the leading structures of the new San Francisco."[12] Despite earlier failures and a limited education, Meyer was able to rise to the expectations of his clients with a handsome, towered structure that could resist both earthquake and fire. As Derleth said, "The structure represents a transition between the methods of building in the old and the new San Francisco."

Royal Globe Insurance Company, 1908
201 Sansome Street
Howells and Stokes, architects; Purdy and Henderson, engineers

In the middle of 1908, construction began on the Royal Globe Insurance Building (figure 13.6).[13] The company had lost its five-story brick and wood building at Pine and Sansome Streets to earthquake and fire. It was not a Class A structure, and it suffered badly, which evidently impressed executives in New York. When they decided to rebuild, they instructed their architects to build a fireproof, earthquake-proof structure.

The Royal Globe Insurance Building's appearance belies its special adaptation to San Francisco's unique problems, because it seems to be a near twin of the company's New York office.[14] Howells and Stokes, the architects who designed the New York building, were not familiar with San Francisco and had never designed a structure to resist earthquakes. The first three stories of the eleven-story San Francisco building were an exact copy of the facade they had designed for the parent company in New York: the same carved Georgia marble, the same bricks and terra-cotta, the same carving over the door of the company's heraldic insignia, executed by Rochette and Parzini of New York. But the interiors of the two buildings tell a different story.

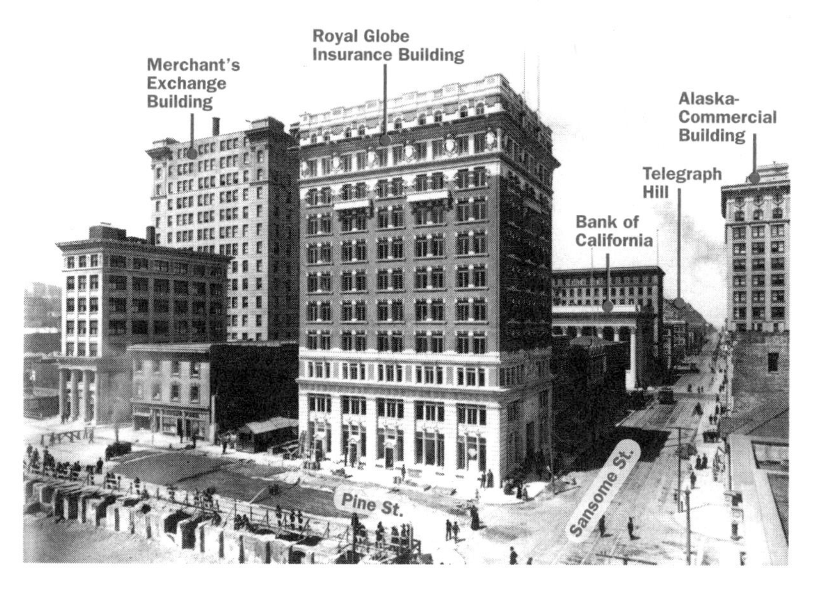

Figure 13.6 (above)
The Royal Globe Insurance Company Building amid other new buildings in downtown San Francisco, c.1910

Figure 13.7 (below)
The Royal Globe Insurance Company Building in construction, 1908. The unusual knee-braced lattice girders contrast with the building's conservative facade.

In articles discussing the unusual safety features of the Royal Globe Insurance Building, the architects are credited with the structural details of the building. But the consulting engineers, Purdy and Henderson, also of New York, must have played an important part as well. Purdy was the structural engineer of the landmark Old Colony Building in Chicago and an acknowledged expert on lateral bracing with several scholarly articles to his credit.[15]

The building was constructed to be very stiff but resilient if subjected to lateral forces from earthquakes.[16] The system devised for the building combined known techniques in new and more effective ways. First, the foundation is extremely solid: the steel-plate columns and angle columns that hold up the superstructure are placed at about sixteen feet on center. Their iron bases rest on I-beam grillages, whose footings are on sand twenty-four inches below the surface of the ground, resting on a bed of piles. The columns are anchored with large bolts through the grillage to its base. The grillages themselves are enclosed in concrete or brickwork, and the floor is reinforced by two crossed layers of steel rods. The steel columns are encased in reinforced concrete. Rods wind around each column, creating "a basketlike network to afford tensile strength in

Figure 13.8
The Royal Globe Insurance Company Building: details of lattice girders with upper and lower knee bracing to resist earthquakes.

every direction." These rods were also threaded into the reinforcing in the concrete floor so that it would perform like a "single strong monolith." This detail is interesting because the architects and engineers were trying to increase continuity, a concept not clearly understood at the time. They used both reinforced concrete and steel to strengthen the building.

Lateral movement in the building is reduced by an advanced system of deep spandrel girders and knee braces (figure 13.7). The design of the steel members resisting lateral drift is the most robust of its era in San Francisco—even heavier than in the Humboldt Savings Bank Building. The deep spandrel girders are reinforced by knee braces or gusset plates on the top and bottom at both ends, which ensures that each story is a rigid box joined to the floor above and below. The overall structure is very rigid compared to other buildings from that period. When inertial forces bend the building in an earthquake, it might yield by bending at the columns, which could be dangerous. But this problem is alleviated by the heaviness of the columns themselves and the masonry infill.

Despite the excellent performance of exterior brick infill walls and decoration in steel-frame buildings in the earthquake, the architects and engineers of the Royal Globe Building decided to reinforce the brickwork and tie it to the steel frame (figure 13.8). Bricks are held onto the framework with bond iron and rods—a version of some of the experiments tried after the earthquake of 1868. This time, the architects and engineers invented a vertical and horizontal web of steel rods that run through all the exterior brick and stonework and attach to the steel frame. Although this feature was used on other structures, this building receives the major notice in the literature. By adopting this straightforward bonding method, Howells and Stokes made a great step forward, providing not only greater safety for the exterior walls of the tall structure, but a safety feature that would become a part of San Francisco's code for thin brick walls in 1914.[17]

Addressing the problem of cornice failures, the architects provided for steel cantilevers attached to the main framework of the building to support the three-foot-deep terra-cotta cornice. While this practice was used immediately following the earthquake, it was not included in the codes until 1909, when the building was in construction.

Many fireproof features protect the building. The floors are of concrete, the partitions are of cement mortar on wire lath and metal furring. The elevators are enclosed in fireproof steel-frame shafts, and the grillages on the shafts are designed with wire glass to eliminate any chance of the shafts acting as chimneys. Wire glass is used on the courtyard windows as well, and the window frames and sashes are made of steel.

Like the Humboldt Savings Bank Building, the Royal Globe Insurance Building has a foundation, a lateral bracing system, reinforced concrete, fire-resistant materials, and an interior designed to resist earthquake damage, but the two buildings are completely different in conception and design. The added elements of cornice attachment and reinforced brick pioneered in the Royal Globe Insurance Building demonstrate the ingenuity of solutions that were invented to meet the problems of construction in earthquake country.

The Phelan Building, 1908[18]
760–784 Market Street
William Curlett, architect; Paul L. Wolfel, engineer

One building that seems especially worthy of praise is the Phelan Building, erected in 1908 (figure 13.9). Former mayor James Phelan, who, depending upon one's view, could be seen as a farsighted urban reformer or an opportunistic bigot, commissioned the architect William Curlett to build a new structure for him on the site of his pre-earthquake building. The triangular ground plan of the structure, on a block of Market Street between O'Farrell and Grant Streets, as well as its sand foundations and its eleven-story height, forced Phelan, Curlett, and the engineer of the building, Paul L. Wolfel of the American Bridge Company of Pittsburgh, to sink deep pile foundations and to bind the steel cage of the building together with particularly deep girders, to adequately absorb and transfer lateral forces.[19] The asymmetrical plan of the structure called for a complicated system of cross bracing that included longitudinal tie-rods similar in function to those in Burnham and Root's pre-earthquake Chronicle Building. An article on the building states that the complicated problem of designing the structure was solved by the architect and engineer working in unison. The last two buildings we examined must also have been team efforts, but here the partnership of architect and engineer is especially noted and favorably described, an interesting comment on what we might see today as an optimal condition for good design.[20]

Figure 13.9
The Phelan Building, 760-784 Market Street, San Francisco, 1908. Photomontage of finished exterior and steel frame in construction. Note the deep spandrel girders, intended to be earthquake-resistant.

Sharon Building, 1912[21]
39–63 New Montgomery Street
George Kelham, architect; Henry J. Brunnier, engineer

One of San Francisco's most famous engineers of well-braced and well-tied buildings was Henry J. Brunnier. He arrived in San Francisco with his wife two weeks after the earthquake, having been sent to California by the New York Edison Company.[22] He quickly learned that many structural engineers in San Francisco were representatives for particular products or manufacturers, but Brunnier wanted to be what he called a "full-service" engineer, without affiliation, and opened his own office in 1908. George Kelham hired him to do the engineering for the Sharon Building (1912), and the two worked together for the remainder of their professional careers. The building was considered a model of responsible engineering in its time and is still home to the office of H. J. Brunnier Associates.

Brunnier's design for the structural system of the Sharon Building called for a heavy steel frame with deep spandrel girders at each floor (figure 13.10). He had been impressed by the old Palace Hotel, which, he wrote, had not been damaged by the earthquake:

> The cement mortar reinforced with steel bands and with the anchors still protruding from the walls indicated that the walls had been thoroughly tied to the wood-frame floors, which, in turn, had acted as a diaphragm.[23]

His employees recalled that he always admonished them to tie their buildings together to act as a unit.[24] He also concurred with the recommendations of the teams that investigated the 1906 earthquake, and in 1955, looking back on his experience, he asserted that structures in earthquake country should be designed to resist a lateral equivalent of 30 to 50 pounds per square foot of exposed area.[25]

When Brunnier calculated the wind load for his building in 1912, he designed it to resist 20 psf wind pressure, the requirement in the 1909 codes. Clearly, the building was designed to be earthquake-resistant. But what Brunnier had not figured in his calculations was the L shape of the plan: this configuration would be hazardous in earthquakes because the legs of the L would move independently, tending to break the joint between them. Although symmetrical configurations were understood in Europe to be superior in earthquakes as early as 1783, no such understanding was present in post-earthquake San Francisco.

Figure 13.10
The Sharon Building, 39–63 New Montgomery, San Francisco, 1912, steel frame with lattice girders. Note the steel framework to hold the cornice in place.

Sather Tower (the Campanile), 1914[26]
University of California, Berkeley
John Galen Howard, architect; Charles Derleth Jr., consulting engineer

UC Berkeley's Sather Tower was the first tower built in the Bay Area after the earthquake and fire of 1906. It was, perhaps, the first building in the world designed to resist earthquake forces using calculations that took acceleration, drift, resonance, and overturning forces into account (figures 13.11a and 13.11b). John Galen Howard, campus architect and chair of the architecture department, had worked in post-earthquake San Francisco. His design for a steel and stone tower was based on the campanile in Piazza San Marco, in Venice. The engineer was Erle L. Cope, and the consulting engineer was Charles Derleth Jr., whose observations on earthquake damage and post-earthquake reconstruction have been a rich resource for this book. Derleth was responsible for all the engineering decisions in the building. His prescience alone might justify the inclusion of this Berkeley building in a chapter on post-earthquake reconstruction in San Francisco. But Derleth's notes offer an even more compelling justification: he used San Francisco buildings, like the Call Building and the Humboldt Savings Bank Building, as models for calculating overturning moments and bracing solutions.

Derleth was not the only professional engineer of his day to calculate the way buildings relate to dynamic earthquake loading. Another was Richard S. Chew. During the discussions of stiff versus flexible, or bending, structural systems at the Structural Association, ideas about dynamic response to earthquake forces began to be discussed. In other words, instead of thinking about a continuous mono-directional wind load pressed on a building, engineers like Chew tried to diagram and calculate the dynamic response of buildings to earthquake force from multiple directions, including the influence of inertial forces.[27] In 1908 Chew prepared a paper for the ASCE in which he diagrammed this type of building movement.[28] Using the classical physical formula of lateral force equaling the building mass times the acceleration against the building, which he saw as a vertical cantilever, and factoring in inertial forces, he created diagrams very similar to modern descriptions of the modes of building vibration (figure 13.12). Drawing on his own observations of earthquake damage, Chew concluded that earthquake stresses, while similar to those caused

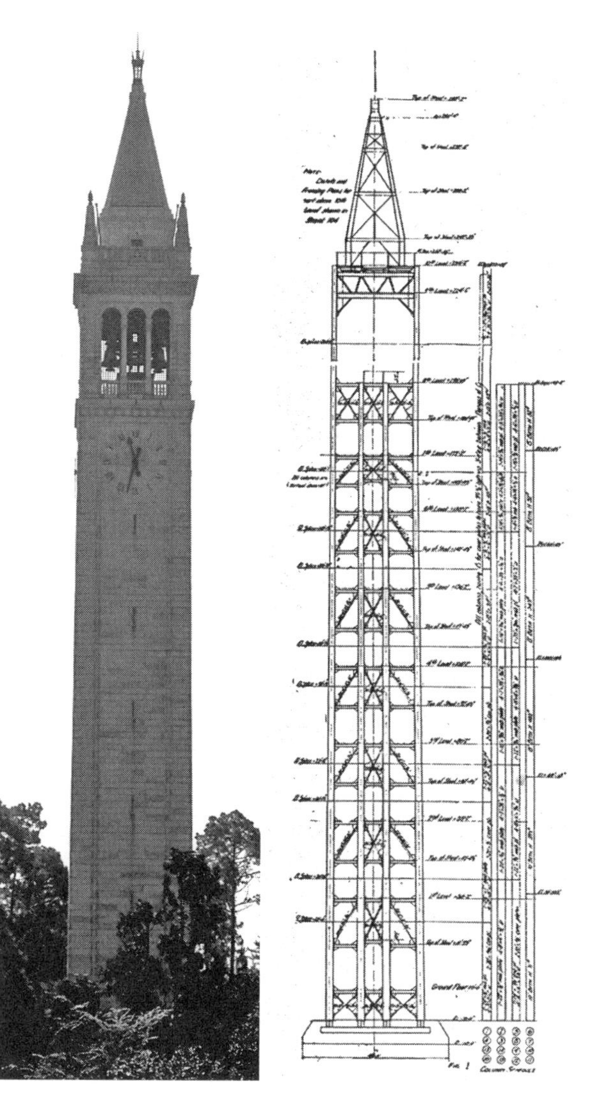

Figure 13.11a
Sather Tower, University of California, Berkeley, 1914: exterior elevation

Figure 13.11b
Sather Tower, Universitiy of California, Berkeley, interior steel frame and details

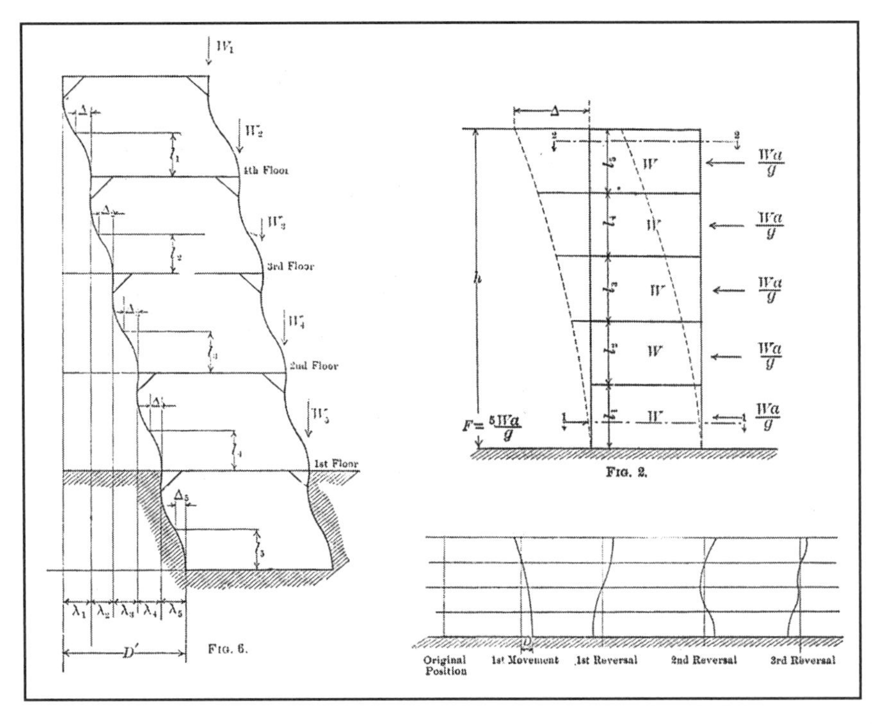

Figure 13.12
Earthquake modes as understood by Richard S. Chew in 1909

by wind, can be significantly increased by multiple quick reversals in motion. He believed all earthquake forces were a direct function of weight, and advised that walls and floors be made as light as possible, and that building frames be highly elastic.

Derleth saw the problems of designing such a high tower in earthquake country, and like Chew, began to think about how to model, analyze, and design it. Buildings must vibrate at a different rate of vibration than earthquakes or risk having their forces and displacements intensified by resonance—the same phenomenon that, when you kick out your feet while swinging, accentuates the trajectory of the swing's curve and thus causes you to travel higher and faster. Derleth understood resonance and, in order to avoid Sather Tower's having the same period of vibration as any earthquake, he designed it to be flexible and to move slowly if shaken. He thought a two-second period of ground motion was impossible in an earthquake, so he designed his tower to move in that rather long period.[29] Intending to design the tall tower with the specified stiffness but also providing back-up flexibility, he installed cross bracing only every other story and made sure all connections were extremely strong. He tested his hypothesis by pounding upon the base of the steel framework and measuring the vibrations at the top of the tower.[30] The exterior stone facing was applied over a reinforced concrete interior that was supposed to help in the uniform loading of the steel frame. Derleth's Sather Tower is an example of how San Francisco's engineers tried to use their observations from the 1906 earthquake in inventive ways in order to build sophisticated, earthquake-resistant structures.

San Francisco City Hall, 1912–1916[31]
Bakewell and Brown, architects; Christopher Snyder, engineer

The need to replace San Francisco's ruined City Hall seemed ever more crucial with the approach of the 1915 Panama-Pacific International Exposition, an event that was expected to draw visitors from around the world. The new structure would be symbolic of San Francisco rising from the ashes, politically cleansed and structurally sound. The basement of the eastern office wing of the old City Hall, which had been serving as a temporary headquarters for city government, was finally vacated and razed. City Hall would be relocated, one of the liabilities of the old site having been "the fact that the ground on the old site is insecure, there having been a lagoon there at one time." A blind design competition was juried by business leaders and architects, including John Galen Howard, Frederick H. Meyer, and John Reid Jr. The firm of Bakewell and Brown was chosen from a field of seventy-three entries.[31]

A whole civic center complex in the Beaux-Arts tradition was envisioned for the new site. Chicago's 1893 World's Fair, with its elegant Beaux-Arts "Great White City," was extremely popular among architects, and it also captured the hearts of aspiring political reformers in the Progressive movement. The aesthetic values of hierarchy, symmetry, and regularity seemed perfectly symbolic of transparent government and reform: city government would be housed in a single building surrounded by secondary buildings created in an identical style, keyed to one another and City Hall: a new library, an opera house, state courts, and an auditorium.

Bakewell and Brown's new, 1912 City Hall was magnificent. The rectangular floor plan, which incorporated hundreds of city offices on five floors organized in corridors around a central rotunda and two light courts, was surmounted by a stately dome. The sources for the plan, the interior detailing, and outside elevations were all French, and the beautifully articulated dome is clearly based on the dome of the eighteenth-century Hotel des Invalides in Paris. The people of San Francisco and their new mayor, James "Sunny Jim" Rolph Jr., loved the look of this new building. (Rolph led the city as mayor from 1912 to 1931, overseeing the reconstruction of the city, the Panama-Pacific International Exposition of 1915, and the opening of the new City Hall in 1916.)

Once the citizens of San Francisco knew that the 1912 City Hall was going to have a steel frame, questions about its structural integrity ceased. But from the reports of the engineer, Christopher Snyder, and the foreman, George Wagner, some interesting and disturbing construction details emerge.[32] Built on wet sand with an abundance of water underneath, the foundation for the structure, designed by Snyder, was the cheapest Wagner had ever laid, costing $40,000, or one-twentieth the cost of the total building, a very low ratio. The foundation, whose design was based on the successes of mat foundations in the 1906 earthquake, was concrete composed of brick aggregate, much of which came from the previous City Hall foundation, crushed and reused.

Snyder's conception of the structure was that it would be "on the lines of a modern office building, as far as the architecture would permit." He stated that "aside from the dome, which was a considerable problem, and the heavy stonework of the exterior walls and cornices, the steelwork and foundations were similar to the ordinary Class A building" (figure 13.13).[33] The dome was a special challenge. The collapse of its predecessor influenced its design in the way the dome was supported, "on four groups of five columns each, latticed together from the second floor to top...These carry girders nine feet deep, sixty feet long." The dome tower was divided into sixteen bays with four circles of columns, "braced radially and circumferentially, like a gas holder frame, with five levels of horizontal

Figure 13.13 (top)
San Francisco City Hall, 1912–1916, with the dome steelwork in construction.

Figure 13.14 (bottom)
San Francisco City Hall with steelwork in construction

bracing to maintain the circular shape against distortion" (figure 13.14). A wind load of 50 pounds per square foot, reduced for curvature and slope, was used in calculating the dome structure. To absorb and counteract inertial forces from the dome, diagonal bracing was placed on the same plane as the floors near the dome, supporting its four bases.

City Hall had a stiffly constructed dome and, hopefully, an appropriate foundation, but buried in Snyder's description of the main portion of the structure was a feature that San Franciscans and their engineers should have—at least—debated. Snyder stated, "No wind was figured on the main structure, nor was any diagonal bracing used below the second floor, in order that the necessary flexibility against earthquakes should be retained."[34] Snyder was putting into practice the idea of a flexible, shock-absorbing first story, which was extremely theoretical in 1912. The Structural Association of San Francisco had discussed this idea immediately after the earthquake of 1906, and it has been debated ever since, with particular vehemence in the 1920s by Brunnier (con) and Nishkian (pro). Chew's ideas of the dynamic forces that acted on structures may also have had an impact. Snyder's concept was revolutionary.

For Snyder the question of earthquake safety was paramount. His decision to include a shock-absorbing first story was tested in the Loma Prieta earthquake seventy-seven years later. According to engineer Eric Elsesser, it probably saved the building from collapse. Snyder's innovation was never mentioned by the architects or the critics of the time who praised the new City Hall, but Snyder himself explained what he had done in the article quoted above that he wrote for his engineering colleagues in *Architect and Engineer.* Arthur Brown Jr., one of City Hall's architects, appears to have been ignorant of the revolutionary character of Snyder's solution.

More Earthquake-Resistant and Fire-Resistant Design in Steel Buildings, 1906–1915

To conclude this section, a few more observations about the design of steel buildings are in order. Buildings being erected throughout San Francisco's downtown shared the assumed earthquake-resistant features of the Humboldt Savings Bank Building, the Royal Globe Insurance Building, the Phelan Building, Sather Tower, and the new San Francisco City Hall. Because stone had performed poorly in earthquakes, terra-cotta replaced it as a favored material for exteriors. When brick was used, either for the facade or for infill, it was most likely reinforced with steel rods or securely tied to steel frames.[35] Reinforced concrete was frequently used for floors, walls, and fireproofing columns.

Earthquake resistance continued to be a major consideration in the structural design of buildings. Though Root died in 1891 and Burnham died in 1912, architect Willis Polk directed their firm in San Francisco with the same commitment to earthquake-resistant design that had been important in the old Chronicle Building. When the newspaper hired Polk to reconstruct the old building and its annex, he strengthened the deep spandrel girders of the annex and rebuilt the steel frame to be "heavy enough to stand future earthquake stresses, [with] much care...being taken in the riveting and anchoring of the great columns and girders."[36]

Also while working for Burnham and Root, Polk designed the elegant First National (Crocker) Bank, the first two floors of which still stand at the intersection of Post, Montgomery, and Market Streets. He declared this building to be "designed to withstand the forces of the elements, including fire and earthquake."[37] Unfortunately, in 1990 it could not withstand "progress" and was cut down to two stories with its upper ten floors replaced by a garden terrace.

Remembering the importance of deep foundations, the architect Henry A. Schulze wrote in 1907 of his new Olympic Club Building, at 524 Post Street, that the foundation went down fifty-five feet, "the deepest of any heretofore undertaken in San Francisco." The steel structure was to be "one of the strongest and safest in the city and capable of resisting the severest stresses that can be imposed upon it."[38]

Another example is Lewis Hobart's Commercial Building (figure 13.15), commissioned as an office building to produce revenue for the California Academy of Sciences and built on the academy's former site. The engineer, Christopher Snyder, working

Figure 13.15
The Commercial Building, 825–833 Market Street, San Francisco, 1908, steel frame with lattice girders

Figure 13.16a (left)
The Alaska-Commercial Building, California Street, San Francisco, 1908 (demolished in 1974)

Figure 13.16b (right)
The steel frame of the Alaska-Commercial Building, with lattice girders as lateral bracing, in construction and nearly complete

Figure 13.17a (above)
The Kohler-Chase Building, 20–26 O'Farrell Street, San Francisco, 1909, steel frame with lattice girders for lateral bracing, in construction

Figure 13.17b (right)
The Kohler-Chase Building's finished facade

for Milliken Brothers Steel, chose to use a system of lattice girders at each floor to stiffen the frame.[39]

Without documentation, we can only guess at the intentions of the many other architects and engineers working in San Francisco. We know that a majority of the engineers investigating the 1906 earthquake had recommended increased lateral bracing. If conspicuous lateral bracing is present in a building, then we might guess that the designers were trying to solve seismic problems. The difficulty with this hypothesis is that architects in cities such as Chicago, as we have seen earlier, designed for wind loads in exactly the same way San Francisco architects designed for earthquakes—with lateral bracing. A number of medium-height San Francisco buildings have outstanding lateral bracing: Henry H. Meyers' and Clarence Ward's Alaska-Commercial Building, 1908 (figures 13.16a, 13.16b); Albert Pissis' White House, 1908; Frederick Meyer's Kohler-Chase Building, with the steelwork designed by Christopher Snyder for Milliken Brothers Steel, 1909 (figures 13.17a, 13.17b); and T. Paterson Ross and A. W. Burgren's Clunie Building (figure 13.18).[40]

Figure 13.18
The Clunie Building, at Montgomery and California, San Francisco, 1908 (later demolished). This construction photograph illustrates the use of deep spandrel girders for lateral bracing.

Class B Buildings: Reinforced Concrete

During this period, the tall steel structures in San Francisco were seen as the most exciting and earthquake-resistant modern building type. However, many architects and engineers considered reinforced concrete to be earthquake-resistant, an appropriate substitute for brick, and particularly adapted to large, fireproof construction. Prior to the 1906 code change, it had been excluded as a load-bearing material, and even after the change it did not have the same status as steel-frame construction: the height of reinforced concrete buildings was limited to 102 feet. Nevertheless, by June 1907, seventy-eight reinforced concrete structures were being built in San Francisco. A writer for the *Architect and Engineer* stated in 1907 that "it is safe to say that there never before has been near as much work of this character going on in any one city at one time."[41]

The quality of workmanship in the field is extremely important in concrete, mistakes and cost-saving substitutions being easily buried. In addition to the size and position of the rebar, the water content of cement and its viscosity are crucial for its strength. Today there is a field test, called a "slump test," in which a cone is filled with newly mixed concrete. When the cone is then inverted, the concrete's ability to retain its shape can be graded.[42] Typical contracts of 1907 make no specification for such a test, and it is likely that no such testing was done. Thus the quality of the concrete in early reinforced concrete structures could vary widely. Similarly, the curing, or drying, of concrete varied from job to job, sometimes with disastrous results. Between 1906 and 1907, two concrete failures occurred, one in the construction of the Hotel Bixby, in Long Beach, California, and the other in a structure for the Kodak Company at Rochester, New York.[43] In the first, the forms were removed too soon, causing collapse and killing several workmen. In the second, a whole series of errors had occurred. The building was not designed properly, the steel was not installed correctly, and some specified reinforcing was omitted. The concrete itself was also bad, containing leaves, sawdust, and other foreign matter.

The brilliant engineer William Hammond Hall, one of those who had analyzed damage caused by the 1906 earthquake, was very concerned about strength and uniformity in reinforced concrete work.[44] An advocate of the material, he eagerly studied many of the 194 new concrete structures that had been built in San Francisco by 1914. His critique provides an entry point for discussing problems with this construction method in which so many people were placing their faith.

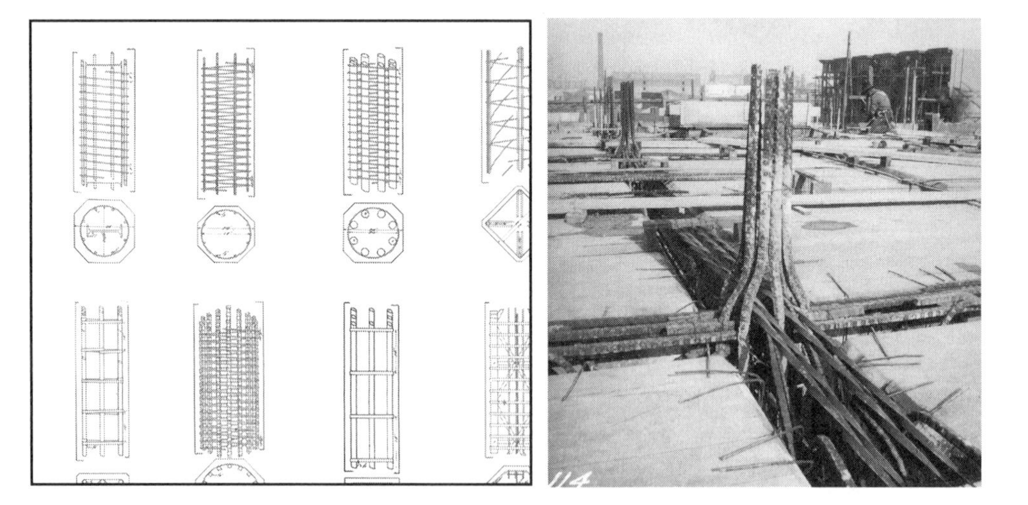

Figure 13.19 (near left) William H. Hall's examples of variation in the rebar in reinforced concrete in San Francisco, illustrating that architects and engineers were experimenting with individual and largely unregulated solutions for reinforcing concrete.

Figure 13.20 (far left) William H. Hall had this photograph of a building in construction taken for his article on San Francisco reinforced concrete. Two different kinds of bars are used: square bars with nubbins to attach to the concrete, and smooth, twisted rods. The rods are "cold twisted," turned on a kind of lath, so concrete will better adhere to them. One of the important ways to ensure strength in concrete is to leave enough room between the rebar, but several in the beam and all the rebar in the column are too close together.

Knowing that every concrete building was being constructed with different kinds, sizes, combinations, and patterns of reinforcement, Hall set out to document this diversity (figures 13.19 and 13.20). The variety of solutions he illustrated shows how young the field was, and the mistakes he found show how easy it is to misunderstand and misuse concrete technology.

In the months following the earthquake and fire in 1906, concrete structures were going up even as code requirements for concrete were being devised and revised. Load tests were being conducted as the new technology evolved, but only in demolition can we see the differences between buildings.[45] For example, the Blake, Moffitt and Towne Building, constructed in 1911 by Willis Polk, had twice the steel reinforcement of its neighbor to the right, Nathaniel Blaisdell's 1907 Golden Gate Building.[46] Today, because it is so hard to examine the rebar in old reinforced concrete work, the "earthquake-proof" material adopted with such enthusiasm and trust presents a unique assessment problem.

The Pacific Building, 1907
801–823 Market Street; Charles Whittlesey, architect

The most noteworthy of the reinforced concrete buildings that arose immediately after the 1906 earthquake was the nine-story Pacific Building, at Market and Fourth Streets, which claimed to be the largest reinforced concrete office building in the world (figure 13.21).[47] The architect of the building, Charles F. Whittlesey of Los Angeles, was one of the leaders of the Mission Revival movement inspired by the missions of California. His railroad station in Albuquerque, New Mexico, was praised by *Architect and Engineer* as "one of the purest examples of modern mission architecture."[48] In San Francisco, however, he designed a huge concrete structure idiosyncratically decorated with rust, green, cream, and yellow tiles and decorative terra-cotta sculpture inspired by the work of architect Louis Sullivan.[49] Whittlesey justified his choice of colors in *Architect and Engineer:*

Figure 13.21
The Pacific Building, 801–830 Market Street, San Francisco, 1907

> The climate of our city is decidedly gray and this is accentuated all about town, especially in the large buildings, by the use of a peculiarly gloomy stone of a disagreeable yellowish gray color that catches and absorbs much of the smudge carried on the winds.[50]

Internally, the structure is completely of concrete, not concrete protecting structural steel columns. Unlike some earlier concrete structures, which depended on cement and forms made in the East, the Pacific Building is built of California cement. According to an article published in *Construction News,* much of the construction method was devised at the job site:

> The reinforcing system has been devised on the spot by the builder, whose ample experience elsewhere in this class of buildings has been brought into service in the present undertaking. All of the major steel rods, three-fourths of an inch in diameter, are of the spiral twist pattern. They are twisted cold, and this process is conducted on the premises, being applied as a test to every rod used.[51]

Using this new technology, stairs and walls could be formed into a reinforced concrete unit, cantilevers were possible without permanent underpinning, and a structure could be erected that would incorporate few burnable surfaces. The building was seen as a cast unit, a series of flexible yet integral wedded parts that could ride out an earthquake. As Whittlesey said:

> In the argument for durability, reinforced concrete is in a class by itself. As far as the elements are concerned, it is practically indestructible. It is by far the most rigid and freest from vibration of any construction known. The steel sinews forming the reinforcement give to the concrete sufficient elasticity to withstand admirably the strains produced by earthquakes, and with the ample bracket connections between columns and floor beams which this method supplies, it would require a greater shock than California has experienced since the coming of the Padres to produce in it any sign of failure. Even though it were strained to the extent of producing cracks, the strength of the structure would be but little impaired, because of the reinforcing metal.[52]

The placement of steel rebar in the concrete and the design of the concrete members are at the heart of the strength and durability of Whittlesey's buildings. And indeed, cross-sections illustrating the placement of reinforcing bars in early concrete structures, as in Whittlesey's Bartell Building, indicate a solid understanding of how to design strong column and floor connections.

Class B and C Brick Buildings

As discussed earlier, many experts condemned brick as a building material, on both experiential and theoretical grounds. Others, citing the many brick buildings that survived the earthquake, argued that blanket condemnations of brick were unjust. Some engineers, architects, and builders attempted to improve upon standard brick construction by reinforcing it with iron, as had been done after the earthquake of 1868. A number of people attempted to popularize patented reinforced brick, and a few significant patents were issued on new systems. Robert W. Gardner, writing about Major Stokes-Roberts of the Royal Engineers, who put up water tanks in reinforced brickwork for the British government, explained that Stokes-Roberts used a light brick wall with telegraph wire reinforcement to withstand the pressure of the water. He incorporated wire ties in the mortar of his bricks and attached the ties to light rods that ran along the outside of the wall and were cemented over. "The result," wrote Gardner, "is practically a reinforced concrete wall. Such a wall could be bulged or twisted by explosion or earthquake, but it could not collapse."[53]

Such a reinforced brick method was never used in California, but an equally interesting patented reinforced brickwork was used at least once. Joseph A. Hofmann developed a patented method dependent upon special hollow bricks that could be placed easily in a wall and reinforced both horizontally and vertically with steel rods (figure 13. 22).[54] Hofmann's method was used in an apartment building on Bush Street in San Francisco and is commemorated by a plaque on the exterior of the building (figure 13.23).

There is no way to know how many of these reinforced brick structures exist in San Francisco. However, it is fairly certain that the practice of laying ordinary

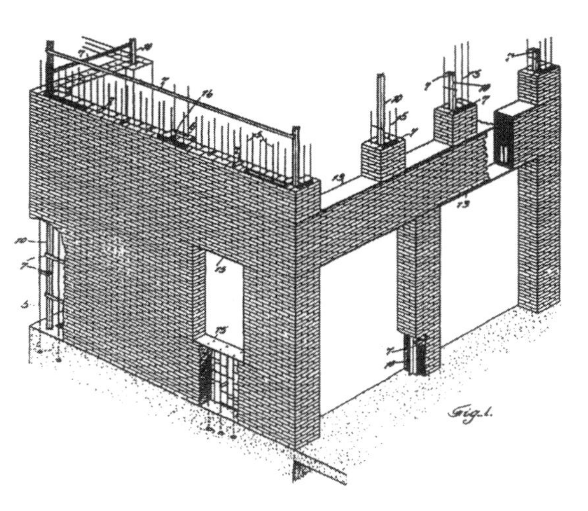

Figure 13.22 (left)
The Joseph A. Hoffmann patent for reinforced brick, July 21, 1908

Figure 13.23 (right)
Apartment building at 1241 Bush Street, San Francisco. On a pilaster on the right side of the facade, a plaque reads: "Reinforced brick construction patented July 1908 by Joseph A. Hofmann/Beasley & Guentier Architects."

bricks on either side of rods or as headers across them became standard for securing brick infill walls to steel-frame buildings of all sizes. This system appears in Class A buildings, like the Levi Strauss Building, as well as in smaller structures, like the apartment building at Post and Jones Streets pictured in figure 13.24, as well as in the Royal Globe Insurance Building.

Figure 13.24a (above, left)
Apartment house, Post and Jones Streets, San Francisco

Figure 13.24b (above, right)
This apartment house has a steel frame and brick walls reinforced with steel rods attached to the frame.

The vast majority of new brick buildings were minimally reinforced. It is possible that the brick buildings constructed between 1868 and 1906 which survived the 1906 earthquake surprisingly well had other earthquake-resistant features. Before the time of steel structures, these brick buildings were the cutting edge of engineering technology and had been built with great attention. The most carefully constructed brick buildings in California had been those in San Francisco. Several engineers writing after the earthquake lamented the poor quality of the new brick structures. In late 1906, P. J. Walker told the Structural Association of San Francisco:

> The first thing that became apparent at that time [after the earthquake and fire] was that more attention would have to be paid to the future construction of brick walls. Unfortunately, I am obliged to confess that so far, with very few exceptions, the disastrously learned lesson has been absolutely disregarded, for today there is no doubt in my mind but that poorer brickwork is being done than before the fire.[55]

Walker mentioned in particular the indiscriminate use of old bricks, the poor construction practices of unscrupulous building contractors, and the use in new construction of the remains of old walls still standing. On the other hand, bond iron, popularized in the 1860s, was stipulated in the codes, and good cement mortar was also prescribed. It is hard to judge the overall quality of the Class C structures. Between 1906 and 1914, 2,699 were built in San Francisco, and two thousand of them still exist.[56] From the outside it is practically impossible to determine whether special reinforcement systems such as Hofmann's are present.

Figure 13.25
Segalas and Plante Building (Class C), Bryant Street, 1911. The positions of bond iron and anchors are illustrated on the building's facade.

Figure 13.26
Excerpt from the original specifications for the Segalas and Plante Building by the architect, Oliver Everett, 1911, with a description of the bond iron to be put into the walls.

Segalas and Plante Building, 1911
Bryant near Third
Oliver Everett, architect

The Segalas and Plante Building, on Bryant near Third Street, is an excellent example of a responsibly designed post-earthquake brick building in San Francisco. Not only the building itself (figure 13.25) but the complete plans and specifications written up by the architect, Oliver Everett, have survived. Every item stipulated in the building codes was included in the construction of the two-story building, and the architect was scrupulous in detailing his design. The foundation and basement floor are reinforced concrete. He included iron girders, anchored into the side walls, to support the brickwork over the shop entrance on the ground floor. To tie the building together, he used bond iron under the joists, as stipulated by code. To tie the walls to the diaphragm, he specified floor joist and roof anchors. Front and rear walls are anchored to the internal joists as well. Everett specified that the brickwork was to be composed of thoroughly cleaned used bricks set with lime and cement mortar in California bond. Even with iron girders, the front and rear walls depend on their own strength to resist in-plane shear. The ground floor still lacked shear strength in the lower part of the facade, there were no metal ties between floors, and the parapet was unbraced. These are the problems that engineers had to face when the building was seismically retrofitted in the 1990s. Although the Segalas and Plante Building would today be classified as unreinforced masonry, and therefore hazardous in earthquakes, it was a good building for its time and should be credited as an example of responsible construction.

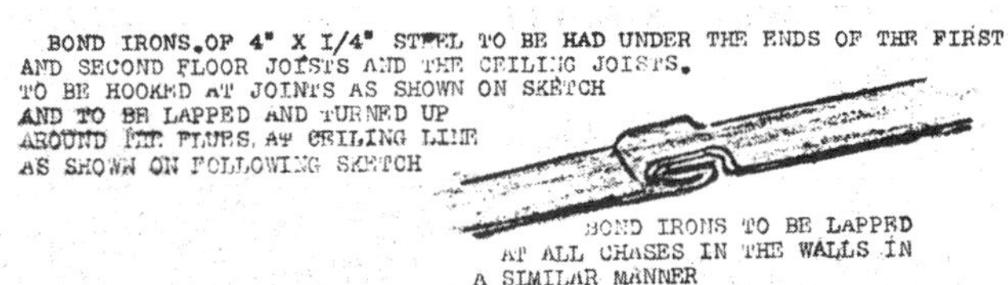
BOND IRONS.OF 4" X I/4" STEEL TO BE HAD UNDER THE ENDS OF THE FIRST AND SECOND FLOOR JOISTS AND THE CEILING JOISTS.
TO BE HOOKED AT JOINTS AS SHOWN ON SKETCH
AND TO BE LAPPED AND TURNED UP
AROUND THE FLUES, AT CEILING LINE
AS SHOWN ON FOLLOWING SKETCH

BOND IRONS TO BE LAPPED
AT ALL CHASES IN THE WALLS IN
A SIMILAR MANNER

118–124 First Street, 1907
Salfield and Kohlberg, architects

118–124 First Street was a Class C brick building constructed in 1907. It was being demolished in 1986 when the photographs in figures 13.27 and 13.28 were taken. The structure was a four-story, L-shaped loft building, with iron columns on the first floor and timber columns from the second to fourth floors. The floors and partitions were all wood. The columns were secured to the beams they carried by top plates, and the beams were strapped together, which was typical of this type of building (see figure 13.29). The beams were secured to the outer, load-bearing walls by square government anchors. Bond iron ran below the joists, placed on the inside edge of the brick wall. As was typical in San Francisco brick buildings, this one was built with American bond, meaning that one wythe of brick was bonded to another by header bricks after several courses. This makes for a very weak wall unless great care is taken with the mortar bonding of the wythes—and in fact the way the wythes cracked apart when the building was being demolished indicates weakness in the walls. Even if iron bonds helped reinforce the structure, the placement of the bond on the outer wythe of bricks just below the joists would seem to minimize its effectiveness. Indeed, the code developed in 1909, two years after this building was constructed, required that the bond be placed in the center of the wall, not on the edge.[57]

Figure 13.27 (above)
Class C Building, 118–124 First Street, San Francisco, 1907. Bond iron placed under joists cascades down demolished wall. The two uppermost pieces are from higher stories. The position of the lowest is clearly visible. The position of this bond iron so close to the edge of the wall negated whatever value it might have had in holding the walls together in an earthquake.

Figure 13.28 (left)
Class C Building, 118–124 First Street, interior, with iron columns on the ground floor and wooden columns on the upper floors. As the building was being demolished, the posts and lintels held together remarkably well.

Figure 13.29 (right)
Interior wooden members and ties, Class C building, Second Street and Folsom. The elements for tying the building together are present, but they are probably not strong enough to save it in an earthquake. (A) is intended to hold the deep joists upright. (B) is a plate on which the second-floor column rests. (C) is an iron or steel bar with ends embedded in the two timbers to hold them together. (D) is a block holding up both sides of the beam. (E) is a plate designed to hold the block and column together as a unit, but the collar is very shallow, making a secure connection nearly impossible.

Lincoln Realty Building, 1908[58]
Willis Polk, architect

The Class C brick Lincoln Realty Building stood east of the Emporium, on what is now the site of Nordstrom's. Originally designed by Sidney Newsom, of Newsom and Newsom, it was to be constructed on the site of Lincoln Grammar School, destroyed in the 1906 earthquake. The site was owned by the city of San Francisco but leased to the developers for thirty-five years, after which it would revert to the city. The developers hoodwinked the city. Because the city had not stipulated that the building should be either steel-frame or concrete, there was a tremendous uproar when the design for a brick building was disclosed. Reporters frantically tried to find something incriminating about the safety of the Class C building. They could not. To make sure the building had a respectable pedigree, Burnham and Co. were given the commission. Project architect Willis Polk altered Newsom's design slightly and the deed was done. The building combined the principles of fire-resistant mill construction with Class C brick building requirements: heavy brick walls on the exterior with huge timbers throughout, metal lath and plaster protecting the staircase, wire-glass skylights, interior sprinklers supplied from rooftop tanks, and standpipes. Although it was fire-resistant and compliant with Class C fireproof design principles, those enraged with the lease agreement called the building a firetrap. When the building was demolished in the 1990s, it became evident that although bond iron and anchors were present, there were no strong ties between its timbers. The protestors had recognized that this was a cheap brick building and that its builders were not among those who were building responsibly.

Special Construction

What did the city government do to assure that the many buildings it constructed after the earthquake and fire were safe? Simply put, earthquake and fire resistance were paid for in some cases, but not in others. It would be hard to conceive of a building more carefully designed than Frederick Meyer's Pumping Station Number 2 at Second and Townsend, with its steel frame, reinforced concrete walls, and fine fireproof design features like Kinnear shutters and wire glass. But the city supervisors, faced with rebuilding so many of San Francisco's firehouses and schools, could not make them all that secure. The locations of the structures dictated the construction level. Schools and firehouses would be made of steel with brick walls in the fire district, but outside the district, wood-frame buildings—"special construction"—predominated.[59]

The special-construction buildings, though not considered adequate today, were built with concern for public safety: they were clad in stucco and partially constructed of fire-resistant materials in order to meet the challenge of fire, and they had wooden frames in order to resist earthquakes. The system was described in 1906:

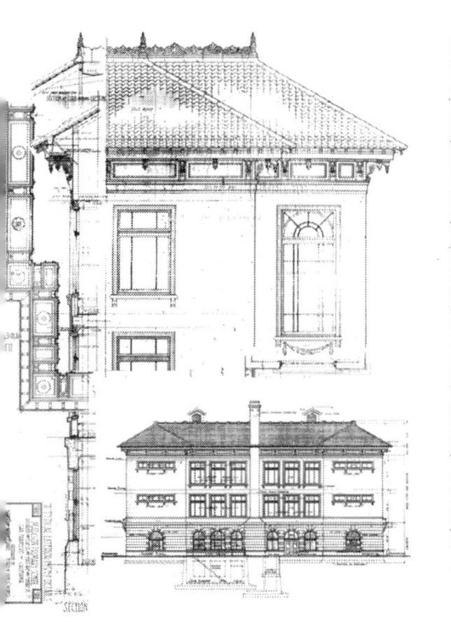

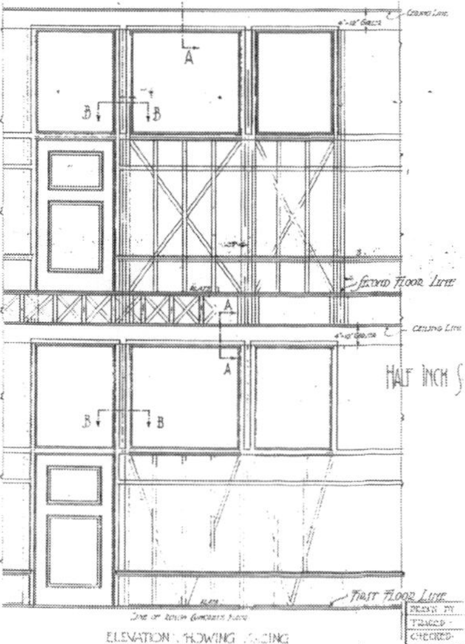

Figure 13.30 (far left)
Madison Grammar School, Clay near Cherry Street, San Francisco, 1909 (demolished c.1977), elevation details of south courtyard facade. Cement stucco was applied to a wooden core to give the exterior a Mission Revival–Italianate look. Though, as the section illustrates, the roof was well-tied, studies in the 1970s showed the concrete work on the first floor was poorly reinforced, jeopardizing the entire building in earthquakes.

Figure 13.31 (near left)
Madison Grammar School, hall elevation of wood studs for partitions. These wall details are the only evidence available that diagonal bracing was present in interior walls. Because of the transom windows, the X-braces would not have contributed much to stiffening the wall.

> A wire-lath and plaster on boards type of building has won the approval of San Francisco city authorities. The new building type has been advocated for schoolhouses and fire engine houses. It consists of a substantial wood-frame structure covered with iron lath and heavy cement mortar on the exterior and lined with iron lath and plaster. The window frames may be covered with sheet metal, and for greater security the window glass may be wired. The advantage of this class of building is its elasticity and the severely tested qualities of frame buildings.[60]

Figure 13.32 (left)
Madison Grammar School, the north-facing Cherry Street facade in construction. This wooden core is the heart of the "special construction" concept. Note that the diagonal siding slopes downward to the right for two bays of windows on the left of the facade; on the right side of the facade, all the siding slopes downward to the left. This was an attempt to use siding for diagonal bracing. Whether there was diagonal bracing between the studs as well is not indicated in the plans.

I attended one of the special-construction grammar schools, Madison Grammar School, on Clay Street, in Pacific Heights (figure 13.32). A photograph of the building in construction illustrates that the wooden sheathing was placed diagonally, instead of vertically, in order to take shear forces in the exterior walls. The studs in the walls were specially constructed with X-bracing to resist lateral movement (figure 13.31). The concept made sense. But in this school, as well as several other special-construction schools, the city architects overlooked problems relating to insufficient shear walls and connections between materials. Still, Madison Grammar School would have been an excellent retrofit candidate. I will never forget standing on the school stage in the 1950s, holding a giant candle and reciting, "Madison Grammar School, forty years old. We wish you many more happy birthdays." It was not to be. Madison School was judged unsafe in the 1970s and demolished.[61]

Firehouses were built in a very similar manner. One finds the same lack of detailing for lateral forces, which is particularly worrisome considering the large first-floor spans that accommodate the fire engines. This lack of careful earthquake-resistant design is especially hard to understand after the prevalence and impact of firehouse failure in the 1906 earthquake.[62]

In 1912, the city erected a number of temporary civic structures to serve the public. One of these was a hospital, and another a stable.[63] In both of these temporary brick buildings, seismic issues seem to have been addressed. They are only one story high and their brick walls are laced with centered bond iron. These temporary buildings seem more carefully designed than the permanent schools and firehouses.

Wood-Frame Residential Construction

Despite all the attention given to the building types discussed above, the vast majority of the structures built between 1906 and 1914 were wood-frame buildings. This is the least documented structural system used in San Francisco, and with few exceptions, we cannot know how frame construction changed as a result of the earthquake. Expecting builders to simply construct wood frames according to the best standard practice of the day, architects rarely included framing diagrams with their drawings, and plans submitted for building permits do not include framing diagrams either. As with other building types, the safety details of wood-frame structures depended upon individual owners, architects, and builders.

Several examples of wood-frame construction pose fascinating questions about Bay Area practice in relation to earthquakes. In the 1901 codes a brief sentence (see chapter 11) admonishes that "angles" should be inserted in exterior and cross walls. These angles are diagonal bracing. When post-earthquake wooden structures are torn down to the studs, diagonal bracing appears on both exterior and interior walls (as in figure 13.33). This diagonal bracing is no longer seen as effective today. But at the time, diagonals were accepted as a lateral resisting feature. It is possible that wooden buildings in the San Francisco Bay Area were traditionally designed to be seismically resistant. In order to prove such a hypothesis, a large sample of wood-frame construction would have to be studied.

Figure 13.33
1911 house demolished to studs, Woolsey Street, Berkeley. This building had diagonal bracing in exterior and interior walls, which may reflect a tradition of earthquake-resistance in wood-frame construction. The diagonals and X braces are placed asymmetrically to ensure the longest possible runs. They were probably not drawn in the plans but were placed by carpenters, hence their ad hoc positioning.

San Franciscans building residences in the Pacific Heights area of the city must have insisted on particularly sturdy buildings; a sample of the houses designed by architects such as Bernard Maybeck, John Galen Howard, Julia Morgan, and the Newsom brothers indicates that all have extensive foundations and sturdy wood frames. Yet a review of the specifications for these houses indicates that stipulations to tie down a chimney or bolt a house to its foundation were rare. Two houses near the corner of Jackson and Locust Streets illustrate the care that was taken in upscale residential construction.

In 1909 Bernard Maybeck designed a house for the Leon L. Roos family, infusing it with both the fun of a charming mock medieval hall and the seriousness of earthquake-resistant detailing.[64] The house was a wedding present to Mrs. Roos from her father, Morris Meyerfield, a partner in the Orpheum Theater Circuit company. Like her father, Elizabeth Leslie Meyerfield was fascinated by the theater, and she had been on the European grand tour. She wanted a theatrical house, and Maybeck obliged. On the exterior, the Roos house was half-timbered and decorated with overscale, medieval-looking carving (figure 13.34). Inside, the entry portico delivered a visitor to the threshold of a great hall roofed in massive timbers, with walls replete with carved redwood paneling, a huge fireplace, mock medieval lighting, and furniture. Putting aside the charm and beauty of the house, an examination of its structure shows that Maybeck and his clients were interested in seismic safety. Sheets of the construction drawings were devoted to structural issues. One of the most telling contains calculations by Maybeck's engineer, Herman Kower, of the dead load of the house. The sheet shows a support column for the basement with knee braces securely tied by bolts to wooden lintels in best engineering practice. Detailed drawings illustrate how posts are to be bolted to the foundation (figure 13.35), and even chimney ties are specified. The Roos house illustrates a conjunction of creative architecture and sound engineering in wood construction.

Designed sometime in 1906 or 1907 and completed by 1908, the house built for my grandfather Oscar Tobriner at

Figure 13.34 (below)
Leon Roos house, 3500 Jackson Street, San Francisco, 1909 (side facade elevation on Locust Street). Note the picturesque but utilitarian chimney tie. The sloping basement of this building is a typical San Francisco design feature. The dead loads on the basement columns were calculated by Maybeck's engineer. All the basement columns are well tied to the upper floor and bolted to the foundations.

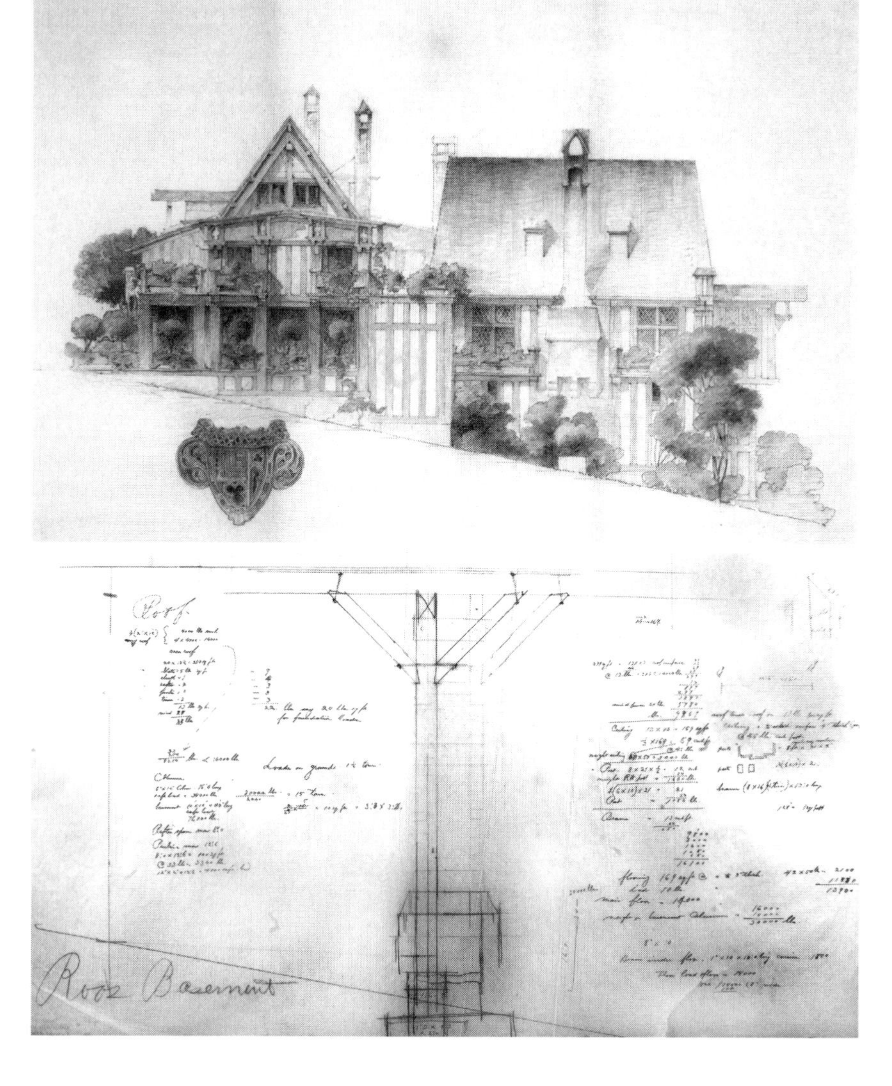

Figure 13.35 (right)
Leon Roos house: foundation detail with bracing

3494 Jackson Street was extremely well planned to resist earthquakes (figures 13.36 and 13.37).[65] The architects were Newsom and Newsom, whose reputation had been made in the creation of large, decorated, Eastlake-style mansions like the Carson House in Eureka. The Tobriner house was built as a three-story shingled cottage with high gabled roofs, but the interior had a Craftsman feel rather than Victorian. Its design was firmly rooted in the Bay Area tradition of shingle houses of the kind built by Ernest Coxhead. The plan was inverted, with the living room at the back, where the view was to the Golden Gate. A massive foundation and a huge, tiered reinforced concrete basement anchored the building on its hillside lot. At least a portion of the house's redwood frame was bolted to the concrete. The basement story on the lowest side of the house was unusually well braced. The fireplace chimney, enclosed in a wood frame, exited at the top of a gable, with very few bricks above the roof level. Floor-to-ceiling diagonal bracing surrounded each aperture, so the back wall looked as thoroughly reinforced as it could be with wood and nails. My grandfather was very proud of the foundation and wall braces and said he had paid so much for the safety of the building that he couldn't complete it without a loan from A. P. Giannini's Bank of Italy (later America). The house, now completely remodeled, still stands.

Figure 13.36 (top)
Basement of Oscar Tobriner house before remodeling, 1982. Like the Roos house, this one is built on a slope. The living room, facing north, is built over a basement more than a story and a half lower than the foundation of the facade of the house. The basement was strengthened to resist lateral loads by (A) diagonal bracing, (B) diagonal siding, (C) diagonal struts, and (D) bridging.

Figure 13.37 (bottom)
Oscar Tobriner house, 3494 Jackson Street, San Francisco, c.1906. The drawing and plan of the original house appeared in the *Overland Monthly* in 1908.

Certainly more examples of similar care in earthquake-resistant design and construction exist in San Francisco, but these few illustrate that some architects, builders, and homeowners were concerned about earthquakes and attempted to make the homes they built as resistant as possible.

The completion of City Hall and the opening of the Panama-Pacific International Exposition in 1915 serve as a fitting end to this cycle in the history of San Francisco. Between 1906 and 1914, 28,507 buildings were constructed in the city. Critics have charged that behind those facades, the new San Francisco was no better built than the old. Certainly there were problems with some city-funded special construction buildings, and some brick buildings were not rigorously constructed either. But how difficult and different such judgments become when we begin to understand the past! Despite the fact that these buildings do not meet today's standards, we can see that architects and engineers of the time made great efforts to build responsibly. San Francisco after 1868 had seen an explosion of interest in earthquake-resistant construction. Sound, seismically resistant buildings like the Montgomery Block, the Palace Hotel, the U.S. Mint, and the Call Building had been built before 1906. Then the earthquake "proved" that tall buildings could be sound, that wood-frame buildings often do fine in earthquakes, that properly tied brick buildings can perform satisfactorily. But just how safe these new buildings really were depended largely on individual owners, architects, engineers, and builders. If they not only followed but exceeded the codes, the newly constructed brick buildings would have some earthquake resistance. If the best practice in reinforced concrete were followed, then the new buildings would probably be strong enough to remain standing in future earthquakes.

The reality of the situation is that individual expertise and personal choice varied greatly and created vast differences in quality. Without complex testing and intrusive structural examinations, the results are hard to judge, but the buildings discussed in this chapter are representative of the care with which many of the thousands of new San Francisco buildings were constructed.

CHAPTER 14

Reality Replaces Myth, 1925–1933

After the fire caused by the 1989 Loma Prieta earthquake had been extinguished, as red and yellow tags fluttered from badly damaged houses and apartments (figure 14.1), a poetic myth began to circulate that the Marina District had been built on rubble dumped there after the 1906 earthquake.[1] That myth grew into a cautionary tale: San Franciscans had built the storybook Panama-Pacific International Exposition of 1915 on this rubble to celebrate the rebuilding of their city. By creating more fill, and building permanent houses on it in the 1920s, they increased their vulnerability to earthquake damage. Charred remains—perhaps debris from 1906?—surfacing after liquefaction in 1989 seemed to confirm the biblical adage that the sins of the fathers are visited on the sons.[2]

It is commonly held that engineers and architects in San Francisco did not construct earthquake-resistant buildings until the 1930s; that they forgot about earthquake danger shortly after 1906, bequeathing to us the fruit of their irresponsibility. As I have tried to show, the situation was far more complex. That complexity can be seen in the 1915 Panama-Pacific International Exposition (PPIE) and continues through the rebuilding of San Francisco before 1933, when a state-mandated code required earthquake mitigation in building design.

Figure 14.1
Apartment building, Beach Street (Marina District), 1989: incipient collapse due to soft-story mechanism on ground floor in the Loma Prieta earthquake

This chapter begins with a brief reconsideration of the PPIE and then addresses the problem of earthquake-resistant design in San Francisco's tall buildings from 1925 to 1933. Like the eyewitnesses who saw the earthquake of 1906, engineers are our eyewitnesses to this history. They need to speak for themselves, which necessitates several long quotations. If you have read this far, you deserve to read for yourself what they wrote.

On Shaky Ground: the Panama-Pacific International Exposition

In 1904, before the Panama Canal was even begun, a San Francisco store owner, R. B. Hale, proposed that the city hold a world's fair to celebrate the canal's completion. Once the idea was presented, other cities vied for the honor of hosting it. In 1911, during the period of post-earthquake reconstruction, Congress chose San Francisco as the site for the fair, which promoters called the Panama-Pacific International Exposition.

The promoters didn't mention earthquakes in their promotional literature, but the engineers working on the site and the architects building temporary structures there did not disregard lessons of the past.[3] For one thing, the decision to build on fill was not as cavalier as we might think. What we now call the Marina District was then called Harbor View, one of the few undeveloped bayside sites of the city. It was eventually chosen as the site of the exposition because it was available and undeveloped. The absentee landlords readily agreed to lease their land to the city for a year. This site lay between two military reservations (the present Fort Mason and the Presidio) whose lands could also be temporarily annexed for the fairgrounds. The PPIE would eventually cover 635 acres, 301 of which were federally owned.[4] One hundred and eighty-four of these acres were under water before the fill. Golden Gate Park had also been proposed as a site for the fair, but the directors were absolutely opposed to its use and Harbor View was chosen by default for the site of the ephemeral exposition.[5]

From the care that was taken in the fill operation, it is clear that San Franciscans had at least taken the dangers of settling to heart, if not those of earthquakes (figure 14.2a, b, and c). In the 1890s, James Fair, who owned water lots along the shoreline of Harbor View, built a seawall around his property with the intention of filling it, but he died before the fill could be placed. Exposition engineers, headed by H. L. Markwart, used the same seawall to hold the fill upon which part of the PPIE (184 acres out of 635) would be built. They stationed two dredges offshore to suck sand from the bottom of the bay and extrude it into a cove behind the seawall through a pipe floating on pontoons. According to Frank

Figure 14.2a
The PPIE site on April 2, 1913, before development, looking north from Pacific Heights. The Marina District is in the background, on the shore of the bay.

Figure 14.2b
The PPIE site in September 1913, after being filled, looking west toward the Golden Gate from Fort Mason.

Figure 14.2c
An anonymous painting of the Panama-Pacific International Exhibition: a bird's-eye view, looking northeast, 1915. The Great White Fleet is shown steaming through the bay

Morton Todd, chronicler of the PPIE, "The engineers were quite finicky about the sort of mud they got."[6] They changed the dredging location until the output suited them and left a gap in the seawall so the heavier sand they were pumping onto the site would squeeze out the "semi-fluid sludge" that made up the natural bottom of the cove.[7]

In the end, the fill—70 percent sand and 30 percent mud—was considered stable. No seismologist of the time suggested to Markwart that he was creating a hazard. Geologists and engineers discussing the hazards of fill in the 1930s barely mention the Marina. The historians Nancy and Roger Olmsted, whose *San Francisco Waterfront: Report on Historical Cultural Resources for the North Shore and Channel Outfalls Consolidation Projects* is the bible for studying San Francisco fills, wrote in 1977 that "the fact that the Marina lands seem solid even today attests to the success of their scheme."[8] Only in the mid-1970s did seismologists begin to understand the danger the Marina posed.

As to the notion of the debris from 1906 emerging in 1989: no report or newspaper notice of the day documents the dumping of debris from the earthquake into the site.[9] No evidence of debris reliably dated to 1906 exists. Geologists studying the site *think* it is possible that debris from 1906 *might* have been dumped on the site, but they offer no proof. If debris exists, it certainly isn't much, since the hydraulic filling was extensive. Rather, documents concur that the major dumping area for the debris from the earthquake was old Mission Bay, south of the present ballpark, a site being intensively developed as I write.[10]

Was there any consciousness of earthquake danger when buildings were constructed for the PPIE? The engineers, who were among the best in the city, all expressed concern about earthquakes. The chief structural engineer of the entire exposition was Henry D. Dewell, whose championing of earthquake-resistant design appears later in this chapter. The structural engineer for Carrere and Hastings' famous 430-foot Tower of Jewels was John D. Galloway. The engineer for Bakewell and Brown's Horticultural Palace was their favorite engineer, Christopher Snyder (the engineer of City Hall). The engineer for Bernard Maybeck's Palace of Fine Arts was M. C. Couchot. The engineer for G. Albert Lansburgh's Motor Transportation Building, Walter L. Huber, wrote seminal papers on earthquakes in the 1920s and 1930s. They all understood the danger of fill in earthquakes and the risk of an earthquake during the one-year run of the exposition. Thousands of piles were driven into the fill to stabilize the buildings. Markwart, writing for *Engineering News,* stated: "It was decided to use piles for the support of all building walls and special loads over the entire site" because they ensured a uniform method of foundation construction, they were more economical or more effective than spread footing, and they afforded *"greater safety in case of earthquake."*[11] Much care was taken in the pile driving, and extensive tests were made to secure proper bearing capacities.

While a few buildings were steel, most were constructed entirely of wood, designed to resist both wind and earthquakes. The Palace of Machinery was purported to be the largest timber structure in the world at the time of its construction.[12] Aside from a small quantity of steel, it was built of Douglas fir imported on schooners from Washington and Oregon. It was a three-aisled building, with each aisle being 101 feet high (figure 14.3). At its center a short transept of three higher (132-foot) aisles intersected the main aisles. The huge timber structure, laterally braced by a series of deep trusses and diagonal trusses, was braced for earthquakes and wind. San Francisco codes stipulated that a wind pressure of 20 pounds per square foot was to be calculated only for buildings higher than 102 feet.[13] It would at first seem that the exhibition's promoters wanted to avoid the cost of extra bracing by making the bulk of the Palace of Machinery one foot shorter than this, but such was not the case. Instead of bracing only buildings 102 feet or taller, the building standards of the exposition used 20 psf as the lateral pressure for all walls up to 150 feet, and 25 psf for anything taller. The words "up to" are critical: "wind" would not affect a three-story building, but earthquakes would (see chapter 5). The inclusion of "wind bracing" for low structures as well as high indicates the same consciousness of earthquakes that appeared in chapter 11.

Figure 14.3
Timber framing of the Palace of Machinery, Panama-Pacific International Exposition, which was claimed to be the largest wooden structure in the world at the time. Extensive X-bracing and huge framing members illustrate a concern for lateral stability in wind and earthquakes.

Markwart, again writing for the *Engineering Record*, recorded no unusual settlement problems. The PPIE buildings, designed to last for just a year, were extremely well built, judging from the blueprints and photographs. When they were demolished and the site was given back to private owners for real estate development, what could not be sold or salvaged was thrown into the site and burned: hence the charred remains that surfaced after the 1989 earthquake.

Until the 1920s, real estate development on the former site of the PPIE, which called for widely separated houses on ample lots, hadn't happened. The detached frame houses planned in the original development, with the right foundations and strong wood construction, probably would have been safe in earthquakes. However, in the mid-1920s and through the 1930s, the streets were replatted by developers into narrow lots upon which small contiguous houses were built. Four-story apartment houses went up at the corners.[14]

In the Loma Prieta earthquake, many of these apartment buildings were badly damaged.[15] Because their ground-floor garages required open space for cars to turn around and park in, they lacked braced, interior walls. The walls between the garage doors were constructed without sufficient shear to resist the intense shaking and lateral forces experienced on the filled soil. The ground floors collapsed because they were less stiff (more flexible than) the stories above. A change in stiffness between one story and another constitutes a discontinuity in the building, in this case called a "soft story." The cumulative loads of a building are concentrated at the base, so if the ground floor is not strong enough, it collapses, as we saw in the "cripple" walls of frame buildings in 1906.

The soft story in these apartment buildings brings up the question of how to treat ground floors in wood and steel. The builders in the Marina may not have been aware of this issue. The term "soft story" only became widely used in the 1970s, after the catastrophic failures in the 1971 San Fernando earthquake made the problem obvious. It has become a red flag for danger in many earthquakes since, like the devastation of hundreds of apartment buildings in the Turkish Kocaeli earthquake of 1999.

When the wood-frame apartment buildings of the Marina District were constructed, diagonal bracing was being "cut in" to wood frames, a practice recommended in the 1920s and incorporated into California codes in 1933. It was in the 1920s that Arthur C. Alvarez, a professor of engineering at the University of California, proposed the insertion of continuous diagonals across studs as a means of bracing wood-frame buildings, providing strength and stiffness.[16] The use of diagonal bracing is demonstrated in the rare example of a framing plan for a modest house in the Berkeley hills designed by Arthur Dudman in 1936 (figure 14.4).[17] Apparently invented by the architect himself, and not described in the codes of his day, the framing plan provides extra stops of wood nailed to ("sistered" to) studs and notched diagonal pieces which cannot easily be dislodged. Experts in earthquake-resistant wood construction would not necessarily approve of this method today; the industry standard for earthquake-resistant bracing in wood includes plywood sheets nailed to mudsills, studs, and top plates, and discourages the use of "cut in" diagonal bracing.

From a post–Loma Prieta vantage point, the seismic hazards created at each step of the Marina's development seem obvious, but they need not be seen as the result of denial, corruption, or incompetence. Instead, each step needs to be seen in context. Remember that earthquakes were still very much a part of public consciousness in 1915, when the PPIE opened. Christopher Snyder's earthquake-resistant City Hall was just being finished. The earthquake-resistant Auxiliary Water Supply System was being completed as well, at enormous cost to the citizens of San Francisco. Seen in light of these projects in other parts of the city, the care taken with the fill for the PPIE and the earthquake-resistant construction of its temporary buildings clearly demonstrate concern for seismic safety. The Marina represents a series of decisions, each one understandable in its own context, that eventually led to a hazardous situation. We can lament this today, but the people of that time, given the weight of their other concerns and their lack of knowledge—particularly in the design of wooden bracing—may not have been able to prevent the danger.

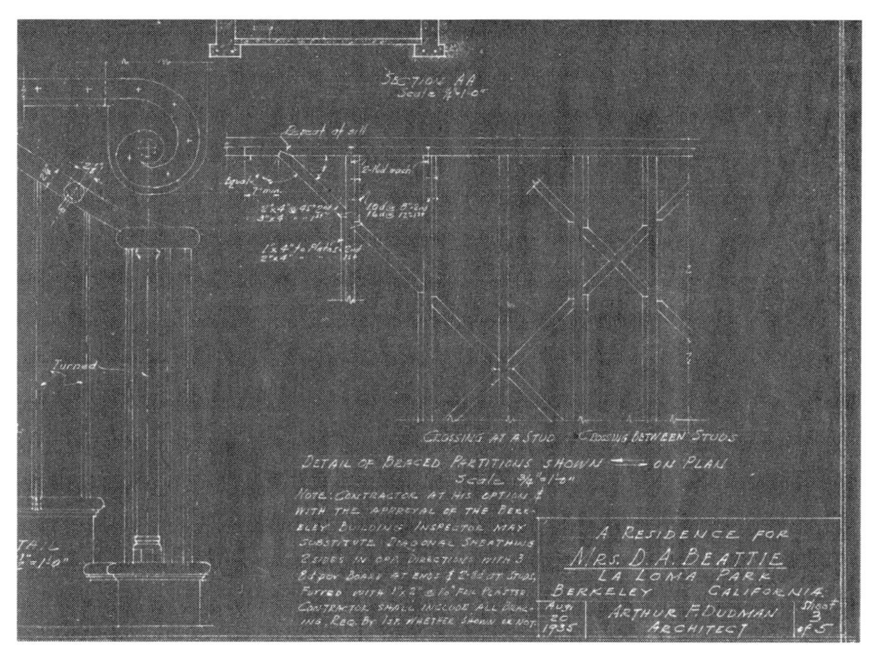

Figure 14.4
Detail of plan for residence for Mrs. D. A. Beattie, La Loma Park, Berkeley, by Arthur F. Dudman, architect, 1935. This detail illustrates bracing for partitions that Dudman apparently invented.

Lateral Forces and the Myth of Denial

In 1925 an earthquake caused substantial damage to what was then the small resort of Santa Barbara, about ninety miles northwest of Los Angeles. The complexity of earthquake mitigation is reflected in discussions after the Santa Barbara earthquake about the wind load strength to which buildings should be built in order to withstand the lateral forces of earthquakes. Henry Dewell, one of the engineers who had worked on the PPIE, was concerned by the extent of the damage in Santa Barbara. He attempted to raise the earthquake consciousness of professionals and lay people:

> The office building or similar structure designed in accordance with today's San Francisco building ordinance is very much weaker than buildings designed immediately after 1906. With respect to safe design against earthquakes, we have not only not progressed, but we have actually seriously regressed.[18]

To make his point, Dewell used quantitative information, starting with the 30 psf wind load requirement in the 1906 code:

> Today the prescribed wind pressure is 15 pounds [psf], a reduction of 50 percent, and the unit stresses for steel increased. Buildings now erected in accordance with the building law are far lighter than those that safely withstood the 1906 shock. San Francisco is not building for safety against earthquakes, and the same statement can be applied to the East Bay cities.[19]

Taking Dewell's comments at face value and looking at the descending wind load requirements in the code, 30 psf in 1906, 20 psf in 1909, 20 psf in 1915, and 15 psf in 1923, we might conclude that engineers were ignoring earthquake danger. But that conclusion would be wrong. We know that some engineers, like Henry Brunnier, Christopher Snyder, and William Day, as well as architects, like George Kelham and Charles Weeks, had continued to struggle with problems of earthquake safety and the quality of construction. After the Santa Barbara quake, Brunnier, who had attempted to build earthquake-resistant buildings throughout his career, joined John G. Little and T. Ronnenberg in urging his colleagues to resharpen their awareness:

> For a number of years after the destruction of San Francisco in 1906, the phrase "Lest we forget" was the universal slogan and warning. But as time rolled on, the people have forgotten and, through ignorance or selfishness, poor construction is creeping in. Let us revive the phrase "Lest we forget" and insist that all buildings be properly designed and carefully inspected and constructed with good materials, honest and careful construction.[20]

A 1925 editorial in the *Engineering News-Record* mused about the post-earthquake wind load level, acknowledging that 30 psf was unrealistically high and poorly enforced, and that some conscientious engineers nonetheless chose to build according to their own seismic standards:

> Perhaps the 30-pound wind load was too high to permit of consistent enforcement. While it was in effect, its requirements were frequently evaded. However, it was unwise to cut the wind load down to 15 pounds with no compensating provision for stiffness or rigidity to lessen earthquake damage. This reduction was a victory for those whose dominating motives were cost saving and competition with other cities for business. But the change has brought a reaction and some builders have been voluntarily using more rigid specifications.[21]

The three most influential engineers in San Francisco in the 1920s and 1930s were Christopher Snyder, L. H. Nishkian, and Henry Brunnier. Snyder and Nishkian attended the Structural Association meetings after the 1906 earthquake and Brunnier saw the ruins on his arrival. None forgot what they had seen, none slavishly followed the code, and none designed buildings for a 30 psf wind load in the 1920s or 1930s. Like Richard Chew and Charles Derleth Jr., they understood that a building had to be both strong and flexible to withstand an earthquake. They tried to find the right balance. Their intuition was correct, but present-day engineers might worry that their buildings could be too flexible, particularly on the ground floor. On the other hand, excellent engineers like Will Day and Felix Spitzer, who built more rigidly, might have designed buildings that were not flexible enough, particularly on rock foundations. The point is that these engineers were doing their best to understand an extremely difficult problem, one which we are still struggling to understand today.

Christopher Snyder, obviously infuriated by the editorial in the *Engineering News-Record*, shot off this reply:

> The editorial "Earthquakes and Building Codes"...seems to me to suggest, by implication at least, that San Francisco designing engineers are unmindful of the earthquake hazard. I believe, on the contrary, that reputable engineers are using as much care in attempting to minimize earthquake damage as they have ever used during the last 25 years. It is true, of course, that we are not building earthquake-proof buildings any more than we are making them fireproof. It would not be practical to do so. Our designs, however, should minimize the hazard, and, insofar as may be, prevent the loss of life or limb and simplify the restoration of damaged structures to their original condition.[22]

According to Snyder, a wind load of 30 psf was unnecessarily high for earthquake-resistant design, an opinion that present-day engineers would share:

> I do not believe in the use of a 30-pound wind load in a country where the actual wind load is only 15 pounds. The principal result is to increase unduly the cost of the building. If the steel frame is honestly designed for the stresses of a 15-pound wind and is otherwise in accord with the building law, the actual resistance is much more, in most cases, due to partition and wall resistance. [Unless diagonal bracing is used] the resistance of the steel frame will not be in evidence until the walls and partitions have sheared horizontally. The ordinary Class A business building is of a type well suited to withstand earthquakes with minimum damage. *My point of view is...not new. I advocated it after observation of the results of the 1906 earthquake, and I was further confirmed in this view by the results of the Santa Barbara earthquake.*[23] [emphasis added]

Snyder's explanation of his idea of earthquake-resistant construction is worth quoting at length:

The ordinary steel-frame office building with stores on the first floor is well adapted to efficient bracing to secure the safety of the occupants and the minimum earthquake damage. From the second floor up, the structure is of relatively low story heights and can be efficiently braced—by diagonal bracing when practical—to make a very rigid box of great weight and, in consequence, great inertia.

The first story is about twice as high as the upper stories and has practically no spandrel walls, being reduced to skeleton piers with glass between. When braced with knees, we have a flexible support, allowing the ground and the first floor to move relatively to the upper portion

Figure 14.5

New, tall buildings in San Francisco as illustrated by John Freeman, in his 1932 *Earthquake Damage and Earthquake Insurance.* The Russ Building (by Kelham with Brunnier as engineer, 1927), prominently labeled, has an extensively braced frame. Also shown:

1. Federal Reserve Bank (Kelham with Brunnier as engineer, 1924)
2. Crown Zellerbach Building (Hyman and Appleton, c.1930)
3. Financial Center Building (Meyer and Johnson, 1927)
5. California Commercial Union (Kelham and MacDonald with Brunnier as engineer, 1923)
6. Matson Building (Bliss and Faville, 1921)
7. Pacific Gas and Electric Building (Bakewell and Brown, 1925, designed for a wind pressure of 15 psf and said to be designed for earthquake resistance)
8. Shell Building (Kelham with Brunnier as engineer, 1929)
9. Standard Oil Building (Kelham with Brunnier as engineer, 1922, said to be braced for earthquakes, designed for 20 psf wind pressure)
10. Hunter-Dulin Building (Schultze and Weaver, with Brunnier as engineer, 1926, said to be braced for earthquakes)
11. Crocker First National Bank (Willis Polk, with Ronnenberg as engineer, 1908, said to be designed to resist earthquakes)

by bending the first-story columns, and so transmitting but little of the ground vibration to the rigid mass above.

Rear walls in the first story should be made of a non-elastic and weak material, such as tile, so as to permit freedom of elastic movement. An earthquake would leave a building of this type entirely or practically uninjured about the second floor; and the walls, partitions, and piers in the first story would be damaged in proportion to the severity of the quake. The occupant would be inconvenienced but little, and the property damage would be largely confined to the first story, where it is most easily repaired.[24]

Figure 14.6
Standard Oil Building, San Francisco. Significant knee bracing on the tower illustrates concern for drift. Freeman's comment about Brunnier's Hunter-Dulin Building describes the Standard Oil Building as well: "This is a steel-frame building braced rigidly against earthquake ('wind pressure') by heavy gusset plates and girders beneath the windows."

L. H. Nishkian also championed the idea of a flexible ground floor because, as he said, "earthquake forces differ from wind forces." The force of an earthquake comes from the foundations of the structure, unlike wind, which blows on the uppermost stories. Stiffness throughout is not the answer, because earthquakes can be stronger than the buildings they shake. Instead, the steel frame should be strong and it should bend, dissipating some of the force of an earthquake. Discussing his Bank of Italy, in San Jose, California, Nishkian says, "The frame and connections were designed to resist the usual 15 pounds per square foot wind pressure." The first story is unusually high, for flexibility in columns:

> Walls [of brick] are comparatively so much stiffer than the structural frame of the building...that the frame will get practically no stress until the walls have failed. All joints should be made strong and flexible, the first story wall columns should be disconnected from the walls in such a manner as to permit a horizontal movement of these columns of about two inches... The building must be thoroughly tied to act as a unit.[25]

Although all of Henry Brunnier's buildings in San Francisco are heavily braced, he didn't adhere to the 30 psf wind load figure either. During the 1920s and 1930s, architect George Kelham designed some of the tallest buildings in San Francisco (figure 14.5). When Brunnier began to design the steelwork for Kelham's huge Standard Oil Company Building, erected at 225 Bush Street in 1922 (figure 14.6), he incorporated spandrel girders and gusset plates in the steel frame of the building. Brunnier first designed the steelwork to resist the code-prescribed wind pressure of 15 psf—but he recalculated the resistance for the building and redesigned his steelwork to be more robust, resisting 20 psf, because he wanted to make sure the building could withstand the wind and earthquake forces he considered possible. Brunnier did not adhere to the idea of a flexible first story.[26]

Engineer Felix Spitzer also seems to have designed his buildings to be strong and stiff but also able to flex. An article on an apartment building he designed on Russian Hill describes the steel structure as being earthquake-proof. The fourteen-story building, constructed by architect-builders Bos and Quandt, had a steel frame designed for a lateral load "far in excess of requirements of the building ordinances" (figure 14.7). The footings for the building were on bedrock, the safest material in the city. The lower stories were stiffened by diagonal bracing, the upper stories by knee braces. All interior connections of the columns to the floor beams were made of heavy top and bottom angles, able to produce "considerable bending movement in the floor beams."[27]

The Brocklebank apartment building, on the northeast corner of Sacramento and Mason Streets across from the Fairmont Hotel, exemplifies socially responsible seismic design for its day (figure 14.8). The apartments were commissioned by Mrs. M.V. Brocklebank MacAdam, who, as she wrote in her memoir, wanted to build an apartment house "which would be a credit to San Francisco and to myself."[28] Mrs. MacAdam wanted to live in the building and create the perfect residence for other well-heeled San Franciscans who also wanted to live on Nob Hill. It was during the Roaring Twenties, and she sunk her entire fortune into the enterprise.

For architect and engineer, she chose the firm of Charles Weeks and Will Day, who in the same year were building the Mark Hopkins Hotel on the site of the old Mark Hopkins mansion we visited in chapter 6. The steel frame for the Mark Hopkins Hotel was strongly braced, as was their steel frame for the Sir Francis Drake Hotel, built in 1928: John Freeman, earlier mentioned in chapter 8, specifically stated that the Sir Francis Drake Hotel had "diagonal bracing for resisting earthquake stress" (figure 14.9).[29] The earthquake-resistant concept of the building Weeks and Day designed for Mrs. MacAdam is based on principles outlined in a University of California report on the Santa Barbara earthquake published in 1925.[30] It is built on rock, the "primary requirement for successful resistance to earthshocks." The structure is of reinforced concrete, not steel. One by one, Weeks and Day addressed the questions that arise when reinforced concrete is used in tall buildings.[31] The walls are poured reinforced concrete, cast integrally, well tied frame, and heavily reinforced with steel. They are substantial, at eight inches thick; codes stipulated only six inches. Continuous shear walls of solid concrete run down through the building, rather than plaster or tile walls, which would have been permitted by the code. The mixing, ingredients, and placement of the reinforced concrete were supervised by hired specialists. In addition, the ratio of reinforced concrete wall to aperture on the exterior of the building is high, giving the building "lateral stiffness." Today the windows look unusually small, perhaps because of this seismic precaution.

Figure 14.8 (left)
Brocklebank apartment house, Nob Hill, 1926. Charles Weeks, architect; Will Day, engineer

Figure 14.9 (below)
Sir Francis Drake Hotel (Charles Weeks, architect; Will Day, engineer). In this plate dated May 10, 1928, from Freeman's *Earthquake Damage and Earthquake Insurance,* the diagonal bracing is clearly visible. The caption reads, "diagonal bracing for resisting earthquake stress."

re 14.7 (above)
eter Gardens, Green and Taylor Streets, Russian Hill, 28. Bos and Quandt, builders; Felix Spitzer, engineer

One would hope that Mrs. MacAdam was rewarded for her care for seismic safety and excellence of design and construction. Unfortunately, during the Depression, the loan on the building was called and she had to vacate. "Added tears are futile," she wrote. "So, with outward calm, I passed through the door which closed upon my little world wherein I had lived in supposed security."[32]

As the engineers' comments on the Brocklebank building prove, by the late 1920s a body of knowledge about the devastating effects of earthquakes and their mitigation was becoming available. The terrible Kanto earthquake and fire of 1923, which leveled Tokyo and killed more than 143,000 people, was a shock to architects and engineers worldwide. Japanese response to the disaster was a lesson for Americans in earthquake mitigation and engineering science. After studies of the damage were completed, the Japanese government adopted a comprehensive seismic code. *Earthquake-Resisting Construction,* by the pioneering Japanese engineer

Tachu Naito, was widely disseminated in California. Naito differentiated between wind and earthquake forces, illustrating his conclusions both mathematically and graphically.[33] Architects were also alerted to earthquake-resistant design by one of their own, Frank Lloyd Wright, whose Imperial Hotel in Tokyo survived the Kanto earthquake of 1923.[34]

As we have seen, the Santa Barbara earthquake of 1925 also played a part in reeducating San Francisco architects and engineers. Building failures had killed thirteen people in the small resort town. Facades and cornices of brick buildings had cascaded into the streets, wrecking the main business district. Santa Barbara was close enough for interested Bay Area architects and engineers to visit, and it hadn't burned: the ruined buildings were there to be seen for weeks after the event.[35] Walter Steilberg, the engineer who worked for Julia Morgan, took hundreds of photographs of the damage. He started to assemble his own book on earthquakes and in the 1930s to patent an earthquake-resistant construction system.[36] Articles by engineers and architects flooded professional journals.

Because of pressure from seismologists, engineers, architects, and concerned local people, two municipal seismic codes became law in 1926, one in Santa Barbara and one in Palo Alto.[37] A nonmandatory seismic section was added to the first edition of the Uniform Building Code, published by the Pacific Coast Building Officials in 1927.[38] Recognizing the higher intensity of ground shaking on filled land, this code had two different lateral load requirements: one for buildings on harder sites, like rock, and one for more malleable soils, like fill. It suggested that buildings constructed on deep, soft soils, like the fill of the Marina District, be designed to resist a lateral force at each floor and roof level equal to 10 percent of the structure's weight at that level. Structures built on firm soils or rock were to be designed for one-third of this force.

After visiting San Francisco, where the 1927 seismic standards were not mandatory, John Freeman wrote:

> Nowhere in the world in this year of grace, 1931, can there be found structural engineers more alert about earthquake hazards, or with better knowledge of how to build structures that will resist them, than some few of those in California, who, in addition to their researches at home, have become well informed of the recent progress in Japan...[Their steel-frame buildings were] braced rigidly against "wind strains" (also against earthquake stress) by horizontal lattice girders as deep as window openings would permit, united to verticals by the deepest practical gusset plates, and the whole, including the floor system, embedded in hard-rock concrete, and curtain walls made integral; also having ornamental work of such a kind and so well-secured that probably the most severe quake would shake off little or nothing upon the heads of people on the sidewalk.[39]

But Freeman pointed out the wide disparity in the buildings he saw:

> Nevertheless, while making a study of local conditions in 1926, the writer found examples within a few hundred yards of each other where on one hand every reasonable precaution appeared to have been taken, while on the other hand there had apparently been simple compliance with the existing easy-going building laws.[40]

Engineers had become ever more responsible for safety in buildings, but whereas architecture was a profession legally recognized and regulated by the state of California, engineering was not. It was the 1925 Santa Barbara earthquake, with its structural failures and the expert reports on them by civil engineers, that brought the profession into prominence and helped set the stage for attaining legal status. With another great disaster, the 1928 collapse of the St. Francis Dam, north of Los Angeles, it became clear that standards for practice and competence in engineering were of paramount importance to the state and the welfare of its citizens. In 1929 California established a State Board of Registration composed of three civil engineers, one of whom had to have a background in structural engineering. With the formation in 1932 of the Structural Engineering Society of California and governmental recognition of their professional status, engineers were positioned to be more influential in framing codes and establishing standards for earthquake-resistant design.[41]

The Long Beach Earthquake of 1933

On March 10, 1933, at 5:45 p.m., a moderate (magnitude 6.3) earthquake struck Long Beach, California, causing $40 million in property damage and killing 115 people.[42] As in the Santa Barbara earthquake eight years earlier, brick buildings performed poorly. The most striking damage was to brick school buildings. Had students been at school during the earthquake, there would have been hundreds of deaths. A coalition of seismologists, engineers, architects, concerned scientists, and citizens' groups lobbied for new state laws to guarantee that new buildings constructed in California would be safe in earthquakes. The time was right. These were the years of the Depression, and people were responding to hard times with public action; the public was aroused and state government was receptive. A 1934 article from the *Engineering News-Record* recaps the results:

> Earthquake-resistant design, considered "too costly" before the Long Beach disaster, is now required by state law; throughout the state, old school buildings are being examined and many condemned; instead of being dimmed by the fog of public apathy that usually closes down again a few months after a disaster, this time the vision has remained clear; and now, more than a year after the earthquake, progress is under way in a dozen lines looking to safer construction.[43]

A sweeping new state law prohibited any construction of brick or concrete block unreinforced masonry buildings. Two more laws established mandatory earthquake-resistant design in California. The Riley Act required all buildings in the state to have a lateral strength equal to 3 percent of their weight. The Field Act established the Office of the State Architect, which was responsible for regulating school construction. This office developed rigorous standards of construction and plan review. It introduced plan review by peers, including structural engineers, which proved crucial as a basic concept in earthquake engineering.

The issue of wind bracing was made moot by the introduction of a new calculation of base shear, which was much more appropriate than lateral wind strengths for determining the effect of dynamic earthquake forces on buildings. Base shear calculations were mandated when California adopted a comprehensive statewide earthquake code in 1933.

Knowledgeable engineers, architects, and lay people no doubt applauded the new provisions. By bringing the discussion of earthquakes into the public forum and advocating structural research, engineers and seismologists opened the door to ongoing studies and observations that would rapidly improve our understanding of earthquakes and seismically resistant design.

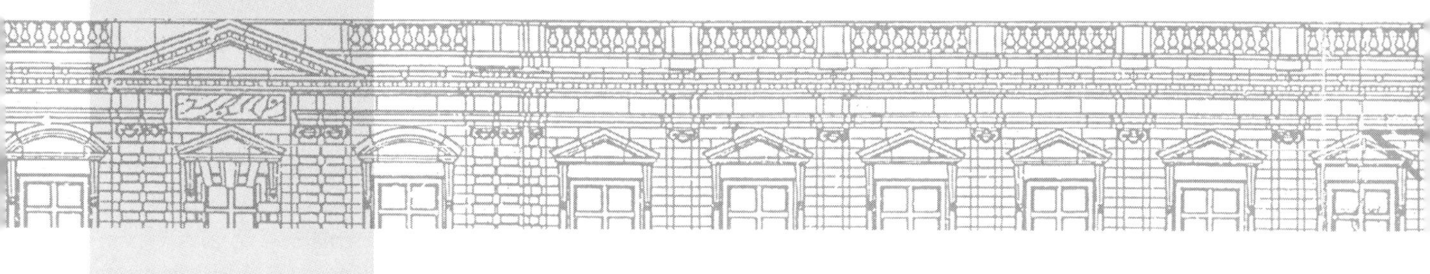

Conclusion

In the end, the history of seismic disaster and reconstruction is the story of the built environment. Not what people say after earthquakes, but what they do; what they have learned, and how they decide to use it to construct safer buildings and cities.

The history of earthquakes and earthquake response is down and dirty. It is not abstract, but physical. The architects and engineers in this book were dealing with objects in the real world: objects with texture, weight, size, and shape. Because they were working before a theory of dynamic response was conceptualized, their attempts to make seismically resistant structures were empirically based. The late Henry J. Degenkolb and the other engineers who rushed around the world to see earthquake damage firsthand understood that engineers had to see the damage for themselves, see the rawness of failure in order to fix it in the future, even with theoretical models available. It is in this same way that we must try to see yesterday's fixes. In this book I have asked how the construction professionals actually built. I have tried to ascertain how they were thinking of earthquake-resistant construction and safety. I have tried to show their intention to construct earthquake-resistant buildings. Their results were tangible, not illusory or theoretical. With knowledge and a modest amount of imagination, it is still possible to see and understand the results of their efforts.

In recounting this story of earthquakes, architecture, and engineering in San Francisco between 1838 and 1933, I hope I have been able to give a glimpse not just of the buildings but of the people who reconstructed their city. These people are not only the Ruefs and the Phelans. They are us: the homeowner making his home safe for his family; the investor building apartments that are safe for her tenants; the businessman who pays extra for earthquake safety; the inventor who deals with innumerable vexing details; the research scientist trying to understand earthquakes; the government official who implements new earthquake policies; the engineer, concerned about earthquake safety, who does his best to ensure it; the architect who realizes that safety is as important as architectural expression; the contractor and subcontractors who understand earthquake safety and build for it.

Disaster is, thankfully, ephemeral. We try to erase it as fast as we can. Even after a small earthquake, the rip in our everyday lives is quickly mended. Books are reshelved, the shards of broken dishes eliminated. "There, now everything is the way it was!" The challenge is to acknowledge the confusion and to realize that the actors in reconstruction after earthquakes are many. The buildings of San Francisco prove that hundreds, perhaps thousands of people, did their best to build safely after the earthquakes of 1865, 1868, and 1906. From the ranks of the engineers alone, a list of previously unknown "heroes" can now be drawn. It is crucial to remember that they did not understand earthquake mitigation as well as we do today, but their intention was to create a safer earthquake-resistant environment. They saw a problem and tried to solve it. They were everyday people whose commitment to their professions, to what was right as they saw it, dictated their choices. We cannot know the entire story of their lives and their choices, but from what they wrote and how they built, we can see what they believed.

About twenty years ago, Frank McClure, a local engineer, told me about a picnic with his family above the University of California at Berkeley. He had seen a building under construction that looked somehow wrong. He went over, examined the building, and found serious errors in construction. He wrote a report to the university about what he had seen, and the errors were corrected. He took his social responsibility as a professional engineer seriously, and he had acted upon it. Many of the engineers, architects, and builders in San Francisco between 1838 and 1933 did the same. The history of earthquakes and reconstruction in San Francisco shows us that every one of us has agency: the power to make a difference in earthquake safety for our families, neighborhoods, and cities.

I have tried to demystify engineering, to uncover the structures of San Francisco buildings, to give engineers and architects their lost histories, and to provide lay people with the knowledge to better ask questions about seismic safety. Many of the engineers and architects of the past did their best to save San Francisco. Will we have the courage and foresight to continue their legacy?

Notes

Introduction

1. Robert S. Yeats, Living with Earthquakes in California: A Survivor's Guide (Corvallis: Oregon State University Press, 2001), 6; Philip L. Fradkin, *The Great Earthquake and Firestorm of 1906: How San Francisco Nearly Destroyed Itself* (Berkeley and Los Angeles: University of California Press, 2005), 239, 242
2. Andrew C. Lawson, et al., *The California Earthquake of April 18, 1906, Report of the State Earthquake Investigation Commission,* vol. 1(Washington, D.C.: Carnegie Institution of Washington, 1908), 434
3. Michele L. Aldrich, Bruce A. Bolt, Alan E. Leviton, Peter U. Rodda, "The 'Report' of the 1868 Hayward Earthquake," *Bulletin of the Seismological Society of America* 76(1), 71-76

Chapter 1

1. "Loma Prieta Earthquake Reconnaissance Report," *Earthquake Spectra,* supplement to vol. 6, May 1990; *Competing against Time, Report to Governor George Deukmejian from the Governor's Board of Inquiry on the 1989 Loma Prieta Earthquake,* State of California, Office of Planning and Research, 1990; Proceedings, "Putting the Pieces Together," National Conference of the Bay Area Regional Earthquake Preparedness Project at the Hyatt Regency San Francisco Airport, October 15–17, 1990
2. Pollio Vitruvius, *Ten Books on Architecture,* trans. Morris Hicky Morgan (Cambridge, Mass.: Harvard University Press, 1914), book 1, chapter 4, 17. Even in his own time, Vitruvius's good advice wasn't taken. Rome itself was founded on an unhealthy site, and nearby were the malarial Pontine marshes.
3. Stephen Tobriner, "Safety and Reconstruction after the Sicilian Earthquake of 1693, the 18th-Century Context," in *Dreadful Visitations, Confronting Natural Catastrophe in the Age of Enlightenment,* Alessa Johns, ed. (New York: Routledge, 1999), 49-50, 52-54
4. Tobriner, *Genesis of Noto, An 18th Century Sicilian City* (Berkeley and London: University of California Press, and Zwemmer Ltd, 1982), 28-31, 40-42, 54-65
5. Tobriner, "A gaiola pombalina: o sistema de construção anti-sismico mais avançado do século XIII," *Monumentos* 21 (Sept. 2004), 164
6. Ibid., 161
7. Tobriner, "Earthquake Damage to Colonial Buildings Reveals Their History," *Newsletter, Center for Latin American Studies* (Winter 2000), 14-15; Roberto Meli, *Ingeriería Estructural de los Edificios Históricos* (Mexico, D.F.: Fundación ICA, 1998), in particular 64
8. Meli, *Ingeniería Estructural,* 60
9. Fray Francisco Palóu, *Historical Memoirs of New California,* ed. Herbert E. Bolton (Berkeley: University of California Press, 1926), 2:129
10. Robert Wallace, ed., *The San Andreas Fault System,* United States Geological Survey (USGS) Professional Paper 1515 (Washington, D.C.: Government Printing Office, 1990); Michael Collier, *A Land in Motion; California's San Andreas Fault* (Berkeley and Los Angeles: Golden Gate Parks Association and University of California Press, 1999)
11. H. E. LeGrand, *Drifting Continents and Shifting Theories* (Cambridge: Cambridge University Press, 1994); Eldridge M. Moores, ed., *Shaping the Earth: Readings from Scientific American Magazine* (New York: W. H. Freeman, 1990)
12. Eric Sandweiss, "Claiming the Urban Landscape: The Improbable Rise of an Inevitable City" in David Harris, *Eadweard Muybridge and the Photographic Panorama of San Francisco, 1850–1880* (Montreal and Cambridge, Mass.: Centre Canadien d'Architecture and MIT Press, 1993), 117-118
13. Zoeth S. Eldredge, *The Beginnings of San Francisco* (San Francisco: Z. S. Eldredge, 1912) 2:751; "Extracts from the Journal of Padre Fray Pedro Font," MS, Bancroft Library, University of California, Berkeley, n.d., n.p.
14. Mariano Guadalupe Vallejo, "Historical and Personal Memoirs Relating to Alta California," trans. Earl R. Hewitt, MS, Bancroft Library, University of California, Berkeley, 19, 238-245: John B. McGloin, *San Francisco: The Story of a City* (San Rafael, Calif.: Presidio Press, 1978), 12-20; and "Presidio of San Francisco, National Historic Landmark District," Historic American Buildings Survey Report, U.S. Department of Interior, National Park Service, Western Region, Division of

National Register Programs, and U.S. Department of Defense, Department of the Army, 1985

15. Erwin Thompson, *Defender of the Gate: The Presidio of San Francisco* (Denver: U.S. Department of Interior, National Park Service, Denver Service Center, 1997), 1:15-16

16. Roger W. Lotchin, *San Francisco 1846–1856: From Hamlet to City* (New York: Oxford University Press, 1974), 7

17. Doris Muscatine, *Old San Francisco: The Biography of a City* (New York: Putnam, 1975), 236-238; Charles Lockwood, *Suddenly San Francisco: The Early Years of an Instant City* (San Francisco: The San Francisco Examiner Division of the Hearst Corporation, 1978), 25; Lotchin, *San Francisco 1846–1856,* 181-183

18. Lotchin, *San Francisco 1846–1856,* 181-183

19. Peter J. McKeon, *Fire Prevention: A Treatise and Textbook on Making Life and Property Safe Against Fires* (New York: Chief Publishing Co., 1912), 1; Edward Atkinson, *The Prevention of Loss by Fire* (Boston: Damrell and Upham, 1900)

20. Herbert E. Bolton, ed., *Anza's California Expeditions* (Berkeley and Los Angeles: University of California Press, 1930), 3:256

21. For one view of the conquest of California, see David Lavander, *California, Land of New Beginnings* (Lincoln and London: University of Nebraska Press, 1972), 136-146.

22. Peter Browning, ed., *San Francisco/Yerba Buena: From the Beginning to the Gold Rush, 1769–1849* (Lafayette, Calif.: Great West Books, 1998), 150-152

23. *San Francisco/Yerba Buena,* 103-109

24. Richard Kagan, *Urban Images of the Hispanic World, 1492–1793* (New Haven: Yale University Press, 2000); D. P. Crouch, D. J. Garr, and A. I. Mundigo, *Spanish City Planning in North America* (Cambridge, Mass.: MIT Press, 1982)

25. Florence Lipsky, *San Francisco La grille sur les collines/The Grid Meets the Hills* (Marseille: Editions Parenthése, 1999), 47-54; Anne Vernez Moudon, *Built for Change, Neighborhood Architecture in San Francisco* (Cambridge, Mass.: MIT Press, 1986), 27-34. For another negative view of O'Farrell's plan see Lotchin, *San Francisco 1846–1856,* 64-65.

26. A. A. Brown et al., *Subsidence and the Foundation Problem in San Francisco* (San Francisco: American Society of Civil Engineers, September 1932), 24-25; James Beach Alexander and James Lee Heig, *San Francisco: Building the Dream City* (San Francisco: Scottwall Associates, 2002), 75; Nancy Elizabeth Stoltz, "Disaster and Displacement: The Effects of the 1906 Earthquake and Fire on Land Use Patterns in San Francisco's South of Market District," master's thesis, University of California, Berkeley (1983), 17; William Sharpsteen, "Vanished Waters of Southeastern San Francisco: Notes on Mission Bay and the Marshes and Creeks of the Potreros and the Bernal Rancho," *California Historical Society Quarterly* 21(2) (1942), 113

27. For an excellent analysis of the nature of San Francisco's growth see Gunther Barth, *Instant Cities: Urbanization and the Rise of San Francisco and Denver* (New York: Oxford University Press, 1975) and James E. Vance Jr., *Geography and Urban Evolution in the San Francisco Bay Area* (Berkeley: Institute of Governmental Studies, University of California, 1964)

28. Lockwood, *Suddenly San Francisco,* 43

29. Bruno Fritzsche, "San Francisco, 1846–1848: The Coming of the Land Speculator," *California Historical Quarterly* 51 (Spring 1972), 17-34; and Nancy Olmsted, Roger Olmsted, and Allen Pastron, *San Francisco Waterfront: Report on Historical Cultural Resources for the North Shore and Channel Outfalls Consolidation Projects* (San Francisco: San Francisco Wastewater Management Program, December 1977), 279-300

30. William T. Sherman, *Memoirs of General W. T. Sherman* (New York: C. L. Webster & Co., 1894), 67

31. Hubert H. Bancroft, *History of California* (San Francisco: A. L. Bancroft and Co., 1884), 6:198

32. For a description of the rubbish see Lotchin, *San Francisco 1846–1856,* 15, 187-188; Stolz, "Disaster and Displacement," 13-14

33. Stoltz, "Disaster and Displacement," 12; Sharpsteen, "Vanished Waters," 113

34. Lotchin, *San Francisco 1846–1856,* 15; Stolz, "Disaster and Displacement," 17

35. Alexander, *San Francisco,* 76-77

36. "The Lake Region Filled In," *San Francisco Real Estate Circular* (February 1873)

37. Lipsky, *The Grid Meets the Hills,* 55-65

38. *Alta California,* January 8, 1863

39. Richard Dillon, *North Beach, the Italian Heart of San Francisco* (Novato, Calif.: Presidio Press, 1985), 41

40. Albert Shumate, *Rincon Hill and South Park, San Francisco's Early Fashionable Neighborhood* (Sausalito, Calif.: Windgate Press, 1988), 42-45

41. *Alta California,* February 2, 1869

42. According to Tousson R. Toppozada and G. Borchardt, "Re-evaluation of the 1836 Hayward and the 1838 San Andreas Fault Earthquakes," *Bulletin of the Seismological Society of America,* 1998), 88, 140-159, the earthquake that supposedly occurred on June 10, 1836, in the Bay Area was actually centered south and was not felt in San Francisco. The June 1838 earthquake was extremely strong and damaged the few houses existent in Yerba Buena. Also see Tousson R. Toppozada, C. R. Reale and D. L. Park, "Preparation of Isoseismal Maps and Summaries of Reported Effects for Pre-1900 California Earthquakes," *Annual Technical Report Fiscal Year 1979–1980* (Sacramento: California Division of Mines and Geology (CDMG), 1980, Open File Report (OFR) 80-15 SAC, 38, and 81-11 SAC, 1980; Don Tocher, "Seismic History of the San Francisco Region," in *San Francisco Earthquakes of March 1957* (San Francisco: California Division of Mines, 1959), Special Report 57, 42. The earthquake of 1838 establishes the beginning of this volume.

43. "Earthquake," *Alta California,* May 16, 1851

44. Ibid.

45. "Singular Freak of Nature," *Daily Alta California,* Nov. 27, 1852

46. Damage occurred throughout the city after the earthquake of February 15, 1856, which was strongly felt at Fort Point ("Violent Shock of Earthquake in this City—Damage to Brick Buildings," *Daily Alta California,* February 16, 1856). Damage was concentrated in areas that would be hit again and again by shaking. Editors of the local press moved quickly to allay fears: A *Daily Alta California* editorial, "The Earthquake—Its Probable Effect upon the Interests of the City," February 18, 1856, states that the damage was only done to frail brick buildings (but the Goodwin Building was new!), the earthquake was the worst in forty years, and people should not worry about quakes in San Francisco. Intensity was RF VIII at San Francisco (Tocher, *Seismic History,* 42), MMI VII (Toppozada, *Annual Technical Report,* 43)

47. The January 9, 1857, earthquake was characterized as a "severe shock." It was felt in the filled area of the city and knocked a frame house at Market and California off its foundations. For an account of this earthquake by George Davidson see California State Earthquake Investigation Commission, *The California Earthquake of April 18, 1906* (Washington, D.C.: Carnegie Institution of Washington, 1908), I:449.

48. On foundations and ironwork see "Specifications of the Different Kinds of Material, Workmanship and Dimensions, April 20, 1853," General Information, Montgomery Block, Deeds Agreements, Folder 2, and "Iron Bill, Iron as per Contract Made and Signed May 12, 1853," General Information, Montgomery Block, Deeds Agreements, Folder 4, Phelan Sullivan Library, The Society of California Pioneers, San Francisco. For earthquake-resistant features see Kenneth Cardwell, "Montgomery Block (Office Building)," Historic American Building Survey Cal-1228 (August 1958); Harold Kirker, *California's Architectural Frontier: Style and Tradition in the Nineteenth Century* (Santa Barbara: Peregrine Smith Inc., 1973), 80-81; O. P. Stidger, "The Montogomery Block: The Realized Dream of Henry W. Halleck," MS, March 1947, Phelan Sullivan Library, The Society of California Pioneers, San Francisco. Also see "Montgomery Block," *San Francisco City Directory* (San Francisco: LeCount & Strong, 1854), 190; Agnes Foster Buchanan, "Some Early Business Buildings of San Francisco," *Architectural Record* 20:1 (July 1906), 25-26; Vernon Aubrey Neasham, ed., "The Montgomery Block, Registered Landmark #80," California Historical Landmark Series (Berkeley: State of California, Department of Natural Resources, 1936); "Wager Halleck, Lincoln's Chief-of-Staff," *California Historical Society* 16(3) (September 1937), 199-200. For iron in brick structures, see chapter 4.

49. "The Earthquake—Its Probable Effect upon the Interests of the City," *Daily Alta California,* February 18, 1856

50. Ibid.

51. Ibid.

Chapter 2

1. Wai-Fah Chen and Charles Scawthorn, eds., *Earthquake Engineering Handbook* (Boca Raton, Fla.: CRC Press, 2003), 29-1

2. John Walker Powell, *An Introduction to the Natural History of Disaster* (University of Maryland: Disaster Research Project, June 30, 1954), 5-13

3. Martyn J. Bowden, J. Eugene Haas, and Robert W. Kates, eds., *Reconstruction Following Disaster* (Cambridge, Mass.: MIT Press, 1977), xxvii. Also see Russell R. Dynes, *Organized Behavior in Disaster* (Lexington, Mass.: D. C. Heath and Co., 1970).

4. George R. Stewart, *Committee of Vigilance: Revolution in San Francisco, 1851* (New York: Ballantine Books, 1964), 89, 103-115. Also see Robert M. Senkewicz, *Vigilantes in Gold Rush San Francisco* (Stanford, Calif.: Stanford University Press, 1985).

5. Elliot Evans, ed., "Some Letters of William S. Jewett, California Artist," *California Historical Society Quarterly* 23(2) (June 1944), 156

6. Frank Soulé, John H. Gihon, M. D., and James Nisbet, *The Annals of San Francisco* (1855; reprint, Berkeley: Berkeley Hills Books, 1998), 230

7. For the fires which follow, see Roger W. Lotchin, *San Francisco 1846–1856: From Hamlet to City* (New York: Oxford University Press, 1974), 174-175; Charles Lockwood, *Suddenly San Francisco: The Early Years of an Instant City* (San Francisco: San Francisco Examiner Division of the Hearst Corporation, 1978), 43-47; and Soulé, *Annals,* 241-242, 274-275, 330-333, 344-347.

8. Lotchin, *San Francisco 1846–1856,* 174; Lockwood, *Suddenly San Francisco,* 43; Soulé, *Annals,* 241-242

9. "San Francisco Fire Department Historical Review 1849–1967," 1967, MS, California Historical Society, San Francisco, 2. There is some disagreement about this date: Archibald MacPhail, in *Of Men and Fire: A Story of Insurance in the Far West* (San Francisco: Fire Underwriters Association of the Pacific, 1948), 48, asserts that Kohler was officially named chief in July 1850. Soulé, in *Annals,* 616, mentions that a citizens' group was formed the day after the fire and Kohler was elected chief. Also see Doris Muscatine, *Old San Francisco: The Biography of a City from Early Days to the Earthquake* (New York: G. P. Putnam's Sons, 1975), 249.

10. Bruce Laurie, "Fire Companies and Gangs in Southwark: the 1840s," in *The Peoples of Philadelphia: A History of Ethnic Groups and Lower-Class Life, 1790–1940,* ed. Allen F. Davis and Mark H. Haller (Philadephia: Temple University Press, 1973), 71-87; Lowell M. Limpus, *History of the New York Fire Department* (New York: E. P. Dutton, 1940); Muscatine, *Old San Francisco,* 248-249

11. "San Francisco Fire Department," 5

12. Theodore H. Hittell, *History of California* (San Francisco: N. J. Stone & Co., 1898), 3:365; Muscatine, *Old San Francisco,* 249-258; Soulé, *Annals,* 618-621

13. Lockwood, *Suddenly San Francisco,* 43

14. Ibid.; Soulé, *Annals,* 274-275

15. William Dallam Armes, "The Phoenix on the Seal," *Sunset* June-July 1906, 113-114

16. Lockwood, *Suddenly San Francisco,* 44; Soulé, *Annals,* 276

17. Soulé, *Annals,* 276

18. Lockwood, *Suddenly San Francisco,* 44; Soulé, *Annals,* 277-278; Armes, in "Phoenix," 114, claims that the fire started from a house, whereas Lockwood and Soulé assert that it started in a bakery. They all agree that the fire began in a building behind the Merchant's Hotel.

19. J. Stanley Rabun, *Structural Analysis of Historic Buildings: Restoration, Preservation, and Adaptive Reuse Application for Architects and Engineers* (New York: John Wiley & Sons, Inc., 2000), 219-220

20. Lockwood, *Suddenly San Francisco,* 44

21. Armes, "Phoenix," 114

22. Lockwood, *Suddenly San Francisco,* 44

23. Armes, "Phoenix," 114; Lotchin, *San Francisco 1846–1856,* 175; Soulé, *Annals,* 290

24. Lockwood, *Suddenly San Francisco,* 45

25. Lotchin, *San Francisco 1846–1856,* 177

26. Soulé, *Annals,* 331

27. Lockwood, *Suddenly San Francisco,* 46

28. Lockwood, *Suddenly San Francisco,* 175-176; Soulé, *Annals,* 329-330

29. Soulé, *Annals,* 339

30. Stewart, *Committee of Vigilance,* 75

31. Ibid., 89

32. Ibid., 89, 103-115. Similar extralegal attempts to stop arson occurred throughout the United States. One of the cruelest took place in New York in 1741 when African Americans were accused of starting fires and in the ensuing panic many innocent people were hanged. Limpus, *History of the New York Fire Department,* 49-53

33. Lotchin, *San Francisco 1846–1856,* 177, 197

34. Ibid., 182-183; Soulé, *Annals,* 340-341; Erwin N. Thompson, *Defender of the Gate, the Presidio of San Francisco, A History from 1846 to 1995* (Denver: National Park Service, NPS-300 digital version, 1997), 42-52

35. C. C. Knowles, *The History of Building Regulation in London 1189–1972* (London: Architectural Press, 1972)

36. For iron buildings and fire, see Cecil D. Elliott, *Technics and Architecture: The Development of Materials and Systems for Buildings* (Cambridge, Mass.: MIT Press, 1994); Frank Marryat, "The Trouble with Iron Houses," in Malcolm E. Barker, comp., *San Francisco Memoirs, 1835–1851: Eyewitness Accounts of the Birth of the City* (San Francisco: Londonborn Publications, 1994), 173-178

37. *Alta California,* December 28, 1849, 2

38. Ibid.; *Alta California,* May 3, 1851, 2

39. Rabun, *Structural Analysis of Historic Buildings,* 219-220

40. Heinrich Schliemann, *Schliemann's First Visit to America* (Cambridge, Mass.: Harvard University Press, 1942), 63-65

41. Soulé, *Annals,* 332

42. Agnes Foster Buchanan, "Some Early Business Buildings in San Francisco," *Architectural Record* 20 (July 1906), 23. For buildings that survived, see John S. Hittell, *A History of San Francisco and Incidentally of the State of California* (1878 and 1888; reprint Berkeley: Berkeley Hills Books, 2000), 3:355.

43. Hittell, *A History of San Francisco,* 170

44. Lockwood, *Suddenly San Francisco,* 20-21, 50; Harold Kirker, "The Parrot Building, San Francisco, 1852," *Journal of the Society of Architectural Historians* 18 (December 1959), 160-161

45. *Daily Alta California,* June 12, 1857

46. Lockwood, *Suddenly San Francisco,* 46; LeCount & Strong, *San Francisco City Directory* (San Francisco: LeCount & Strong, 1854), 213

47. Lockwood, *Suddenly San Francisco,* 48; LeCount & Strong, *SF Directory,* 214

48. Ibid., 194-195; Rodger C. Birt, "The San Francisco Album and its Historical Moment: Photography, Vigilantism, and Western Urbanization," in *San Francisco Album: Photographs of the Most Beautiful Views and Public Buildings* (San Francisco: Frankel Gallery, 1999), 99-122

49. Buchanan, "Some Early Business Buildings," 19

Chapter 3

1. "Earthquakes," March 6, 1864, Bancroft Scraps, vol. 29: Earthquakes, 14, Bancroft Library, University of California, Berkeley.

2. Philip L. Fradkin, *Magnitude 8: Earthquakes and Life along the San Andreas Fault* (New York: Henry Holt, 1998), 76

3. For brick buildings in 1868 see "The Resources of Vallejo, no. XV, Fires," October 24, 1868, Bancroft Scraps, 48. For the steady increase of brick buildings see *San Francisco Directory for the Year Commencing December, 1860* (San Francisco: Henry G. Langley, 1860), 65-66, and *San Francisco Directory for the Year Commencing December, 1864* (San Francisco: Henry G. Langley, 1864).

4. "Another Earthquake," December 20, 1863, Bancroft Scraps, 14. See also Toppozada 1979, 46.

5. "Earthquakes," March 6, 1864, Bancroft Scraps, 14. Intensity registered at RFVII+ (Tocher, "Seismic History," 42), M 5.9 (Toppozada 1979, 46), MMI 6-7 (Toppozada 1981, 46).

6. "Another Earthquake," May 21, 1864, Bancroft Scraps, 15. This earthquake is dated May 21, 1864, in Toppozada 1979, 46-47, with intensity at MMI VI (Toppozada 1981, 47).

7. Tocher, "Seismic History," 42

8. Walter L. Huber, "San Francisco Earthquakes of 1865 and 1868," *Bulletin of the Seismological Society of America* 20(4) (December 1930), 263

9. Tocher, "Seismic History," 42; Toppozada 1981, 53. MMI VIII intensity observed in San Francisco.

10. B. E. Lloyd, *Lights and Shades in San Francisco* (1876; reprint Berkeley: Berkeley Hills Books, 1999), 320-321

11. Mark Twain, *Roughing It* (New York: Harper & Row, 1899), 2:138-139

12. For a more detailed discussion of seismology, see Bruce A. Bolt, *Earthquakes* (New York: W. H. Freeman and Co., 1993), 115-136. Other good introductions include Matthys Levy and Mario Salvadori, *Why the Earth Quakes: The Story of Earthquakes and Volcanoes* (New York and London: W. W. Norton and Co., 1995) and Robert S. Yeats, *Living with Earthquakes in California, A Survivor's Guide* (Corvallis: Oregon State University Press, 2001).

13. Huber, "San Francisco Earthquakes," 261-262

14. Ibid., 261

15. Lloyd, *Lights and Shades,* 321

16. William Hammond Hall, "Some Lessons of the Earthquake and Fire," I, *San Francisco Chronicle,* May 19, 1906; Zoeth S. Eldridge, "Location of an Ancient Laguna," *San Francisco Chronicle,* May 22, 1906; T. L. Youd and S. N. Hoose, *Historic Ground Failures in Northern California Triggered by Earthquakes,* USGS Professional Paper 993 (Washington, D.C.: Government Printing Office, 1978), 24-59; George F. Whitworth, ed., *Subsidence and the Foundation Problem in San Francisco, a Report of the Subsoil Committee of the San Francisco Section* (San Francisco: American Society of Civil Engineers, September 1932), 9, 15-29

17. Thomas Rowlandson, "Notes and Comments on the Late Earthquake," December 9, 1865, Bancroft Scraps, 22

18. See note 1 of this chapter; also California State Earthquake Investigation Commission, *The California Earthquake of April 18, 1906* (Washington, D.C.: Carnegie Institution of Washington, 1908), 449; "The Earthquakes," November 7, 1865, "Earthquake Phenomena," November 1865, and Rowlandson "Notes and Comments," Bancroft Scraps, 19-22.

19. Harold Kirker, *California's Architectural Frontier: Style and Tradition in the Nineteenth Century* (Santa Barbara: Peregrine Smith Inc., 1973), 74-75

20. William Craine and George Cofran to Charles James, "U.S. Custom House Damage Report," October 17, 1865, MSS 2005/26 c, Bancroft Library, University of California, Berkeley

21. Ibid.

22. Ibid.

23. United States, Custom House, San Francisco Records, 1847–1912, MSS C-A 169 Pt. III, Bancroft Library, University of California, Berkeley

24. For a brief survey of iron in construction see Cecil D. Elliott, *Technics and Architecture: The Development of Materials and Systems for Buildings* (Cambridge, Mass.: MIT Press, 1994), 68-69, 463. For iron rods in medieval and early Renaissance buildings see Luciano Patetta, "Le 'catene' come scelta progettuale negli edifici tra XIII e XV secolo," *Saggi in Onore di Renato Bonelli,* edited by Corrado Bozzoni, Giovanni Carbonara and Gabriella Villetti in *Quaderni dell'istituto di Storia dell'Architettura,* Nuova Serie, Fasc. 15-20 (1990-1992), 233-242. For iron in Brunelleschi's buildings, see Howard Saalman, *Filippo Brunelleschi. the Buildings* (University Park, Penn.: Pennsylvania University Press, 1993), 40-48. For iron in Renaissance and Baroque domes, see Elwin C. Robison, "St. Peter's Dome: The Michelangelo and Della Porta Designs," Proceedings, International Symposium on Domes from Antiquity to the Present, Mimar Sinan University, Istanbul, 1988, 101-110, and Joseph Connors, "Borromini's S. Ivo alla Sapienza: The Spiral," *The Burlington Magazine* 138, October 1996, 680. For the Louvre, see Richard W. Berger (with Roland Mainstone), *Palace of the Sun, the Louvre of Louis XIV* (University Park, Penn.: Pennsylvania University Press, 1993), 65-67. For iron in earthquake-resistant construction, see Stephen Tobriner, "A History of Reinforced Masonry Construction Designed to Resist Earthquakes: 1755–1907," *Earthquake Spectra* 1:1 (Nov. 1984), 125-149; Tobriner, "La Casa Baraccata: Earthquake-Resistant Construction in 18th Century Calabria," *Journal of the Society of Architectural Historians* 42(2) (May 1983), 131-138; and M. Comerio, M. Green, and Tobriner, "Reconnaissance Report on the Umbria-Marche, Italy, Earthquakes of 1997," *EERI Special Earthquake Report* (December 1997), 5.

25. "A Genuine Earthquake," October 9, 1865, Bancroft Scraps, 4

26. "The Earthquake Yesterday," October 22, 1868, Bancroft Scraps, 33

27. After surveying the damage and stating the building was safe to repair and reoccupy, James Ballenti, J. E. Nutting, and George Cofran made the comment about the building being "well ironed" in a report dated Oct. 26, 1868. *Alta California,* Oct. 27, 1868

28. "A Genuine Earthquake"

29. *San Francisco Bulletin,* October 10, 1865

30. *Daily Alta California,* August 31, 1866

31. *Daily Alta California,* October 10, 1865

32. Order 725, Sec. 9 (August 21, 1866), The Consolidation Act, or Charter of the City and County of San Francisco; with Other Acts Specially Relating to San Francisco, and of General Interest Therein; and the General Orders of the Board of Supervisors

33. Perry Byerly, "History of Earthquakes in the San Francisco Bay Area," in *Geologic Guidebook of the San Francisco Bay Counties* (San Francisco: California Division of Mines, 1951), 151; Bolt, *Earthquakes,* 277

34. Lloyd, *Lights and Shades,* 322; *Alta California,* October 22, 1868; Bolt, *Earthquakes,* 277

35. Huber, "San Francisco Earthquakes," 265-266

36. "The Earthquake Yesterday"

37. Huber, "San Francisco Earthquakes," 266, 270

38. State Earthquake Investigation Commission, *The California Earthquake,* plate 146

39. "The Earthquake Yesterday." Any doubt that the Custom House had been reinforced after 1865 is ended by quotes like this one: "Had it not been for the anchors and ties put into the Custom House after the shock of the 1865 earthquake, it is probable that a portion of the building would have fallen to the ground."

40. Ibid.

41. "United States Engineers Condemn the Custom-House Building," *Daily Morning Chronicle,* October 24, 1868

42. "Meeting of the Committee on Earthquakes," November 25, 1868, Bancroft Scraps, 63

43. Albert Shumate, *A San Francisco Scandal: The California of George Gordon* (Spokane: Arthur H. Clark Co., 1994), 220-223

44. "Meeting of the Committee on Earthquakes"

45. Ibid.

46. Rowlandson blames the committee in *A Treatise on Earthquake Dangers, Causes and Palliatives* (San Francisco: Dewey & Co. Publishers, 1869), 3-7. In addition, George Davidson blames the committee chairman for the suppression of the report in his 1908 letter to the Seismological Society: see William H. Prescott, "Circumstances Surrounding the Preparation and Suppression of a Report on the 1868 California Earthquake," *Bulletin of the Seismological Society of America* 72(6) (December 1982), 2389-2393.

47. "Earthquake Commission," Bancroft Scraps, 64

48. David Farquharson's Bank of California is described in "The Earthquake Yesterday": "The balustrade of the Bank of California is being removed. The main coping of the building, which is bolted down the walls for six feet, will stand. The whole building is clamped and tied together, so that it could not be wrecked except [if] it was thrown bodily over. The building, the ornamental work on the roof excepted, has sustained no injury, but it is the intention of the Director to take down the chimneys and fire wall between it and the building in the rear, so as to leave nothing for chance in the future to displace. The present chimneys will be replaced by wrought iron."

49. "Earthquake Commission"

50. "Earthquake Commission"

51. "Earthquake Philosophy, and Earthquake-Proof Building," *The Mining and Scientific Press* 17(19) (November 7, 1868), 296

52. Prescott, "Circumstances," 2390

53. "Earthquake Philosophy," 296

54. "Earthquake Commission"

55. State Earthquake Investigation Commission, *The California Earthquake,* 434

56. Harry O. Wood to George D. Louderback, March 26, 1947, MS, George Davis Louderback papers, 1900–1956, Bancroft Library, University of California, Berkeley

57. Michele Aldrich, Bruce Bolt, et al., "The 'Report' of the 1868 Hayward Earthquake," *Bulletin of the Seismological Society of America* 76(1) (February 1986), 71-76

Chapter 4

1. The Committee on Instruction to the Regents of the University of California, Committee file 1868–1879, University Archives, CU-1, box 2, folder 11, Bancroft Library, University of California, Berkeley

2. Ibid.

3. John S. Hittell, *The Resources of California,* 3d ed. (San Francisco: A. Roman and Co., 1867), 41-42

4. Ibid. See chapter 13, note 65.

5. Committee on Instruction to the Regents

6. "The University Building and Earthquakes," *San Francisco Bulletin,* March 22, 1869

7. "California Architecture," October 24, 1868, Bancroft Scraps, vol. 29: Earthquakes, 44, Bancroft Library, University of California, Berkeley

8. "Architecture Adapted to Earthquakes," February 19, 1869, Bancroft Scraps, 70-72

9. Vitruvius Frazer, "Earthquake-Proof Buildings—How to Construct Them—Suggestions by a Professional Architect," January 6, 1869, Bancroft Scraps, 69

10. Stephen Tobriner, "Bond Iron and the Birth of Anti-Seismic Reinforced Masonry Construction in San Francisco," *The Masonry Society Journal* 5(1) (Jan.-June 1986), 12; Tobriner, "A History of Reinforced Masonry Construction Designed to Resist Earthquakes: 1755–1907," *Earthquake Spectra* 1(1) (Nov. 1984), 129

11. Tobriner, "A History of Reinforced Masonry Construction," 129; C. W. Pasley, *Observations on Limes, Calcareous Cements, Mortars, Stuccos, and Concrete,* 2d ed. (London: J. Weale, 1847)

12. John Kelly, Construction of Seawalls, U.S. Patent 74,547, February 18, 1868. See also U.S. Patent Office, *Annual Report of the Commissioner of Patents*, vol. 1375 (Washington, D.C.: Government Printing Office, 1868), 157, 613.

13. Building Ordinance no. 291, sec. 130 (May 8, 1901), *General Orders and Ordinances of the Board of Supervisors of the City and County of San Francisco*

14. William H. Foye, Improved Submarine Foundation, U.S. Patent 92,033, June 29, 1869

15. Ibid.

16. Case file 12345, *Pacific Submarine Company v. United States,* General Jurisdiction Case Files, Record Group 124, Records of the United States Court of Claims 1884, National Archives, Washington D.C. Also see note 33.

17. "A New Invention," *Morning Daily Call,* December 19, 1872

18. B. S. Alexander, "Earthquake-Proof Buildings," *Daily Alta California,* November 23, 1870, 1

19. "The Foye System of Binding Buildings," *Mining and Scientific Press* 24 (June 1, 1872), 345

20. U.S. Patent Office, *Official Gazette of the United States Patent Office,* vol. III (Washington D.C.: GPO, 1873), 423

21. Jules Touaillon, Improvement in Buildings, U.S. Patent 99,973, February 15, 1870. For early concepts of base isolation see David Stevenson, "Notice of Aseismic Arrangements, Adapted to Structures in Countries Subject to Earthquake Shocks," *Transactions of the Royal Scottish Society of Arts* 7 (1868), 565, and David A. Stevenson, "Earthquake-Proof Buildings," *Nature* 32 (1885), 316. Milne called base isolation "free foundation" and intimated that the Stevensons had gotten the idea from another seismological pioneer, Mallet. See John Milne, "On Construction in Earthquake Countries," *Minutes of Proceedings of the Institution of Civil Engineers* 83 (1886), 282-283.

22. *San Francisco Newsletter,* May 7, 1870, 3. Also *The San Francisco Directory for the Year Commencing December, 1869* (San Francisco: Henry S. Langley, 1869), 18: "The building will be of iron and brick, with thoroughly-braced wooden frames...the iron and brick will be secured with anchors, placed vertically and horizontally...so as to render the building comparatively secure against earthquakes."

23. *San Francisco Directory for the Year Commencing December, 1869,* 18

24. Ibid.

25. Ibid., 18-19

26. Ibid., 19

27. Circuit Court of the United States of the Ninth Circuit in and for the District of California, "Testimony of William H. Foye," *Pacific Submarine and Earthquake Proof Wall Company vs. P. H. Canavan, Joseph G. Eastland, and Charles E. McLane* (San Francisco, 1874–1877), 15

28. Harold Kirker, *California's Architectural Frontier: Style and Tradition in the Nineteenth Century* (Santa Barbara: Peregrine Smith, 1973), 95; *The San Francisco Directory for the Year Commencing December, 1876* (San Francisco: Henry G. Langley, 1877), 23. The Palace Hotel was so well constructed that it was extremely difficult to demolish for the erection of a new building after the earthquake and fire. "The difficulty that the wreckers are having in razing the Palace Hotel in San Francisco does not bear out the reports which are so freely circulated concerning inferior construction in that city...The bricks were laid in a mortar composed of lime and cement, with strips of Norway iron and steel cables at frequent intervals. The mortar is too hard to be scattered by the cleaning machines. The supports include 100 pillars weighing seven tons each. The walls are first dynamited and then pulled down by hoisting engine—1½ in. steel cables—equivalent to the strength of 10,000 men. Notwithstanding this, the strips of wall remain unbroken on the ground and require picks and sledges to loosen the bricks." ("High Class Brick Work," *Mining and Scientific Press,* April 20, 1907) The engineer John Blume told me in personal correspondence that he thought his father had been in charge of demolishing the Palace.

29. "Board of Supervisors—Report of the Earthquake Committee," *San Francisco Chronicle,* October 24, 1868

30. *Report of the Committee Appointed to Investigate the Affairs of the San Francisco City Hall Commission* (Sacramento: T. A. Springer, 1874), 4

31. Stephen Williams and John Wright, "Instruction and Suggestions to Architects," adopted June 23, 1870, *San Francisco Municipal Reports for the Fiscal Year 1871–72, ending June 30, 1872* (San Francisco: Board of Supervisors, 1872), 419; *San Francisco Municipal Reports for the Fiscal Year 1888–89, ending June 30, 1889* (San Francisco: Board of Supervisors, 1889), 769

32. Ibid.

33. Basin Research Associates, *San Francisco Main Library Project: Archaeological Monitoring and Architectural Documentation, Site of the Former City Hall Completed in 1897,* vol. 1 (San Francisco: Basin Research Associates, December 1994), 1, 27-29

34. A. B. Mullet to J. F. Morse, November 2, 1868, Entry 7, Box 19, Folder 2, RG 121, National Archives, 279

35. Ibid., 279-280

36. A. B. Mullet to J. F. Morse, October 26, 1868, Entry 7, Box 19, Folder 2, RG 121, National Archives, 132

37. A. B. Mullet to W. P. C. Stebbins, September 20, 1869, Entry 7, Box 25, Folder 1, RG 121, National Archives, 389 1/2

38. Ibid. For similarities to Foye's construction methods, see Foye, Improved Submarine Foundation, patent.

39. Stephen Tobriner, "South Hall and Seismic Safety at the University of California in 1870," *Chronicle of the University of California* 1:1 (Spring 1998), 13-23

40. "Architecture Adapted to Earthquakes," February 19, 1869, Bancroft Scraps, 71

41. John Cotter Pelton, "San Francisco: Its Position in Architectural and Constructive Development," in *Modern San Francisco, 1907–1908* (San Francisco: Western Press Association, 1908). After the 1906 earthquake and fire had reduced Farquharson's Nevada Block in San Francisco to ruins, Pelton, a famous local engineer, saw the remains of one of the "T" columns embedded in brick and remarked that this form of skeletal construction predated skeleton construction in Chicago by almost a decade.

42. Tobriner, "South Hall," 13-23

43. *Morning Call,* July 17, 1873

Chapter 5

1. Oscar Lewis, *San Francisco: Mission to Metropolis* (San Diego: Howell-North Books, 1980), 166–184; William Issel and Robert W. Cherny, *San Francisco 1865–1932: Power, Politics, and Urban Development* (Berkeley: University of California Press, 1986), 80-106, 118-160

2. The text draws on the development of tall buildings in the East. In particular, Sarah Landau and Carl Condit, *Rise of the New York Skyscraper 1865–1912* (New Haven: Yale University Press, 1996), 5-15

3. Quoted by Donald Freeman, *Historical Building Construction, Design, Materials and Technology* (New York and London: Norton, 1995), 8

4. Richard M. Levy, "The Professionalization of American Architects and Civil Engineers, 1865–1917" (Ph.D. dissertation, University of California, Berkeley, 1980); Gerald L. Geison, ed., *Professions and Professional Ideologies in America* (Chapel Hill, N.C.: University of North Carolina Press, 1983)

5. William Barclay Parsons, *Engineers and Engineering in the Renaissance* (Cambridge Mass.: MIT Press, 1968); William H. Wisely, *The American Civil Engineer 1852–1974: The History, Traditions and Developments of the American Society of Civil Engineers* (New York: American Society of Civil Engineers, 1974), 1-30; Levy, "Professionalization"; Henry A. Schulz, "Architectural versus Engineering Requirements," *American Builders Review* 4(3) (September 1906), 270

6. Levy, "Professionalization," 229-277

7. "The Chronicle's New Home," *San Francisco Chronicle,* June 22, 1890, 2

8. Ibid.

9. Ibid.

10. Ibid.

11. Gray Brechin, *Imperial San Francisco: Urban Power, Earthly Ruin* (Berkeley, Los Angeles: University of California Press, 1999), 177

12. Robert Livingston Schuyler, ed., *Dictionary of American Biography* Supplement II, "Charles Strobel" (New York: C. Scribner's Sons, 1940), 22:638. See also Michael Corbett, *Splendid Survivors* (San Francisco: California Living Books, 1979), 85.

13. Ralph Peck, "History of Building Foundations in Chicago, Structural and Civil Engineering Design," in William Addis, ed., *Studies in the History of Civil Engineering* vol. 12 (Brookfield, Vt.: Ashgate Publishing, 1999), 105-158

14. For a discussion of the engineering decisions and the details of construction see: "The Spreckels Building, San Francisco," *Engineering Record* (April 9, 1898), 412-414, and *Engineering Record* (April 16, 1898), 433-436; Frank Soulé, "Foundations as Affected by Earthquakes," *American Builders Review* 4(3) (September 1906), 256; John D. Galloway, "Notes on the Construction of Steel Frames for Buildings," *American Builders Review* 4(3) (September 1906), 266

15. Soule, "Foundations," 257

16. "Economic Forces Prove Stronger Than Earthquakes," *Architectural Record* 88(6) (December 1940), 82

17. Corbett, *Splendid Survivors,* 85

18. Forell/Elsesser Engineers, Inc., *Final Report: Seismic Damage Evaluation and Upgrade Recommendation for U.S. Court of Appeals and Post Office, San Francisco, California* (San Francisco: June 1990), 1.1-1.5 and Appendix E

19. "Earthquake and Fire. Concerning the Fire Resistance of Building Materials," Wire Glass Manufacturers, n.d., 2

20. "The Whittell Building," *American Builders Review* 4(3) (March 1907), 71-78

21. Ibid., 71

22. Christopher H. Nelson, "Classical California: The Architecture of Albert Pissis and Arthur Brown, Jr.," Ph.D. dissertation, University of California, Santa Barbara, 1986, 146-149

23. In a November 1988 interview with the author, Henry J. Degenkolb discussed the Flood Building. He discussed the ground floor's vulnerability in earthquakes. For rivets, see chapter 8, notes 21, 22.

24. *San Francisco Municipal Reports for the Fiscal Year 1888–89, ending June 30, 1889* (San Francisco: Board of Supervisors, 1889), 764-798

25. Ibid., 764; also see 783-784

26. Ibid., 784-786

27. Ibid., 786

28. Compare the frame of City Hall exposed after the earthquake with the structure of the dome of the Federal Building (1905) by Henry Ives Cobb in Chicago, reproduced in John Zukowsky, ed., *Chicago Architecture 1872–1922: Birth of a Metropolis* (Munich: Prestel-Verlag, 1987), 113, fig. 12

29. "Earthquake," *San Francisco Call,* June 21, 1897; and "Incidents of the Shocks Ashore and at Sea," *San Francisco Call,* April 1, 1898, 2. Both earthquakes (June 20, 1897 and April 1, 1898) cracked terra-cotta plates on the new dome.

30. Henry J. Cowan, *Science and Building: Structural and Environmental Design in the Nineteenth and Twentieth Centuries* (New York and London: John Wiley and Sons, 1978), 23-30, 61-68; Carl W. Condit, *American Building: Materials and Techniques from the First Colonial Settlements to the Present* (Chicago: University of Chicago Press, 1968), 79-130; Zukowsky, *Chicago Architecture 1872–1922,* 39-55; Carl W. Condit, *American Building Art: The Nineteenth Century* (New York: Oxford University Press, 1960), 103-162; Stephen P. Timoshenko, *History of the Strength of Materials* (New York: Krieger, 1953), 184-186; C. T. Purdy, "The Steel Construction of Buildings," *Bulletin of the University of Wisconsin* 1 (1894), 49

31. For examples of the many analyses for wind and earthquake forces, see Robins Fleming, *Wind Stress in Buildings* (New York:

John Wiley & Sons, 1930) or James Ambrose and Dimitry Vergun, *Simplified Building Design for Wind and Earthquake Forces* (New York: John Wiley & Sons, 1980), 37-71. For the history of the concept of wind pressure, see Carl W. Condit, "The Wind Bracing of Buildings," *Scientific American* (February 1974), 92-105.

Carl Condit ("The Two Centuries of Technical Evolution Underlying the Skyscraper," in Lynn S. Beedle, ed., *Second Century of the Skyscraper* [New York: Van Nostrand Reinhold, 1988], 14-15), after discussing early contributions of Robert Hooke, John Smeaton, and Jean Charles Borda, expresses confusion over why "it was more than 150 years before scientific investigations began to bear fruit in iron-framed building" in the United States: "At this stage we reach a mystery, and I am sorry that I cannot yet unravel it. By the end of the Civil War we know that the necessity for bracing in an iron-framed building was at least recognized" but no evidence of this can be found in structures themselves. As late as 1893, Adler and Sullivan's Stock Exchange Building in Chicago has no wind bracing. Condit sees Quimby as "the voice of authority in wind bracing" and credits the 1892-3 ASCE *Transactions* as the first professional discussion of the subject. Also see Landau and Condit, *Rise of the New York Skyscraper,* 23, 160-167; Donald Freeman, *Historical Building Construction, Design, Materials and Technology* (New York and London: Norton, 1995) 80-82; Cowen, *Science and Building,* 61-68

32. *Transactions,* American Society of Civil Engineers 27 (September 1892), 224-225

33. Ibid., 236

34. Ibid., 238

35. Ambrose and Vergun, *Simplified Building Design,* 37-71

36. *Transactions,* 227

37. Ibid., 228

38. Ibid., 234-235

39. William H. Birkmire, *The Planning and Construction of High Office Buildings,* 3d ed. (New York: John Wiley & Sons, 1903), 205-206

40. "A Lively Shake-Up, the City Rocked by Two Earthquakes," *San Francisco Chronicle,* March 26, 1884, 3

41. Ibid.

42. "A Slight Earthquake," *San Francisco Chronicle,* March 1, 1888; "Heavy Earthquake, Plastering of Buildings Cracked at Biggs," *San Francisco Chronicle,* April 29, 1888

43. Don Tocher, in "Seismic History of the San Francisco Region" (*San Francisco Earthquakes of March 1957;* [San Francisco: California Division of Mines, 1959, Special Report 57], 43), estimates the 1889 quake at RF VII in San Francisco. The 1890 quake was strong in Monterey (RF VIII or IX). See also *San Francisco Call,* April 25, 1890.

44. "The Quake is Over," *Morning Call,* April 23, 1892

45. "Why They Fell, Faulty Construction More to Blame than the Temblors," *San Francisco Call,* April 24, 1892

46. Ibid.

47. "Tall Buildings and Earthquakes," *San Francisco Call,* April 1, 1898

48. "Damage from Earthquakes Was Exaggerated," *San Francisco Chronicle,* April 23, 1892; "By the Way," *Morning Call,* April 24, 1892

49. "The Earthquake Bogie," *San Francisco Call,* April 23, 1892

50. "After the Event," *San Francisco Chronicle,* April 23, 1892

51. Tocher, "Seismic History," 43; "Earthquake," *San Francisco Call,* June 21, 1897

52. "Navy-Yard Severely Damaged," *San Francisco Call,* April 1, 1898, 1

53. "Incidents of the Shocks Ashore and at Sea," *San Francisco Call,* April 1, 1898, 2

54. Ibid.

55. "The Earthquake and the Untruth," *San Francisco Call,* April 1, 1898, 6

Chapter 6

1. The most comprehensive source for the earthquake is the State Earthquake Investigation Commission's two-volume report, *The California Earthquake of April 18, 1906* (Washington, D.C.: Carnegie Institution of Washington, 1908)

2. Bruce A. Bolt, "The Focus of the 1906 California Earthquake," *Bulletin of the Seismological Society of America* 58, 1968, 457-471; David J. Wald, Hiroo Kanamori, Donald V. Helmberger, Thomas H. Heaton, "Source Study of the 1906 San Francisco Earthquake," *Bulletin of the Seismological Society of America* 83, 1993, 981-1019; Wai-Fah Chen and Charles Scawthorn, *Earthquake Engineering Handbook* (Boca Raton, Fla.: CRC, 2003), Ch. 9. It was formerly believed the epicenter was near Olema. For the reaction of Daly City officials to a group that wants to place a plaque marking the Daly City epicenter see Ryan Kim, "Daly City Official Unmoved by Quake Notoriety," *San Francisco Chronicle,* April 11, 2004.

3. For the effects on ships, see State Earthquake Investigation Commission, *The California Earthquake.*

4. State Earthquake Investigation Commission, *The California Earthquake,* II:3

5. Frank Neumann and Harry O. Wood had not yet created MMI in 1906, so Wood, feeling the Rossi-Forel scale wasn't specific enough to the damage in San Francisco, invented

the following scale for grading intensity of shaking: Grade A, Very violent (MMI X), which did not occur in San Francisco. Grade B, Violent (MMI VIII+), comprises fairly general collapse of brick and frame buildings when not unusually strong; serious cracking of brickwork and masonry in excellent structures; wavelike folds in paved and asphalt-coated streets; destruction of foundation walls by undulation of the ground; breaking of sewers and water pipes; and lateral displacement of streetcar tracks. Grade C, Very strong (MMI VII-VIII), comprises badly cracked brickwork and masonry with occasional collapse; some gable walls thrown down; frame buildings lurched on weak underpinning; considerable cracking of foundation walls; and general destruction of chimneys and of brick, masonry, or cement veneer. Grade D, Strong (MMI VI-VII), comprises general but not universal fall of chimneys; also cracks in brickwork, masonry, foundation walls, and curbing. Grade E, Weak (MMI VI), comprises occasional fall of chimneys and damage to plaster, partitions, plumbing, and the like. See Harry O. Wood, "Distribution of Apparent Intensity in San Francisco," in State Earthquake Investigation Commission, *The California Earthquake,* 224-225.

6. Carl W. Stover and Jerry L. Coffman, "Seismicity of the United States, 1568–1989" (Revised), U.S. Geological Survey Professional Paper 1527 (Washington, D.C.: Government Printing Office, 1993); http://neic.usgs.gov/neis/eq_depot/usa/1906_04_18.html

7. *San Francisco,* directed by W. S. Van Dyke (1936; Burbank, Calif.: Warner Home Video, 1990); Philip L. Fradkin, *The Great Earthquake and Firestorms of 1906: How San Francisco Nearly Destroyed Itself* (Berkeley: University of California Press, 2005)

8. Karl Steinbrugge, *Earthquakes, Volcanoes, and Tsunamis: An Anatomy of Hazards* (New York: Skandia America Group, 1982), 224

9. Gladys Hansen, "The 1906 Numbers Game," "The San Francisco Numbers Game," *California Geology* 40, Dec. 12, 1987, 271-274; Gladys Hansen, Frank Quinn, and F. E. McClure, "The 1906 Numbers Game," *Seismological Research Letters* 58(1) (Seismological Society of America, 82nd Annual Meeting," 1987, abstract only

10. Dan Kurzman, *Disaster! The Great San Francisco Earthquake and Fire of 1906* (New York: William Morrow, 2001), 248; James Dalessandro, *1906* (San Francisco: Chronicle Books, 2004), 14, 359

11. For the Messina earthquake, see Enzo Boschi et al., *Catalogo dei forti terremoti in Italia dal 461 a.C.al 1980* (Rome: Istituto Nazionale di Geofisica, 1995), 462-473; Chen and Scawthorn, *Earthquake Engineering,* 1-2, Table 1.1

12. There is an ongoing debate about the number of casualties in the Messina-Calabria earthquake. According to Japanese and Italian investigators Omori and Rizzo, the casualties were above 100,000. This figure was challenged as excessive by Davidson, who cited an upper estimate of around 82,000 for the major cities of Messina and Reggio without including the smaller cities in Calabria. In the recent catalogue of earthquakes in Italy, Boschi et al. cite no total number of casualties but estimate that 60,000 people died in Messina and its territory, which they calculate to have been 42 percent of the population. They also state that 21 percent of the population of Reggio died, about 12,000. Their total is therefore 72,000. But estimates very widely, with Wakabayashi near the top with 120,000 dead. The 98 percent destruction of houses was advanced by Omori and accepted by Davidson. For a discussion of fatalities in relation to Japanese damage estimates see Gregory Clancey, "Art Nation/Earthquake Nation: The Cultural Economy of Japanese Seismicity," Symposium: Architecture and Modern Japan, Columbia University, October 21, 2000, 15-16. See also Charles Davison, *Great Earthquakes* (London: T. Murby & Co., 1936), 203; Fusakichi Omori, "Preliminary Report on the Messina-Reggio Earthquake of December 28, 1908," *Bulletin of the Imperial Earthquake Investigation Committee in Foreign Languages* 3(1), 1909; Minoru Wakabayashi, *Design of Earthquake-Resistant Buildings* (New York: McGraw Hill, 1986), tables 1-2; Chen and Scawthorn, *Earthquake Engineering,* 1-2, Table 1.1; Boschi et al., *Catalogo dei forti terremoti,* 462-473. The lower mortality rate in Calabria may relate to the reinforcing system, called the *casa baraccata,* used in many masonry buildings constructed after the 1783 earthquake and still extant in 1908. See chapters 1 and 4.

13. Gladys Hansen has been working to establish an accurate total of the victims of the 1906 earthquake and fire for decades. The only comprehensible tally that she has published appears in "The San Francisco Numbers Game," *California Geology* 40, 271-274. In that article she lists the death toll by category. In her list are victims of heart problems (80) and suicides (80). She estimated 1,498 deaths in both the earthquake and fire. Since 1987 her estimates of deaths have climbed steadily but no article or scientific statement supporting the ever growing number of fatalities has appeared. Following the suggestion of Donald Cheu of the Governor's Earthquake Task Force, Hansen is now defining earthquake death as "an immediate fatality resulting from an earthquake or an earthquake-caused injury or illness that becomes fatal within a period of one year following the earthquake." This definition is not helpful for researchers attempting to estimate deaths directly caused by building failure. In 1986 Hansen, along with her son Richard, established the San Francisco Earthquake Research Project under the auspices of the California Academy of Sciences with engineers Frank McClure and Henry Degenkolb, seismologist Bruce Bolt, and others as an advisory council. I discussed the pilot study with the late Henry Degenkolb and the late Bruce Bolt. It consisted of ten blocks, one of which was in the South of Market. Of the 203 buildings studied, 146 were wood and 23 deaths—all the deaths in this area--occurred in them. What killed these people is still a question: collapse of the buildings themselves? chimney failures? neighboring chimney failures? foundation or cripple-wall failures? falling interior furniture? On the larger issue, does the fatality figure for the pilot study indicate that wooden buildings were hazardous in the earthquake? How does it relate to fatality figures as a whole? Wood as a material is no guarantee of earthquake safety. Safety depends on how the wooden structure is constructed and where. Quality of construction is impossible to know. Bolt was unconvinced by Hansen's fatality figures for the earthquake as a whole and didn't endorse future studies. If more of this kind of research is attempted, figures must be checked, results understandably reported and interpreted, and totals confirmed by outside observers. Since Hansen is claiming a governmental cover-up or conspiracy, it is her responsibility to prove her charge. Modern authorities estimate the number of deaths throughout California from the earthquake to have been

"2000+" (Chen and Scawthorn, *Earthquake Engineering*, 1-2, Table 1.1).

14. California Code of Regulations, Sec. 1626 (2001)

15. The eyewitness accounts I am using are drawn largely from the San Francisco *Argonaut*. Between 1926 and 1927, the *Argonaut* published accounts collected by the Subcommittee on History of the Committee of Fifty. The Subcommittee on History, headed by UC Berkeley Professor Henry Morse Stephens, asked those who had experienced the earthquake to write an account and send it to Stephens, who was to collect and publish the thousands of responses. His work on the collection is well documented. He gave several lectures about the reminiscences on the East Coast but never published the collection. When he died in 1919, the material passed to another committee member, John S. Drumm. The many boxes of accounts had no home and were shuttled to various storage areas with no one taking an interest in them. A small part of the collection surfaced again in the late 1920s to appear in the *Argonaut*, after which the remains of the collection were lost. Was this some kind of conspiracy to suppress valuable earthquake material that might damage the reputation of San Francisco, as some authors have implied? I think not. When Stephens died, the History subcommittee's collection was as good as dead too. With no one to protect or nurture it, it was probably destroyed. Denial and conspiracy didn't destroy the collection, indifference did. Further, years later, when authors writing about the earthquake and fire stumbled on the *Argonaut* articles, they kept the source a secret. The importance of footnotes and bibliography is that they not only record the origin of the information but help to conserve the source itself. I found the *Argonaut* by chance in the 1980s. Malcolm Barker also found it by chance, cited it, and has published many accounts from it in his *Three Fearful Days, San Francisco Memoirs of the 1906 Earthquake and Fire* (San Francisco: Londonborn Publications, 1998). Unfortunately, the eyewitness reports are not always reliable. Frank Louis Ames (Ibid., 75) wrote that he was in the Palace Hotel when the quake hit and saw the clock tower of the Chronicle Building "waver." This would have been impossible, since the clock tower had burned the year before. I have done my best to corroborate the accounts published in this chapter.

    For the beginnings of the history collection, see Charles Derleth Papers, MS, The Bancroft Library, University of California, Berkeley, for the following articles: "Citizens Give History to City; Report of the History Committee Unfolds Undertaking of Magnitude," November, 24, 1906; "Compiling History of Earthquake," *Berkeley Gazette*, April 12, 1907; "Preparing Story of the Disaster; Progress of Earthquake History Outlined by Professor Stephens," March 29, 1908. Also see Henry Morse Stephens, Correspondence and Papers, 1890-1910, Cartons 1, 4, 5, and 7, Bancroft Library, University of California, Berkeley.

16. *Argonaut*, May 1, 1926, 4

17. *Argonaut*, May 15, 1926, 4

18. *Argonaut*, July 10, 1926, 5

19. John Ripley Freeman, *Earthquake Damage and Earthquake Insurance* (New York: McGraw-Hill, 1932), 332

20. Fred G. Plummer to G. K. Gilbert, October 10, 1906, MS, George Davidson Papers, Bancroft Library, University of California, Berkeley

21. Horace D. Dunn, July 19, 1906, MS, George Davidson Papers, Bancroft Library, University of California, Berkeley

22. *Argonaut*, July 10, 1926, 5

23. E. I. de Laveaga, diary entry, from the personal collection of Connie de Laveaga

24. Virtual tours are just that. I have tried to be as accurate as possible, using the sources available, the accounts, photographs, and Sanborn Insurance maps.

25. We know that Wright was interested in earthquake-resistant buildings and was responsible for the specifications for the 1871 City Hall, South Hall at UC Berkeley, and the state's school for the deaf, dumb, and blind, in Berkeley, all of which had earthquake-resistant features. Whether or not he used the same techniques when he designed the Hopkins Mansion can never be known, but it is the belief of the author that Wright did employ similar methods. See Norman J. Ronneberg, "The Bittersweet Life of John Wright," in *John Wright (1830–1915), Grandfather of West Coast Architecture* (Victoria, B.C.: Maltwood Art Museum and Gallery, 1990).

26. Using photographs as evidence is always a problem, because the date and time (and sometimes the place) are difficult to establish. In the case of the earthquake and fire, we know from insurance records that photographs were important documents, establishing that a building had not collapsed in the earthquake. If the building stood, and had not lost its walls, then it was covered by fire insurance. A traffic in fake photographs must have existed, but I have yet to see a faked photograph which convincingly "repairs" building damage. I have seen photographs retouched to make them appear more dramatic. Gladys Hansen has stated that the "Burning City" panorama has been altered. She contends the tower of the Hall of Justice fell in an early aftershock and was reinserted in this picture. Actually, the superstructure of the tower never fell in the earthquake or the fire. Only the tiny cupola at the top slumped over, either because of aftershocks or fire, although the masonry cladding did. In other photographs before and after the earthquake, the tower looks out of place in relation to the rest of the Hall of Justice because its shaft is aesthetically so discordant with the facade behind. That may explain its strange look in the photograph. In a similar case, we might say Selby's shot tower was faked because it did not survive the fire. However, there are other photographs of it standing as the fire burns around it, proving it survived the initial shock and aftershocks only to fall for an unknown reason. One would also have to question the motive of retouching the tower of the Hall of Justice. The photograph records a lot of other damage. Why try to retouch a small cupola at the top of the tower of the Hall of Justice? I believe that this photograph and the others I have used are reliable. Beware though: in this chapter I have altered a photograph of the Chronicle Building to recreate its appearance on the morning of April 18.

27. Bricks from the California Hotel fell on the Bush Street fire station next door, injuring Captain Dennis Sullivan. An apparently trustworthy eyewitness states that the cupola of the hotel fell onto the fire station. This seems unlikely, because the round plan of the brick turret would probably have prevented

it from collapsing on the fire station. However, the tall chimney on the side of the turret in a pre-1906 photograph (see chapter 8) appears to be extremely dangerous. It is this huge chimney which hit the fire station. Because the chimney is on the southeastern side of the hotel, it would not be visible from the Hopkins mansion. By analogy, damage on the southeast and south sides of all blocks would not appear here. Photographs like this one give impressions of damage, but obviously much more evidence is needed for conclusive proof of damage or lack of it.

28. For Temple Emanu-El, see Ruth Hendricks Willard and Carol Green Wilson, *Sacred Places of San Francisco* (San Francisco: Presidio Press, 1985), 57; Harold Kirker, *California Architectural Frontier,* 3d ed. (Salt Lake City: Gibbs M. Smith, 1986), 76.

29. For the column, see Michael Corbett, *Splendid Survivors* (San Francisco: California Living Books, 1979), 238; United States Geological Survey Bulletin 324, *The San Francisco Earthquake and Fire of April 18, 1906, and Their Effects on Structures and Structural Materials*, ed. Grove Karl Gilbert et al. (Washington, D.C.: Government Printing Office, 1907), 90; Bernice Scharlach, *Big Alma, San Francisco's Alma Spreckels* (San Francisco: Shotwell Associates, 1999), 16.

30. Corbett, *Splendid Survivors,* 161; A.L.A. Himmelwright, *The San Francisco Earthquake and Fire of 1906* (New York: Roebling Construction Company, 1906), 172, 175; USGS (Gilbert et al.), *San Francisco Earthquake and Fire,* 43-44, 103-104

31. "Chronicle Tower Will Be Rebuilt," *San Francisco Chronicle,* November 9, 1905

32. *Argonaut,* April 23, 1927, 5. Although there was disagreement about how badly the stones on the upper floors of the Call Building had been displaced, observers agreed that some damage had been done: Frederick Meyer wrote that the 13th floor had distorted eye-bar frames due to swaying ("The Humboldt Bank Building," *American Builders Review* 5[1], October 1906, 1).

33. Himmelwright, *San Francisco Earthquake and Fire*, 63-66; USGS (Gilbert et al.), *San Francisco Earthquake and Fire,* 35

34. Ibid.

35. Maurice C. Couchot, "A Pictorial Record of the Results of Earthquake and Fire in San Francisco," *Engineering Record* 53 (May 5, 1906): 577-578; Himmelwright, *San Francisco Earthquake and Fire,* 66; USGS (Gilbert et al.), *San Francisco Earthquake and Fire,* 35

36. Corbett, *Splendid Survivors,* 85

37. USGS (Gilbert et al.), *San Francisco Earthquake and Fire,* 42-43, 96

38. *Argonaut,* August 28, 1926, 5; November 13, 1926, 5; November 27, 1926, 5

39. *Argonaut,* November 13, 1926, 5; November 27, 1926, 5

40. *Argonaut,* November 20, 1926, 5; November 27, 1926, 5

41. "New and Handsome Business Block," *San Francisco Chronicle,* January 12, 1896

42. Gladys Hansen and Emmet Condon, *Denial of Disaster* (San Francisco: Cameron and Co., 1989), 34

43. "Burdette's Building Is Intact amid Ruins," *San Francisco Call,* June 18, 1906

44. Corbett, *Splendid Survivors*, 102

45. Himmelwright, *San Francisco Earthquake and Fire,* 124

46. I examined the Atlas Building shortly after the Loma Prieta earthquake and took numerous photographs to document its condition.

47. Himmelwright, *San Francisco Earthquake and Fire,* 179, 184; USGS (Gilbert et al.), *San Francisco Earthquake and Fire,* 42

48. USGS (Gilbert et al.), *San Francisco Earthquake and Fire*, 41. The building was judged "substantial" with exterior damage in displacement of stones on facades, internal damage to tiles and finishes, some attributed to dynamiting, but not a collapse hazard.

Chapter 7

1. This is an oral account and may not be wholly accurate. I present it here as an example of the stories that San Francisco families have that bring the city's history to life. Hopefully, the text blends a fascination with the historic event of the fire itself with an analysis of what occurred. According to my father, Justice Mathew Tobriner (1904–1982), his father, Dr. Oscar Tobriner (1874–1960), was staying with the Samuel Leszynsky family at 3046 Jackson Street. The address is corroborated by the May 1906 *San Francisco Directory.* Because of the earthquake and fire, the only other source is the 1905 *Directory,* which lists Samuel Leszynsky at 2928 Pacific Ave., an address which also appears under his name in the May 1906 *Directory.* According to both my father and my cousin Beth Davis, Mrs. Leszynsky's reaction to the sad spectacle of an exhausted man pushing a woman in a wheelchair on April 18 was, "Look at those poor people! Oh my God, that's my daughter!" My grandfather was staying with his wife's family, rather than with his Tobriner relatives, who had a house at 2820 Clay Street. They all must have evacuated together. At the time, my grandfather's office was on the eighth floor of the Mutual Savings Bank Building at 708 Market Street, across from the Call Building. It survived the earthquake but burned in the fire.

2. James Cornell, *The Great International Disaster Book* (New York: Charles Scribner's Sons, 1976), 279-280. Prior to 2005, the most costly disaster in American history was Hurricane Agnes, in 1972, and the most deadly disaster was a hurricane that hit Galveston, Texas, in 1900 (Cornell, 27). Hurricane Katrina, which devastated the Louisiana coast in 2005, will probably prove to be the most costly disaster in U.S. history to date. For the 1906 fire as the largest urban fire in the United States, see

C. Scawthorn and T. D. O'Rourke, "Effects of Ground Failure on Water Supply and Fire Following Earthquake: The 1906 San Francisco Earthquake" (paper presented at the Second Annual U.S.-Japan Workshop on Liquefaction, Large Ground Deformation and Their Effects on Lifelines, Grand Island and Ithaca, New York, September 26-29, 1989).

3. "Reports of the Fire Officers of the San Francisco Fire Department on the Fire of 1906: Copies of the Reports in the Possession of Battalion Chief Fred J. Bowlen," MS, Bancroft Library, University of California, Berkeley, 1

4. Ibid. Captain J. R. Mitchell stated that 51 fires had started after the earthquake (experiences of Captain J. R. Mitchell, Engine No. 3 in "Reports"), but the exact number is still in question. According to the *Argonaut,* July 23, 1927, 5, acting Fire Chief John Dougherty reported 52 fires immediately after the earthquake.

5. According to "Reports of the Fire Officers," the fires were of diverse origins: for example, electrical and chimney fires (Experiences of J. J. Conlon, Battalion Chief, Ninth District); drugstore fires (Experiences of Captain G. F. Brown, Engine Company No. 2, and His Men, Station, Pine near Larkin); tall industrial chimney collapses (Experiences of Captain Schmidt on April 18, 1906). Other factors include carelessness (see "Ham and Eggs," in Fred J. Bowlen, "Outline of the History of the San Francisco Fire," MS, Library of the California Historical Society, San Francisco, 12-13); arson (the Alcazar fire, Bowlen, 21; or the fire at Sixth and Harrison, Bowlen, 24).

6. A. M. Hunt, "The Effect of the Earthquake and Fire on the Power and Lighting Situation," *Journal of Electricity, Power and Gas* 35 (May 5, 1906), 4

7. See Edwards account in *Argonaut*, October 23, 1926, 5

8. *Argonaut,* November 6, 1926, 5

9. "Equipment of Fire Departments," *Insurance Engineering* 11(1) (January 1906), 38-60

10. Doris Muscatine, *Old San Francisco: The Biography of a City from Early Days to the Earthquake* (New York: G. P. Putnam's Sons, 1975), 258

11. "San Francisco Fire Department Historical Review 1849–1967," 1967, MS, California Historical Society, San Francisco, 7

12. *San Francisco Municipal Reports for the Fiscal Year 1866–1867, ending June 30, 1867* (San Francisco: Joseph Winterburn and Co., 1867), 203 and attached lists

13. Ibid., 9-10

14. *San Francisco Municipal Reports for the Fiscal Year 1872–1873, ending June 30, 1873* (San Francisco: Board of Supervisors, 1873), 129

15. *San Francisco Municipal Reports for the Fiscal Year 1879–1880, ending June 30, 1880* (San Francisco: Board of Supervisors, 1880), 196-197

16. Ibid., 196, 199

17. *San Francisco Municipal Reports for the Fiscal Year 1898–1899, ending June 30, 1899* (San Francisco: Board of Supervisors, 1899), 511-516

18. *San Francisco Municipal Reports for the Fiscal Year 1904–1905, ending June 30, 1905* (San Francisco: Board of Supervisors, 1907), 206-220; *San Francisco Municipal Reports for the Fiscal Year 1905–1906, ending June 30, 1906, and Fiscal Year 1906–1907, ending June 30, 1907* (San Francisco: Board of Supervisors, 1908), 64-73; S. Albert Reed, *The San Francisco Conflagration of April, 1906: Special Report to the National Board of Fire Underwriters Committee of Twenty* (New York: The Committee, May 1906), 6

19. O'Rourke and Scawthorn, "Effects of Ground Failure"

20. *Report on the City of San Francisco, California* (New York: The National Board of Underwriters Committee of Twenty, October 1905), 63

21. There were 38 engine companies in San Francisco. See *Report on the City of San Francisco*, 27. The companies housed in brick buildings were 1, 4, 5, 7, 9, 12, 15, 17, 18, 19, and 35. The wood-frame buildings were occupied by companies 11, 14, 16, 20-23, 25, 27-32, and 36-38. See Fred J. Bowlen, "A Summary of the Introductory Paragraphs," MS, Library of the California Historical Society, San Francisco, 2-5

22. For jammed doors and stampeding horses, see Bowlen, "Reports" and "A Summary," as well as *Argonaut* (July 23, 1927). The data that emerges is that stampedes or runaway horses hampered engine companies 4, 6, and 19, while doors jammed in numbers 5, 7, and 9. The *Argonaut* states that of the total of 45 firehouses in the city, 15 were temporarily disorganized by the earthquake.

23. *Argonaut,* Sept. 11, 1926, 5

24. "San Francisco Fire Department," 12

25. Building collapses trapped inhabitants, and fire personnel were delayed in rescue operations. See "Reports of the Fire Officers": Experiences of Captain C. J. Cullen, Engine Company No. 6, and His Men, Station 62 South Street Near Seventh; Experiences of Captain J. Conniff and His Men from April 18th to April 20th; Experiences of Captain T. J. Murphy, Engine 29, and His Men; Experiences of F. Nichols, Truck No. 4 Captain, 1648 Pacific; also see *Argonaut* (August 6, 1927), 5.

26. Fred J. Bowlen, "The San Francisco Earthquake and Fire 1906," MS, California Historical Society, 1934, 5

27. For the over 23,000 broken water connections, see Herman Schussler, *The Water Supply of San Francisco, California, Before, During and After the Earthquake of April 18, 1906, and the Subsequent Conflagration* (New York: M. Brown Press, 1906), 34, and Marsden Manson et al., *Report, The Sub-Committee on Statistics to the Chairman and Committee on Reconstruction* (San Francisco, April 24, 1907), 1

28. *Report on the City of San Francisco*, 64

29. Bowlen, "A Summary," 9-10; *San Francisco Municipal Reports for the Fiscal Year 1904–1905*, 212

30. *Report on the City of San Francisco, California,* 24; Edwin Duryea Jr., "Suggestion from a Preliminary Report of the Sub-Committee on Water Supply and Fire Protection, Committee of Forty on Reconstruction of San Francisco, on Present Condition of Water System Serving San Francisco, and Notes on the Particular Features in which this System Fails to Give an Efficient Fire Protection to this City," May 18, 1906,

MS, Stanford Library; O'Rourke and Scawthorn, "Effects of Ground Failure"

31. Bowlen, "A Summary," 6

32. For water used to save the Mint, see Bowlen, "Outline," 10A. In another case, water used to save the territory south of Bryant, west of Eighth, south of Harrison, and west of Eleventh was supplied by the United Railroads powerhouse (Bowlen, "Outline," 50).

33. Bowlen, "Outline," 52-53

34. Bowlen, "A Summary," 7

35. O'Rourke and Scawthorn, "Effects of Ground Failure"

36. Bowlen, "Outline," 2-3

37. Fires controlled by noon included blazes at London and Paris Streets; Clement Street; 22nd and Mission; 2021 California; Golden Gate Avenue and Buchanan; Fulton and Octavia; Hayes and Laguna; between 22nd and 23rd on Harrison; at O'Farrell and Taylor; Pacific and Leavenworth; 17th Avenue near Clement; China Avenue near London Street; Davis between Pacific and Broadway; Front Street between Vallejo Street and Broadway; Clay between Walnut and Laurel; Geary near Divisadero; Ashby and Waller; Waller and Masonic; Oak near Stanyan; Polk between Bush and Pine; Bay at Powell; Bay and Kearny; Dupont between Broadway and Pacific; Schrader between Waller and Frederick; Davis between Clay and Commercial. See Bowlen, "Outline," 2-7. For an excellent overview of the progression of the fire, see S. Albert Reed, *The San Francisco Conflagration.*

38. Bowlen, "Outline," 13-14

39. Ibid., 12-13

40. *LeCount & Strong's San Francisco City Directory for the Year of 1854* (San Francisco: San Francisco Herald Office, 1854), 213-214; *San Francisco Municipal Reports for the Fiscal Year 1874–1875*, 828-829

41. John P. Young, *San Francisco: A History of the Pacific Coast Metropolis* (San Francisco: S. J Clarke Publishing Company, 1912), II, 569-570. Also see "Order Defining the Fire Limits and Regulating the Construction of Buildings in the City and County of San Francisco," Order No. 1617, approved March 2, 1881, San Francisco, 1881.

42. "Order No. 1,752: To Define the Fire Limits of the City and County of San Francisco, and Making Regulations Concerning the Erection and Use of Buildings in Said City and County," approved December 28, 1883, *General Orders of the Board of Supervisors of the City and County of San Francisco* (San Francisco: Board of Supervisors, 1884), 160-161

43. "Order No. 1,917: To Define the Fire Limits of the City and County of San Francisco, and Making Regulations Concerning the Erection and Use of Buildings in Said City and County," approved June 21, 1887, *General Orders of the Board of Supervisors of the City and County of San Francisco* (San Francisco: Board of Supervisors, 1888), 128-172

44. *California Architect and Building News* 10(11) (November 15, 1889), 141, and 10(12) (December 15, 1889), 150

45. Building Ordinance No. 645, Sec. 96 (February 5, 1903), *General Orders and Ordinances of the Board of Supervisors of the City and County of San Francisco* (1904)

46. Committee on Fire-Resistive Construction, *The Baltimore Conflagration: Report of the Committee on Fire-Resistive Construction of the National Fire Protection Association* (Chicago: National Fire Association, 1904), 5

47. Ibid.

48. USGS (Gilbert et al.), *San Francisco Earthquake and Fire,* 146

49. Bowlen, "Outline," 9

50. *Report on the City of San Francisco*

51. Building Ordinance No. 291, Sec.18 (May 8, 1901), *General Orders and Ordinances of the Board of Supervisors of the City and County of San Francisco* (1904)

52. Robert K. Mackenzie, *The San Francisco Earthquake and Conflagration, April, 1906* (Liverpool: Liverpool Post and Mercury Printing Works, 1907), 8

53. I am grateful to Dr. Stephen L. Quarles, Wood Durability Advisor, Agriculture and Natural Resources, University of California, for discussing the qualities of redwood with me. See Stephen L. Quarles, Laurence G. Cool, and Frank C. Beall, *Performance of Deck Board Materials Under Simulated Wildfire Exposures*, Seventh International Conference on Wood Fiber-Plastic Composites (and Other Natural Fibers), May 19-20, 2003, Madison, Wisc. (Madison, Wisc.: Forest Products Society, 2003), 89-93, and Vytenis Babrauskas, *Ignition Handbook, Principles and Applications to Fire Safety Engineering, Fire Investigation, Risk Management and Forensic Science* (Issaquah, Wash.: Fire Science Publishers, 2003), 1073-1077.

54. For the Mint fight, see Bowlen, "Outline," l0A

55. For the Post Office firefight, see *Argonaut,* December 18, 1926, 5, and December 25, 1926, 5.

56. William Bronson, *The Earth Shook, the Sky Burned* (1959; reprint, New York: Crown, 1986), 54

57. *Argonaut,* August 13, 1927, 6

58. *Argonaut,* December 4, 1926, 5

59. See Gordon Thomas and Max Morgan Witts, *The San Francisco Earthquake* (New York: Stein and Day Publishers, 1971), in particular pp. 88-95, for a scathing evaluation of Funston's seizure of the city.

60. Bowlen, "Outline," 40, 59, 61; *Argonaut,* August 13, 1927, 6

61. See Captain LeVert Coleman's statement in *Earthquake in California, April 18, 1906, Special Report of Maj. Gen. Adolphus W. Greely*, U.S. War Department Annual Report of the Secretary of War (Washington, D.C.: Government Printing Office, 1906). For dynamiting and the burning of Chinatown, see Thomas and Witts, *San Francisco Earthquake,* 137, and *Argonaut,* August 13, 1927, 6.

62. Bowlen, "Outline," 42-43

63. Ibid., 41, 59-63. People would have defended their property if given the chance. Russian Hill was defended by citizen firefighters who, although they had been ordered to leave

the area, eventually saved a few houses on Green Street (Bowlen, "Outline," 40-41). Citizen fighters led by Van Wych successfully fought the fire at Greenwich Street and saved a store by using vinegar to fight the fire. On Russian Hill the house of O. D. Baldwin was saved by citizen firemen. The Robert Louis Stevenson home was saved by volunteers from the Bohemian Club. Citizens saved their homes on Telegraph Hill (ibid., 59-61). For post office employees and employees of Folgers and for the defense of the Kohl, see *Argonaut* (August 13, 1927), 6.

64. "San Francisco Conflagration Report," *Coast Review* 72(3) (September 1907), 635

Chapter 8

1. The major investigating groups were: the California State Earthquake Investigation Commission, composed of Andrew C. Lawson, G. K. Gilbert, H. F. Reid, J. C. Branner, A. O. Leuschner, George Davidson, Charles Burkhalter, and W. W. Campbell (Andrew C. Lawson et al., *The California Earthquake of April 18, 1906, Report of the State Earthquake Investigation Commission,* 2 vols. [Washington, D. C.: Carnegie Institution of Washington, 1908]; Japanese government investigators Omori and Nakamura ("World's Greatest Seismologist Says San Francisco Is Safe," *San Francisco Call,* August 5, 1906, 4-8; Fusakichi Omori, "Notes on the Comparison of the Faults of the Three Earthquakes of Mino-Owari, Formosa, and California," in *The California Earthquake,* Lawson et al., 146; Omori, "Preliminary Note on the Cause of the California Earthquake of April 18, 1906," in *The California Earthquake of 1906,* David S. Jordan, ed. (San Francisco: A. M. Robertson, 1907), 281-318; Omori, "Preliminary Note on the Seismographic Observation of the San Francisco Earthquake of April 18, 1906," *Bulletin of the Imperial Earthquake Investigation Committee* 1 (January 1907), 26-43; "Dr. Nankamure [sic] on the Effects of Severe Earthquakes," *San Francisco Chronicle,* June 19, 1906, 14).

2. Engineering reports comprised the U.S. Geological Survey, composed of Grove Karl Gilbert, Richard Lewis Humphrey, John Stephen Sewell, and Frank Soulé (Gilbert et al., eds., *The San Francisco Earthquake and Fire of April 18, 1906, and Their Effects on Structures and Structural Materials,* USGS Bulletin 324, Washington, D.C.: Government Printing Office, 1907); San Francisco Subcommittee on Statistics, composed of Marsden Manson, Edwin Duryea Jr., Virgil G. Bogue, James W. Reid, and C. H. McKinstry (Manson et al., *Report, the Sub-Committee on Statistics to the Chairman and Committee on Reconstruction,* San Francisco, April 24, 1907); the general committee of the American Society of Civil Engineers composed of Edwin Duryea Jr., C. D. Marx, Franklin Riffle, Arthur L. Adams, and William W. Harts ("The Effects of the San Francisco Earthquake of April 18th, 1906, on Engineering Constructions," in *Transactions of the American Society of Civil Engineers* 59 [December 1907], 208-329); the Structural Association of San Francisco with the following officers: C. B. Wing, president; W. J. Miller, vice-president; Lewis A. Hicks, second vice-president; L. E. Hunt, third vice-president, and Charles Derleth Jr., secretary-treasurer, and 141 members (papers published in *Engineering Supplement to the American Builder's Review,* June 30, 1906-July 1907)

3. S. Albert Reed, *The San Francisco Conflagration of April, 1906: Special Report to the National Board of Fire Underwriters Committee of Twenty* (New York: The Committee, May 1906) and *Report of the Committee of Five to the "Thirty-Five Companies" on the San Francisco Conflagration, April 18-21, 1906,* A. R. Hosford, Chairman (1906).

4. A. L. A. Himmelwright, *The San Francisco Earthquake and Fire of 1906* (New York: Roebling Construction Company, 1906)

5. Charles Derleth Jr., *The Destructive Extent of the California Earthquake: Its Effect upon Structures and Structural Materials within the Earthquake Belt* (San Francisco: A. M. Robertson, 1907)

6. John R. Freeman, *Earthquake Damage and Earthquake Insurance* (New York: McGraw-Hill, 1932)

7. See Philip L. Fradkin, *The Great Earthquake and Firestorms of 1906: How San Francisco Nearly Destroyed Itself* (Berkeley: University of California Press, 2005), "Selected Readings," 395-402, and Hansen and Condon (note 8) for numerous book citations. The best book written on the San Francisco earthquake and fire is Frank W. Aitken's *A History of the Earthquake and Fire in San Francisco; An Account of the Disaster of April 18, 1906, and Its Immediate Results* (San Francisco: The Edward Hilton Co., 1906). Aitken's lucid and well-considered descriptions of both the fire and the earthquake, the trustworthiness of his information, and the 284 illustrations he includes (many by him) make this the classic nonengineering text on the twin disasters.

8. William Bronson, *The Earth Shook, the Sky Burned* (Garden City, New York: Doubleday, 1959); Eric Saul and Don DeNevi, *The Great San Francisco Earthquake and Fire, 1906* (Millbrae, Calif.: Celestial Arts, 1981); Gladys Hansen and Emmet Condon, *Denial of Disaster: The Untold Story and Photographs of the San Francisco Earthquake and Fire of 1906* (San Francisco: Cameron, 1989)

9. Duryea et al., ASCE *Transactions* 59, 264-329

10. Lawson et al., *California Earthquake,* 233-241; USGS (Gilbert et al.), *San Francisco Earthquake and Fire,* 56

11. Lawson et al., *California Earthquake,* 220-227 and accompanying atlas maps 4, 17, 18, 19

12. Ibid., 239-240

13. Ibid., 239

14. Ibid., 238-239

15. Ibid., 238

16. Ibid., 232-233

17. USGS (Gilbert et al.), *San Francisco Earthquake and Fire,* 143

18. "The Effects of the San Francisco Earthquake," ASCE *Transactions* 59, 233

19. Ibid. and USGS (Gilbert et al.), *San Francisco Earthquake and Fire,* 105, 144

20. USGS (Gilbert et al.), *San Francisco Earthquake and Fire,* 57, 105-107, 144; F. A. Koetitz, "The Tower of the Union Ferry Depot, San Francisco," *American Builders Review* 6(1) (January 1907), 1-11

21. H. C. Vensano, "A Suggestion for Earthquake Design in High Building Construction, *American Builders Review* 4(3), September 1906, 267: "A good example of this destruction of rivets at the first floor level can be seen in the James Flood Building, where nearly all the rivets connecting the spandrel girders at that level have sheared off." Himmelwright says nothing of this problem (Himmelwright, 49), and the building was quickly repaired.

22. C. F. Weiland, "Safe Structural Treatment of Building Frontages," *American Builders Review* 5(2) (November 1906), 67-69

23. Duryea et al., ASCE *Transactions* 59, 233

24. Maurice C. Couchot, "A Pictorial Record of the Results of Earthquake and Fire in San Francisco," *Engineering Record* 53 (May 5, 1906), 577-578; Himmelwright, *San Francisco Earthquake and Fire,* 187; USGS (Gilbert et al.), *The San Francisco Earthquake and Fire,* 41 and 45.

25. Himmelwright, *San Francisco Earthquake and Fire,* 121; and USGS (Gilbert et al.), *San Francisco Earthquake and Fire,* 45, 103

26. Weiland, "Safe Structural Treatment," 68

27. Himmelwright, *San Francisco Earthquake and Fire,* 213-222

28. USGS (Gilbert et al.), *San Francisco Earthquake and Fire,* 90-91

29. Charles M. Coleman, *P. G. and E. of California: The Centennial Story of Pacific Gas and Electric Company, 1852–1952* (New York: McGraw-Hill, 1952), 240

30. Sara H. Boutelle, *Julia Morgan, Architect* (New York: Abbeville Press Publishers, 1988), 45

31. Himmelwright, *San Francisco Earthquake and Fire,* 231-233; USGS (Gilbert et al.), *San Francisco Earthquake and Fire,* 34

32. Himmelwright, *San Francisco Earthquake and Fire,* 70

33. Duryea et al., ASCE *Transactions* 59, 322

34. For a discussion of the problem of URMs and retrofitting strategies in relation to past earthquakes, see Rutherford and Chekene Consulting Engineers, "Seismic Retrofitting Alternatives for San Francisco's Unreinforced Masonry Buildings: Estimates of Construction Cost and Seismic Damage, Prepared for the City and County of San Francisco Department of City Planning," May 1990, 5.1-5.22.

35. Duryea et al., ASCE *Transactions* 59, 230-231; Gretchen Smith and Robert Reitherman, *Damage to Unreinforced Masonry Buildings at Stanford University in the 1906 San Francisco Earthquake* (Redwood City: Scientific Service, December 1984)

36. Duryea et al., ASCE *Transactions* 59, 223, 225, 230

37. Rutherford and Chekene, "Seismic Retrofitting Alternatives," 3.4

38. Duryea et al., ASCE *Transactions* 59, 230

39. *Argonaut,* June 5, 1926, 4; July 12, 1926; July 10, 1926, 5; August 7, 1926, 5

40. *Argonaut,* September 11, 1926, 5, for the account

41. *Argonaut,* June 26, 1926, 4

42. *Argonaut,* May 1, 1926, 4

43. Duryea et al., ASCE *Transactions* 59, 326

44. *Argonaut,* June 26, 1926, 4

45. *Argonaut,* July 3, 1926, 5

46. Ibid.

47. *Argonaut,* October 9, 1926, 5; October 16, 1926, 5

48. See *Argonaut,* July 10, 1926; January 1, 1927; April 23, 1927, 5, for examples of collapsed brick buildings.

49. Beth Israel and Pike's Memorial were both newly completed and generally seem to have been poorly designed and built with "weak and flimsy framing, insufficient bracing and poor mortar." USGS (Gilbert et al.), *San Francisco Earthquake and Fire,* 26

50. The Girls' High School collapse, according to Humphrey, was due to "lack of proper tie between the floor and roof timbers and the walls and the poor quality of the mortar." USGS (Gilbert et al.), *San Francisco Earthquake and Fire,* 26

51. The Majestic Theater (Ninth and Market): USGS (Gilbert et al.), *San Francisco Earthquake and Fire,* 40, and Himmelwright, *San Francisco Earthquake and Fire,* 197; Tivoli, ibid., 227; Colonial Theater, ibid., 195; Bell Theater, ibid., 233; Jackson Brewery, ibid., 236

52. R. L. Humphrey thought the partial collapse of Frederick Meyer's Hahnemann Hospital was due to "bad design, the roof trusses butting against the walls and the floor timbers resting upon the walls without adequate ties. The poor quality of the mortar permitted a ready disintegration of the brick-veneered walls, although some band iron [sic] had been used for the purpose of strengthening them." USGS (Gilbert et al.), *San Francisco Earthquake and Fire,* 27

53. "Chimneys collapsed most generally, breaking about halfway up, and destroyed in part at least the structures upon which they fell...It is quite evident that brick stacks and similar tall structures built of brick or stone without reinforcement against flexure, or without being guyed, are unsuitable for use in countries liable to earthquake shock. They should be constructed either of steel, guyed, or if self-supporting, of steel or reinforced concrete": USGS (Gilbert et al.), *San Francisco Earthquake and Fire,* 57. Investigators like Sewell were surprised that the two brick stacks of the U.S. Mint were intact, and credited their survival to the thickness of the masonry: USGS (Gilbert et al.), *San Francisco Earthquake and Fire,* 95. But he was of the opinion that "nearly all [smokestacks] suffered more or less damage": ibid., 112. The team from ASCE concurred: Duryea et al., ASCE *Transactions,* 236. Also see L. Wagoner's comments on reinforcement for tall brick chimneys (ibid., 272) and Himmelwright, *San Francisco Earthquake and Fire,* 242. Among those that failed were the Valencia Street power plant: USGS (Gilbert et al.), *San Francisco Earthquake and Fire,* 57.

54. Selby's shot tower survived the initial earthquake and aftershocks (see Hansen, *Denial,* 13) but evidently succumbed

to the fire. Post-fire photographs don't show it, and observers don't include it in their accounts of successful chimneys.

55. Smith and Reitherman, *Damage,* 14-15; USGS (Gilbert et al.), *San Francisco Earthquake and Fire,* 112-114

56. This assertion that Freeman makes seems to be borne out by several photographs of the Jackson Square area in the Bancroft Library.

57. Freeman, *Earthquake Damage,* 319-358. Freeman includes the following brick buildings as "Examples of Resistance to the San Francisco Quake of 1906": U.S. Appraiser's Building, Palace Hotel, Phelan Building, Girls' High School, Spring Valley Building, Crocker Building, New Chronicle Building, California Electric Works, National Ice Company Building, Black Point Pumping Station, Dunham Carrigan & Hayden Co. Building, Grant Building, St. Mary's Cathedral, California School of Mechanical Arts, Wells Fargo Building, Kohl Building, Atlas Building, U.S. Mint, Grace Cathedral, St. Francis Church, Clarendon Heights Pumping Station, Montgomery Block, Folger Building, and Lowry Building.

58. Himmelwright, *San Francisco Earthquake and Fire,* 28; USGS (Gilbert et al.), *San Francisco Earthquake and Fire,* 154

59. Fred G. Plummer to G. K. Gilbert, October 1906, Andrew C. Lawson Papers, Bancroft Library, University of California, Berkeley

60. For background on these materials, see chapters 4 and 6.

61. USGS (Gilbert et al.), *San Francisco Earthquake and Fire,* 149

62. Ibid., 151

63. James R. Tapscott, letter to Mrs. I. J. Tapscott, April, 22, 1906, MS, Bancroft Library, University of California, Berkeley

64. Himmelwright, *San Francisco Earthquake and Fire,* 179, 184

65. Stephen Tobriner, "South Hall and Seismic Safety at the University of California in 1870," *Chronicle of the University of California* 1(1), 14

66. Duryea et al., ASCE *Transactions* 59, 325

67. Ibid., 266-267

68. Ibid., 326

69. USGS (Gilbert et al.), *San Francisco Earthquake and Fire,* 57

70. Duryea et al., ASCE *Transactions* 59, 229

71. Freeman, *Earthquake Damage,* 356

72. G. A. Wright, "A Few Notes on Construction Suggested by the Great Fire, *American Builders Review* 4(1), July 1906, 177. "It is a question worthy of consideration as to whether the balloon-frame building, properly braced and fire stopped, is not preferable to the prevalent method of building up story by story, especially when we consider the twisting effect which some earthquakes seem to possess."

73. Although recent evaluations have asserted that wood-frame buildings did quite well in the quake. See Karl Steinbrugge, *Earthquakes, Volcanoes, and Tsunamis: An Anatomy of Hazard* (New York: Skandia America Group, 1982), 303. For chimney damage, see Himmelwright, *San Francisco Earthquake and Fire,* 15, and USGS (Gilbert et al.), *San Francisco Earthquake and Fire,* 111.

74. Clemens Max Richter, "Autobiography and Reminiscences," MS, Bancroft Library, University of California, Berkeley, October 1922

75. USGS (Gilbert et al.), *San Francisco Earthquake and Fire,* 110

76. For the opinion that 95 percent of chimneys in San Francisco fell, see Himmelwright, *San Francisco Earthquake and Fire,* 15.

77. Lawson et al., *The California Earthquake,* vol. 1, 354-357

78. *Argonaut,* September 18, 1926, 5

79. Stephen Tobriner, "The Failure of San Francisco's City Hall in the Earthquake of 1906," Domes from Antiquity to the Present, Proceedings of the IASS-MSU Symposium (Istanbul: Mimar Sinan University), 1988, 733-742. Newton J. Tharp's testimony appears in *San Francisco Municipal Reports for the Fiscal Year 1907-8, ending June 30, 1908* (San Francisco: Board of Supervisors, 1909), 743-749. For the condition of the dome, see pages 746-748.

80. Tharp, *San Francisco Municipal Reports*

81. Ibid., 747

82. San Francisco Main Library Project, Archaeological Monitoring and Architectural Documentation Site of the Former City Hall Completed in 1897 (Project #8920.00/9014.00), Basin Research Associates, San Leandro, Calif., 1994, 1, 19-21

83. At the time of the earthquake, the cost of steel-frame and fireproofing seldom exceeded 27 percent of the cost of the structure. See "All American Cities in Danger," *Ohio Architect and Builder* 9 (June 1907): 17.

84. Reed, *San Francisco Conflagration,* 2

85. *Report of the Special Committee of the Board of Trustees of the Chamber of Commerce of San Francisco on Insurance Settlements Incident to the San Francisco Fire* (San Francisco: Chamber of Commerce, November 13, 1906), 20; *Argonaut,* July 23, 1927, 5

86. "Majority Builds Poorly," *American Builders Review* 4(3) (September 1906), 250

87. Horace D. Dunn, "Statement Relative to the Effects of the Earthquake in San Francisco, Ca. July 19, 1906," MS, Davidson Papers, Bancroft Library, University of California, Berkeley

88. "Owners Blamed for Poor Construction," *Architect and Engineer* 9(3) (July 1907), 70

89. A. M. Hunt in Freeman, *Earthquake Damage,* 190

90. Harry O. Wood to George D. Louderback, February 12, 1947, MS, George Davis Louderback papers, 1900–1956, Bancroft Library, University of California, Berkeley

91. Steinbrugge, *Earthquakes, Volcanoes, and Tsunamis,* 224

92. "Majority Builds Poorly," *American Builders Review* 4(3). Also see F. W. Fitzpatrick, "The Red Monster," *Architect and Engineer* 7 (January 1907), 57.

93. Ibid., Fitzpatrick

Chapter 9

1. Judd Kahn, *Imperial San Francisco: Politics and Planning in an American City, 1897–1906* (Lincoln: University of Nebraska Press, 1979), 133-135, 154-158; Walton Bean, *Boss Ruef's San Francisco* (Berkeley: University of California Press, 1952), 121-123; Christopher M. Douty, *The Economics of Localized Disasters: The 1906 San Francisco Catastrophe* (New York: Arno Press, 1977), 98; J. Eugene Haas, Robert W. Kates, Martyn J. Bowden, *Reconstruction Following Disaster* (Cambridge, Mass.: MIT Press, 1977), 1-23
2. Kahn, *Imperial SF,* 133-135; Bean, *Boss Ruef,* 121-123
3. Kahn, *Imperial SF,* 184-185; Douty, *Economics,* 98, 101 (The legal holiday was to last until June 3, 1906.)
4. Bean, *Boss Ruef,* 162
5. Frank W. Aitken and Edward Hilton, *A History of the Earthquake and Fire in San Francisco* (San Francisco: Edward Hilton Co., 1906), 180; Douty, *Economics,* 87
6. Aitken, *History of Earthquake,* 181-182
7. William Bronson, *The Earth Shook, the Sky Burned* (Garden City, N.Y.: Doubleday, 1959), 118. For food distribution see Douty, *Economics,* 106-121; for tents see Douty, 126
8. Douty, *Economics,* 120-121
9. Kahn, *Imperial SF,* 136; Bean, *Boss Ruef,* 123
10. "Grand Jury Inquires into Shack Theaters," *San Francisco Chronicle,* February 16, 1907. For Ruef's manipulation of post-earthquake recovery, see Bean, *Boss Ruef,* 127-144; Douty, *Economics,* 105-106. Two scandalous incidences of giving preference to one company over another for a payoff were the Home Telephone franchise and the United Railroad's trolley ordinance. For the climate of reconstruction in relation to business see Martyn J. Bowden, "Reconstruction Following Catastrophe: The Laissez-Faire Rebuilding of Downtown San Francisco after the Earthquake and Fire of 1906," *Proceedings of the Association of American Geographers,* 1970, vol. 2, 22-26.
11. General Adolphus W. Greely in *San Francisco Relief Survey* (New York: The Russel Sage Foundation, 1913), 82-83; Bronson, *Earth Shook,* 122. Also see A. W. Greely, *Reminiscences of Adventure and Service* (New York: C. Scribner's Sons, 1927), 219-224.
12. Bronson, *Earth Shook,* 127; Douty, *Economics,* 129-136. There are several websites devoted to the investigation and preservation of the "earthquake shacks." The most informative are "Refugee Cottages," Presidio of San Francisco, National Park Service, Golden Gate National Recreational Area, www.nps.gov/prsf/history/1906eq/cottages.htm and Western Neighborhoods Project—1906 Earthquake Refugee Shacks, www.outsidelands.org/shacks.html.
13. "Will Erect Model Homes for Laborers," *San Francisco Chronicle,* May 26, 1906; for a comprehensive view of the housing program, see Douty, *Economics,* 309-338.
14. Bronson, *Earth Shook,* 127
15. Aitken, *History of Earthquake,* 183; Douty, *Economics,* 176-178
16. Aitken, *History of Earthquake,* 184-186
17. "Building Right Over the Debris to Save Time," *San Francisco Chronicle,* May 11, 1906
18. Frank Morton Todd, *Eradicating Plague from San Francisco* (San Francisco: The Citizens' Health Committee, March 31, 1909), 38-39
19. Ibid., 30
20. Ibid., 39
21. Ibid., 65, 73, 84, 103, 126-150
22. Ibid., 183
23. Fred J. Bowlen, "The San Francisco Earthquake and Fire of 1906," MS, California Historical Society, 1935), 10-11; Douty, *Economics,* 225-226
24. "Tearing Down Ruined Walls," n.d., Derleth papers, MS, Bancroft Library, University of California, Berkeley; Bowlen, "San Francisco Earthquake," 10-11
25. "Army Will Withdraw on the First of July," June 13, 1906, and "Regular Troops Will Quit Duties To-Day," July 2, 1906, Derleth papers; "Everybody Willing to Aid in Cleaning the City," *San Francisco Chronicle,* February 16, 1907; "Thousands Await the Call for Cleaning Up To-day," *San Francisco Chronicle,* March 3, 1907
26. Bronson, *Earth Shook,* 170
27. Louis J. Stellman, "What Becomes of San Francisco Scrap Iron and Pipe," *San Francisco Chronicle,* May 26, 1907, 9; E. O. Ritter, "Steel an Important Factor in the Rebuilding of San Francisco," *Architect and Engineer* 10(2) (September 1907), 65-66
28. "Steam Cars to Carry Debris," *San Francisco Examiner,* April 27, 1906; "Ocean Shore Railroad Gridirons Burnt District," *San Francisco Examiner,* May 5, 1906; "Free Removal of Debris," *San Francisco Chronicle,* May 6, 1906; illustration of ocean dumping operation, *San Francisco Examiner,* May 2, 1906; *San Francisco Municipal Reports for the Fiscal Year 1907-8, ending June 30, 1908* (San Francisco Board of Supervisors, 1909), 708
29. "Lad Crushed by Palace Walls," *San Francisco Call,* December 1, 1906
30. Ibid.
31. "Falling Wall Kills Workmen," *San Francisco Call,* May 29, 1906
32. "Disregarded Warning and Are Struck by Debris," *San Francisco Call,* February 28, 1908
33. "Collapse of Wall Kills a Laborer," *San Francisco Call,* May 30, 1907
34. "Comparatively Few Men Have Been Killed in Rebuilding," *San Francisco Call,* July 19, 1908

35. Stephen Tobriner, *The Genesis of Noto: An Eighteenth-Century Sicilian City* (Berkeley and London: University of California Press and A. Zwemmer, 1982); José-Augusto França, *Une Ville des Lumières, La Lisbonne de Pombal* (Paris: SEV-PEN, 1965), 125-134; Margarida Alçada, dir., *monumentos* 21 (September 2004); Maria Helena Ribeiro dos Santos, *A Baixa Pombalina, Passado e Futuro* (Lisbon: Livros Horizonte, 2000)

36. Howard Saalman, *Haussmann: Paris Transformed* (New York: George Braziller, 1971); Donald J. Olsen, *The City as a Work of Art* (New Haven: Yale University Press, 1986), ch. 4; Anthony Sutcliffe, *Paris: An Architectural History* (New Haven: Yale University Press, 1993), ch. 6; Joan Draper, "Paris by the Lake: Sources of Burnham's Plan of Chicago," in John Zukowsky, ed., *Chicago Architecture 1872–1922* (Munich: Prestel-Verlag, 1987), 107-120

37. Daniel H. Burnham and Edward H. Bennett, *Report on a Plan for San Francisco* (San Francisco: Sunset Press, 1905); Kahn, *Imperial SF,* 82-86

38. Kahn, *Imperial SF,* 95; for Burnham on the reconstruction of San Francisco, see Herman Scheffauer, "City Beautiful: San Francisco Rebuilt," *Architectural Review* 20 (July-December 1906), 3-8, 86-94.

39. Marsden Manson didn't believe that firebreaks were effective (see M. Manson, H. D. H. Connick, T. W. Ransome, W. C. Robinson, *Reports on an Auxiliary Water Supply System for Fire Protection for San Francisco. California,* Board of Public Works, 1908, 60-61).

40. Bean, *Boss Ruef,* 124-127

41. William Issel and Robert W. Cherny, *San Francisco, 1862–1932: Politics, Power, and Urban Development* (Berkeley: University of California Press, 1986), 126-130

42. *San Francisco Bulletin,* April 29, 1906

43. *San Francisco Chronicle,* April 27, 1906

44. "For Chinatown on Old Site," *San Francisco Chronicle,* May 11, 1906

45. F. W. Fitzpatrick, "Fears Another Wooden City," *Architect and Engineer* 12(3) (January 1907), 77

46. Much of the steel in the Emporium was recycled from its own ruins (Ritter, "Steel an Important Factor," 65-66, and *Iron Age,* May 1907), but not the dome (Fred T. Huddart, "Largest Steel Dome in United States," *Architect and Engineer* 13(2), June 1908, A5).

47. "High Class Brickwork," *Construction News* (February 2, 1907)

48. The cistern under the courtyard of the original Palace Hotel appears still to be partially intact under the present building: see plans on loan to the Environmental Design Archives, University of California, Berkeley.

49. For Hale Brothers reconstruction see Michael R. Corbett, *Splendid Survivors: San Francisco's Downtown Architectural Heritage* (San Francisco: California Living Books, 1979), 94.

50. Thus far the blueprints from the San Francisco Building Inspection Department have yielded twenty-five examples of post-earthquake buildings whose foundations include pre-earthquake walls. These plans are stored in the Environmental Design Archives, University of California, Berkeley.

51. "Fairmont as New City Hall," Polk Scrapbook, College of Environmental Design Archives, University of California, Berkeley

52. "New Pacific Union Club," Polk Scrapbook, College of Environmental Design Archives, University of California, Berkeley

53. The Subtreasury Building before the earthquake was four stories and was reduced in height in the reconstruction in 1906. "Sub-treasury in Close Quarters," *San Francisco Chronicle.* June 6, 1906, 8

54. For the most famous buildings that still exist, see Corbett, *Splendid Survivors,* 29. A complete catalog of all the pre-1906 survivors should be compiled.

55. For the history of the Rialto project, see Sewall Bogart, *Lauriston: An Architectural Biography of Herbert Edward Law* (Portola Valley, Calif.: Alpine House Publications, 1976), 16-33, 37, 68.

56. F. H. Meyer, "Fireproofing of the Structural Parts or Skeleton Frame of a First-Class Building," *Engineering Supplement to the American Builders Review* 1(3) (July 7, 1906), 7

57. The area around the boilers was demolished and had to be extensively rebuilt, and since witnesses claim the building was not dynamited, a gas explosion might account for what appears to have been an explosion on its eastern side. The Sub-Committee Report of the Underwriters Adjusting Bureau for the Rialto Building, September 22, 1906 (Corporate Archives, CIGNA Service Company, 1600 Arch Street, Philadelphia, PA), includes statements from four witnesses that the building was not dynamited. It is possible that the four were lying, but the adjuster believed them—they were probably telling the truth. Gilbert, however, seems certain that the building was damaged by an explosion, which he attributes to dynamite (Gilbert et al., eds., *The San Francisco Earthquake and Fire of April 18, 1906, and Their Effects on Structures and Structural Materials*, USGS Bulletin 324, Washington, D.C.: Government Printing Office, 1907). A gas explosion could have caused similar damage.

58. The drawings for the Rialto Building on loan to the Environmental Design Archives, University of California, Berkeley, from the San Francisco Public Library show damage on all facades with instructions for repair. The schedule for the replacement of columns and girders shows that there was extensive damage in the eastern part of the structure.

59. "Dome Will Once More Be Lighted, Battered City Hall Will Be Brilliant on Night of April 18th," April 6, 1907, and "City Hall Is Not All Ruins; East Wing and Hall of Records Will Soon Be Ready for Occupancy," April 1907, Derleth papers

60. The reconstruction is featured in *Sunset* 18 (April 1907); 19(5) (September 1907); 20(6) (April 1908); and 21(7) (November 1908).

61. Albert W. Whitney, *On Insurance Settlements Incident to the 1906 San Francisco Fire* (1906; reprint Pasadena: California Institute of Technology, 1972), 11

62. Marquis James, *Biography of a Business* (Indianapolis & New York: Bobbs-Merrill, 1942), 212-229

63. Henry R. Gall and William George Jordan, *One Hundred Years of Fire Insurance* (Hartford: Aetna Insurance Company, 1919), 118-120; James, *Biography*, 168

64. A. R. Hosford et al., *Report of the Committee of Five to the "Thirty-five Companies" on the San Francisco Conflagration, April 18–21, 1906* (New York: Mail and Express Job Print, 1907), 27

65. *Report of the Special Committee of the Board of Trustees of the Chamber of Commerce of San Francisco on Insurance Settlements Incident to the San Francisco Fire* (San Francisco: Chamber of Commerce, November 13, 1906), 40

66. Ibid., 18

67. Robert K. MacKenzie, *The San Francisco Earthquake and Conflagration, April, 1906* (Liverpool: Liverpool Daily Post and Mercury Printing Works, 1907), 20-21

68. Julius Kahn, "The San Francisco Disaster—Honest and Dishonest Insurance," Thursday, June 28, 1906, in *Speeches of Julius Kahn of California in the House of Representatives* (Washington, D.C.: Government Printing Office, 1906)

69. *San Francisco Chronicle,* April 23, 1906

70. MacKenzie, *San Francisco Earthquake*, 23

71. Ibid., 10-11; *Report of the Special Committee,* 23

72. George C. Pardee, "Proclamation of the Governor, June 2, 1906," Pardee papers, Bancroft Library, University of California; MacKenzie, *San Francisco Earthquake*, 11

73. Whitney, *On Insurance Settlements*, 30-31; MacKenzie, *San Francisco Earthquake,* 18-19; Frederick L. Hoffman, *Earthquake Hazards and Insurance* (Chicago: Spectator Co., 1928), 128-129

74. *Report of the Special Committee,* 37

75. Whitney, *On Insurance Settlements*, 37-41

76. *Second Decennial Edition of the American Digest: A Complete Digest of all Reported Cases from 1906 to 1916*, vol. 13 (St. Paul: West Publishing Co., 1920), 421

77. *Pacific Reporter, July 5-August 16, 1909*, vol. 102 (St. Paul: West Publishing Co., 1909), 811-812

78. MacKenzie, *San Francisco Earthquake,* 11-12

79. George Brooks, *The Spirit of 1906* (San Francisco: California Insurance Company, 1921), 57-59

80. For the Fireman's Fund recovery see William Bronson, *Still Flying and Nailed to the Mast* (Garden City, NY: Doubleday, 1963), 81-110.

81. For the continuing struggle with insurance companies to reduce their rates because portions of the system were working, see the following articles: "Cut in Insurance Soon to Be Made," *San Francisco Call,* August 10, 1912; "Fire Insurance Men Agree to Reduce Rates," *San Francisco Call,* October 5, 1912; "Fire Insurance Premiums Reduced," *San Francisco Call,* October 12, 1912; "8 Percent Insurance Reduction," *San Francisco Examiner,* July 25, 1915; "City's Fight on Insurance Men Is Over," *San Francisco Examiner,* September 12, 1915; *Engineering News* 74 (April 12, 1915), 39-40.

82. Frederick Meyer, "The Humboldt Bank Building," *American Builders Review* 5(1) (October 1906), 3

83. Committee on Fire Prevention, *Report of the City of San Francisco* (New York: National Board of Underwriters, 1910)

Chapter 10

For interesting studies of other building codes see: Roger H. Harper, *Victorian Building Regulations* (London: Mansell Publishing, 1985); John P. Comer, *New York City Building Control, 1800–1941* (New York: Columbia University Press, 1942); and C. C. Knowles and P. H. Pitt, *The History of Building Regulation in London, 1189–1972* (London, Architectural Press, 1972).

1. "Buildings to Be Fireproof," *San Francisco Call,* May 4, 1906; "Committee of Forty Names Committees," *San Francisco Chronicle,* May 5, 1906; "To Plan Work of Reconstruction," *San Francisco Chronicle,* May 7, 1906. The committee also included several supervisors: James L. Gallagher, Daniel G. Coleman, George F. Duffy, A. N. Wilson, L. A. Rea, Charles Boxton, and Max Mamloch.

2. "Building Laws Committee Completes its Labors," *San Francisco Chronicle,* May 22, 1906

3. "No Decision Reached Regarding Fire Limits," *San Francisco Chronicle,* May 12, 1906

4. Ibid.

5. Ibid.

6. "Height Limit of Class A Buildings Not Fixed," *San Francisco Chronicle,* May 16, 1906

7. Nancy Elizabeth Stoltz, "Disaster and Displacement: The Effects of the 1906 Earthquake and Fire on Land Use Patterns in San Francisco's South of Market District," Master of City Planning thesis, University of California, 1983, 38-41; R. A. Burchell, *The San Francisco Irish, 1848–1880* (Manchester: University Press, 1979), 38-41, 46-47

8. "Mass Meeting Protests against Ordinance Extending Fire Limits," *San Francisco Chronicle,* May 28, 1906, 2

9. "Supervisors Reverse Committee's Action," *San Francisco Chronicle,* May 29, 1906, 1

10. "Mass Meeting," *Chronicle,* May 28, 1906, 2

11. "Supervisors Reverse," *Chronicle,* May 29, 1906

12. "New Roof Ordinance in San Francisco," *Coast Review* 69(3) (March 1906), 160; Building Law of the City and County of San Francisco, Sec. 5, 102 (1906)

13. S. Albert Reed, *The San Francisco Conflagration of April, 1906: Special Report to the National Board of Fire Underwriters Committee of Twenty* (New York: The Committee, May 1906), 28

14. Ibid., 9

15. Peter Collins, *Concrete: The Vision of a New Architecture* (London: McGill-Queen's University Press, 2004), 56-75. Maurice Couchot voiced a typical judgment of the performance of concrete in "How the Earthquake Affected Certain Buildings, *Architect and Engineer of California* 5(1), May 1906: "Reinforced concrete has but few examples in the calamity but as few as they were, they all stood the earthquake and the few that were through the fire came out hardly injured." Octavius Morgan was enthusiastic: in "A Los Angeles Architect's Impressions of the San Francisco Earthquake and Fire," *Architect and Engineer of California* 5(1), May 1906, he said, "I believe the full steel-frame and the reinforced concrete buildings to be as nearly earthquake-proof as possible."

16. Concrete was not accepted in the San Francisco codes, although Ernest Ransome was one of the pioneering developers of the new technology. See Ernest Ransome and Alexis Saurbrey, *Reinforced Concrete Building* (New York: McGraw-Hill, 1912), 3-6. For concrete's appearance in city codes across the nation, see "The Structural Design of Buildings," *Cement Age* 1(6) (Nov. 1904), 221.

17. Edwin Duryea et al., eds., "The Effects of the San Francisco Earthquake of April 18th, 1906, on Engineering Constructions," *Transactions of the American Society of Civil Engineers* 59 (December 1907), 233-234; Grove Karl Gilbert et al., eds., *The San Francisco Earthquake and Fire of April 18. 1906, and Their Effects on Structures and Structural Materials,* USGS Bulletin No. 324 (Washington, D.C.: Government Printing Office, 1907), 33, 150

18. Duryea et al., *ASCE Transactions,* 230-233; Gretchen Smith and Robert Reitherman, *Damage to Unreinforced Masonry Buildings at Stanford University in the 1906 San Francisco Earthquake* (Redwood City, Calif.: Scientific Service, December 1984)

19. "Brickmakers' Opposition to Re-enforced Concrete," *Architect and Engineer* 5(1) (May 1906), 68

20. "Blunders Delay the Building Ordinance: Supervisors Listen to Much Argument on Merits of Brick and Concrete," *San Francisco Chronicle,* June 19,1906

21. Ibid.

22. Building Ordinance No. 645, Sec. 97 (February 5, 1903), *General Orders and Ordinances of the Board of Supervisors of the City and County of San Francisco*

23. *Building Law of the City and County of San Francisco,* Sec. 32A (1906); Building Ordinance No. 1008, Sec. 75 (December 22, 1909), *General Orders and Ordinances of the Board of Supervisors of the City and County of San Francisco*

24. Tobriner, "La casa baraccata: un sistema antisimico nella Calabria del xviii secolo," in C. Latina, ed., *Per Costruire in Laterizio, antologia di saggii dalla rivista ufficiale* (Rimini, 1999), 203-209

25. See chapter 11

26. *Building Law of the City of Boston, Being Acts of 1907* (October 1907), Ch. 550, Sec. 18

27. James Cornell, *The Great International Disaster Book* (New York: Charles Scribner's Sons, 1976), 259-261

28. *Building Law of...San Francisco,* Sec. 310, 311 (1906)

29. Two hundred men employed on the 11th and 12th floors of the World Building continued to work at their desks in spite of the smoke because they believed in the building's fireproof qualities: "Fireproof Buildings," *San Francisco Call,* September 23, 1895. The horrible Triangle Fire in 1911 showed the necessity for exits in tall buildings while highlighting the need for fire protection in New York's tenements and sweatshops: David von Dreble, *Triangle: The Fire that Changed America* (New York: Atlantic Monthly Press, 2003); Leon Stein, *The Triangle Fire* (New York: A Carroll & Graf/Quicksilver Book, 1962)

30. *Building Law of...San Francisco,* Sec. 310A (1906)

31. J. L. Van Arum, "Some Engineering Lessons of the San Francisco Disaster," *Architect and Engineer of California* 10, August 7,1906, 52-55, distrusts plaster-lath. Maurice Couchot, in "How the Earthquake Affected Certain Buildings," has this to say: "The only protection that can be said to protect steel frame thoroughly is to have it absolutely incased in concrete... Concrete has stood best of any material, as was the case in Boston and Baltimore...Common brick has failed more than I ever expected. Terra-cotta for floors was a failure. Plaster partitions, either hollow or solid, have failed."

32. Gladys Hansen and Emmet Condon, *Denial of Disaster* (San Francisco: Cameron and Company, 1989), 138. Condon states that the code was finally corrected in 1984, but "by then the damage was done. Primarily because of this one construction defect the city, in recent years, has experienced numerous multi-building fires."

33. Building Ordinance No. 645, Sec. 18 (February 5, 1903)

34. *City of New York, Code of Ordinance* (New York: Bureau of Buildings, July 17, 1917), Art. 22, Sec. 473

35. *The Building Law of...San Francisco,* Sec. 92-99 (1906)

36. "Building Ordinance Goes to the Printer," *San Francisco Chronicle,* June 12, 1906, 2

37. Duryea et al., *ASCE Transactions* 59 (December, 1907), 327

38. Building Ordinance No. 1008 (December 22, 1909)

39. "State Tenement House Law" (May 29, 1915), *General Orders and Ordinances of the Board of Supervisors of the City and County of San Francisco,* 107-128

40. Eugene E. Schmitz, "Lest We Forget," *The Observer* 1(19) (April 22, 1916), 283

Chapter 11

1. Stephen Tobriner, "The History of Building Codes to the 1920's," in *Proceeding of the Structural Engineers Association of California 1984 Convention* (Monterey, Calif.: October 18–20, 1984), 47-58; Wai-Fah Chen and Charles Scawthorn, eds., *Earthquake Engineering Handbook* (Boca Raton, Fla.: CRC Press, 2003), ch. 11

2. Carl-Henry Geschwind, *California Earthquakes: Science, Risk & the Politics of Hazard Mitigation* (Baltimore: Johns Hopkins University Press, 2001), 26-27, 31; Gladys Hansen and Emmet Condon, *Denial of Disaster* (San Francisco: Cameron and Co., 1989), 137-144

3. Charles Derleth Jr., *The Destructive Extent of the California Earthquake, Its Effect Upon Structures and Structural Materials within the Earthquake Belt* (San Francisco: A. M. Robertson, 1907), originally published as a chapter in a collection of essays edited by David Starr Jordan, *The California Earthquake of 1906* (San Francisco: A. M. Robertson, 1907); "Discussion on San Francisco Earthquake," *Transactions of the American Society of Civil Engineers* 59 (December 1907), 311-323

4. Derleth, *Destructive Extent,* 122

5. Building Ordinance No. 645, Sec. 67 (February 5, 1903), *General Orders and Ordinances of the Board of Supervisors of the City and County of San Francisco*

6. *The Building Law of the City and County of San Francisco*, Sec. 110 and 293 (1906)

7. Building Ordinance No. 1008, Sec. 233 (December 22, 1909), *General Orders and Ordinances of the Board of Supervisors of the City and County of San Francisco*

8. "Comments by the California State Board of Architects," *The Brickbuilder* 15(5) (May 1906), 102

9. Derleth, *Destructive Extent,* 122

10. No reinforcement was stipulated for parapet walls in the 1903 code enforced in 1906 (Building Ordinance No. 645 Sec. 220, February 5, 1903). No reinforcement specifically labeled for parapets appeared in 1906 codes *(The Building Law,* Sec. 294, 1906), but under Sec. 293, "Cornices, Belts, Gutters and Others Appendages," iron brackets were stipulated, and in Sec.110 cornices are supposed to be tied to the frame of the structure, although again parapets are not specifically mentioned. Parapets are dealt with specifically in San Francisco in Building Ordinance No. 1008, Sec. 38 (December 22, 1909).

11. A. L. A. Himmelwright, *The San Francisco Earthquake and Fire* (New York: Roebling Construction Co., 1906), 21

12. William H. Hall, quoted in "Building Materials in the San Francisco Earthquake," *Municipal Engineering* 31 (July 1906), 172

13. Maurice C. Couchot, "The Well-Built Steel Frame Buildings Shown to Resist Earthquake and Fire in San Francisco Catastrophe," *Engineering Record* 53(18) (May 5, 1906), 578

14. Derleth, *Destructive Extent,* 62

15. Derleth, *Destructive Extent,* 63

16. C. H. Snyder, "Steel Construction and Design," *Engineering Supplement to the American Builders Review* 1(8) (August 11, 1906), 7-8

17. J. D. Galloway, "Notes on the Construction of Steel Frames for Buildings," *Engineering Supplement to the American Builders Review* 1(8) (August 11, 1906), 10-11

18. Derleth, *Destructive Extent,* 62

19. Derleth, *Destructive Extent,* 62-63

20. Galloway, "Notes on the Construction," 11

21. *Argonaut,* April 23, 1927, 5. Although there was a disagreement about how badly the stones on the upper floors of the Call Building had been displaced, observers agreed that some damage had been done: Frederick Meyer wrote that the thirteenth floor had distorted eye-bar frames due to swaying ("The Humboldt Bank Building," *American Builders Review* 5[1] [October 1906], 1).

22. Himmelwright, *San Francisco,* 269

23. H. C. Vensano, "A Suggestion for Earthquake Design in High Building Construction," *Engineering Supplement to the American Builders Review* 1(8) (August 11, 1906), 12-13

24. Eric Elsesser, personal communication with author

25. *The Building Law*, Sec. 50, 66-70 (1906); *The Building Code of the City of New York*, Sec. 134-140 (April 12, 1906); *Proposed Building Ordinance of the City of Chicago,* Sec. 252, 516-569 (1905, rev. June 27, 1910)

26. *The Building Law*, Section 69, 33 (1906)

27. Galloway, "Notes on the Construction," 11; also see Edwin Duryea Jr., et al., "The Effects of the San Francisco Earthquake of April 18th, 1906, on Engineering Construction," *Transactions of the ASCE* 59 (December 1907), 211, 316.

28. Richard Michael Levy, "The Design of Tall Frames in America" (paper presented before the Mohawk-Hudson Chapter of the ASCE, n.d.), 29; J. Charles Rathbun, "Wind Forces on a Tall Building," *Transactions of the ASCE* 105 (1940), 1-41

29. *The Building Law*, Sec. 41-48, 255 (1906)

30. Cecil D. Elliott, *Technics and Architecture* (Cambridge, Mass.: The MIT Press, 1992), 376-378

31. Derleth, *Destructive Extent,* 66, 70, 123, 124

32. Ibid., 52-53

33. Duryea, "The Effects of the San Francisco Earthquake," 211

34. Derleth, *Destructive Extent,* 55-56

35. T. H. Skinner, *Engineering Supplement to the American Builders Review* 1(13) (Sept. 15, 1906), 10

36. F. C. Davis, "Anchoring Masonry," *Engineering Supplement to the American Builders Review* 1(13) (Sept. 15, 1906), 6-7

37. Stephen Tobriner, "Bond Iron and the Birth of Anti-Seismic Reinforced Masonry Construction in San Francisco," *Masonry Society Journal* 5(1) (Jan.-June 1986), G12-18; P. J. Walker, "Notes on Brickwork," *Engineering Supplement to the American Builders Review* 6(1) (Jan. 7, 1907), 31

38. Derleth, *Destructive Extent,* 126. Also see unheeded advice on securing a building to its foundation: Joseph W. Rowell, "Amendments to Fire Laws," *San Francisco Call,* May 20,1906.

39. *The Building Law,* Sec. 262 (1906)

40. *City of New York Code of Ordinances, Chapter 5, Building Code,* Sec. 472 (July 17, 1917)

41. Derleth, *Destructive Extent,* 127

Chapter 12

1. San Francisco Board of Public Works, *Reports on an Auxiliary Water Supply System for Fire Protection for San Francisco, California,* report prepared by Marsden Manson, H. D. H. Connick, T. W. Ransom, and W. C. Robinson (San Francisco, 1908), 7

2. *Excerpts from San Francisco Municipal Reports for Fiscal Year 1905–06, ending June 30, 1906 and Fiscal Year 1906–07, ending June 30, 1907* (San Francisco: Board of Supervisors, 1908), 783-784

3. Ibid., 783

4. Edwin Duryea Jr., "Suggestion from a Preliminary Report of the Sub-Committee on Water Supply and Fire Protection, Committee of Forty on Reconstruction of San Francisco, on Present Condition of Water System Serving San Francisco, and Notes on the Particular Features in which this System Fails to Give an Efficient Fire Protection to this City," May 18, 1906, MS, Stanford Library, 7

5. *Reports on the Water Supply of San Francisco, California, 1900–1908* (San Francisco: Board of Supervisors, 1908), 5-6; Ted Wurm, *Hetch Hetchy and Its Dam Railroad* (Berkeley: Howell-North Books, 1973), 9-16; also see Robert W. Righter, *The Battle over Hetch Hetchy: America's Most Controversial Dam and the Birth of Modern Environmentalism* (New York: Oxford University Press, 2005), 29-44.

6. Mel Scott, *The San Francisco Bay Area: A Metropolis in Perspective* (Berkeley: University of California Press, 1959), 64

7. Ibid.

8. Wurm, *Hetch Hetchy,* 17-22; *Reports on the Water Supply,* 97-123

9. *Reports on the Water Supply,* 10

10. Wurm, *Hetch Hetchy,* 17-22; *Reports on the Water Supply,* 21-22; Righter, 45-65

11. *Reports on the Water Supply,* 97-123

12. *Excerpts from San Francisco Municipal Reports for Fiscal Year 1905–06,* 784

13. Charles Derleth Jr., "Some Effects of the San Francisco Earthquake on Water Works, Streets, Sewers, Car Tracks and Buildings," *Engineering News* 55(20) (May 17, 1906), 550-552

14. Herman Schussler, *The Water Supply of San Francisco, California, Before, During and After the Earthquake of April 18, 1906, and the Subsequent Conflagration* (New York: M. Brown Press, 1906), 34

15. Duryea, "Suggestion from a Preliminary Report," 8

16. Ibid.

17. Ibid., 10

18. San Francisco Board of Supervisors, *San Francisco Municipal Reports for the Fiscal Year 1903-1904, ending June 30, 1904* (San Francisco, 1905), 414-418

19. San Francisco Board of Public Works, *Reports on an Auxiliary System,* plate 4. This plate outlines high-pressure auxiliary systems throughout the United States.

20. Edward F. Croker, *Fire Prevention* (New York: Dodd, Mead, 1912), 217

21. Ibid., 227

22. "What Shall We Do for Credit?" *Insurance Engineering* 12(1) (July 1906), 7

23. "Our Need of Fire Protection," *San Francisco Chronicle,* August 19, 1907; John Kenlon, *Fires and Firefighters* (New York: George H. Doran Co., 1913), 206. Kenlon provides an example of rate reductions outside of San Francisco.

24. Charles Derleth Jr. Collection, "Underwriters Agree to Shaughnessy's Plans," July 14, 1907, in Newspaper Clipping IV (Bancroft Library, University of California, Berkeley)

25. Ibid., "The Auxiliary Fire System; Continued Delay in Agreement upon the Details," October 2, 1907

26. San Francisco Board of Public Works, *Reports on an Auxiliary System.* The authors meticulously describe the auxiliary water system.

27. W. H. Ticknor, "San Francisco High-Service Fire System: A General System—High-Pressure Gravity Supply and Auxiliary Pumps—Earthquake Proof," *Insurance Engineering* 24(3) (Sept. 1912), 129

28. The pumping stations were both situated on rock, and since they required chimneys, these were specified for an earthquake acceleration of 6.2 feet per second and made of reinforced concrete. See *Engineering News* 74 (April 12, 1915), 308

29. Ticknor, "San Francisco High-Service Fire System," 139-140. For the continuing struggle with insurance companies to reduce their rates because portions of the system were working, see the following articles: "Cut in Insurance Soon to Be Made," *San Francisco Call,* August 10, 1912; "Fire Insurance Men Agree to Reduce Rates," *San Francisco Call,* October 5, 1912; "Fire Insurance Premiums Reduced," *San Francisco Call,* October 12, 1912; "8 Percent Insurance Reduction," *San Francisco Examiner,* July 25, 1915; and "City's Fight on Insurance Men Is Over," *San Francisco Examiner,* September 12, 1915.

30. "Lack of Water Menace to the City, Says San Francisco Fire Chief," *San Francisco Gazette,* October 20, 1907

31. Derleth, "Some Effects," 552

32. Wurm, *Hetch Hetchy,* 22

33. Ibid.

Chapter 13

1. *San Francisco Municipal Reports for the Fiscal Year 1905-6, ending June 30, 1906, and Fiscal Year 1906-7, ending June 30, 1907* (San Francisco: Neal Publishing Co., 1908), 481; *San Francisco Municipal Reports for the Fiscal Year 1907-8, ended June 30, 1908* (San Francisco: Neal Publishing Co., 1909), 797; *San Francisco Municipal Reports for the Fiscal Year 1908-9, ended June 30, 1909* (San Francisco: Neal Publishing Co., 1910), 632; *San Francisco Municipal Reports for the Fiscal Year 1909-10, ended June 30, 1910* (San Francisco: Neal Publishing Co., 1911), 709; *San Francisco Municipal Reports for the Fiscal Year 1910-11, ended June 30, 1911* (San Francisco: Neal Publishing Co., 1912), 1025; *San Francisco Municipal Reports for the Fiscal Year 1911-12, ended June 30, 1912* (San Francisco: Neal Publishing Co., 1913), 971; *San Francisco Municipal Reports for the Fiscal Year 1912-13, ended June 30, 1913* (San Francisco: Neal Publishing Co., 1915), 564; *San Francisco Municipal Reports for the Fiscal Year 1913-14, ended June 30, 1914* (San Francisco: Neal Publishing Co., 1916), 367

2. Carl-Henry Geschwind, *California Earthquakes: Science, Risk & the Politics of Hazard Mitigation* (Baltimore: Johns Hopkins University Press, 2001), 20-42

3. *San Francisco Municipal Reports 1905-6 and 1906-07,* 481

4. Ibid. Also see "Numerous Owners and Contractors Courting Arrest; Building Law Is Boldly Violated," *San Francisco Call* (May 11, 1906), 2

5. Building law scandals surfaced in the press regarding theaters not meeting city code requirements and temporary structures which, though meant to be razed after a few months, became permanent dwellings. See notes, chapter 10.

6. The drawings for the Humboldt Bank Building are held in the Environmental Design Archives, College of Environmental Design, University of California, Berkeley. Christopher Snyder's personal papers were donated to Stanford University. While his books were saved, his calculations and specifications for individual buildings are said to have been discarded. Snyder is listed in John William Leonard, *Who's Who in Engineering; A Biographical Dictionary of Contemporaries,* 2d ed. (New York: Who's Who, 1925), 1953. The best profile of Snyder as an engineer and a person was given by Michael V. Pregnoff in his oral history (Stanley Scott, interviewer; *Michael V. Pregnoff and John E. Rinne, Connections: The EERI Oral History Series* [Oakland, Calif.: Earthquake Engineering Research Institute; 1996], 11-17). According to Pregnoff, who worked in Snyder's office in 1923, "Before opening his office [around 1910], C. H. Snyder worked as a sales engineer for Milliken Steel Company, and in those days, years ago, he would come to the architect and say, 'I will give you the layout and structural steel sizes and you will give the steel contract to my company.' That's the way they did it in those days." Pregnoff recounted that there were very few structural engineers in San Francisco, the best offices being those of Henry Brunnier, C. H. Snyder, L. H. Nishkian, E. L. Cope, and Austin Earl. Smaller, one-man offices included R. S. Chew and Gus Saph. I had the opportunity to interview Michael Pregnoff in 1986. I did not yet have all the articles Snyder published from as early as 1906 about his philosophy of earthquake-resistant design (See Snyder's long exposition about seismically resistant design in chapter 14, when Pregnoff was in the office) but showed Pregnoff the article about City Hall. He told me that he had not known about the article, and that Snyder had not discussed the issue with him until they designed the Opera House in 1928. According to Pregnoff, Snyder was influenced by the work of R. S. Chew to include a lateral force of 10 percent of gravity in his calculations for horizontal force. But tellingly, Pregnoff said in his oral history that "many engineers thought, like Snyder, "If a building is designed for 30 to 50 pounds per square foot of resistance to wind, it's good enough" (*Connections,* 14). Perhaps he had not understood what Snyder meant by wind loads. Pregnoff did say (perhaps because Pregnoff and I had talked earlier), "I cannot elaborate freely on the prevalence or nonprevalence of seismic design in the '20s" (*Connections,* 17). There is absolutely no doubt from Snyder's articles that he was extremely interested in seismic design right after the 1906 earthquake through the 1920s. I cannot explain the discrepancy between what Snyder wrote and Pregnoff remembered. In Snyder's obituary, Arthur Brown Jr., architect of City Hall, wrote that Snyder had "little sympathy with the vague experiments of some of the innovators…in engineering" and stressed his simple, straightforward engineering solutions. It seems that Brown was ignorant of Snyder's innovative engineering solutions. See *Proceedings, American Society of Civil Engineers,* October 1938, 64(8), part 2, 1887-1888

7. "Dr. Nakamura on the Effects of Severe Earthquakes," *San Francisco Chronicle* (June 19, 1906), 14

8. "Humboldt Building of Concrete," *Architect and Engineer* 5(2) (June 1906), 70

9. Frederick Meyer, "The Humboldt Bank Building," *American Builders Review* 5(1) (October 1906), 3

10. Drawings for the Humboldt Bank Building can be found in the Environmental Design Archives, University of California, Berkeley. Meyer made changes in the details of the elevators and stairwells after the earthquake. No firedoors appeared on drawing 108, dated March 31, 1906. The first revision, dated June 20, 1906, illustrates a firedoor over the elevators only. But in a later revision, dated December 22, 1906, Meyer added four firedoors, which would effectively enclose both the elevators and the stairwells. On drawing 119, dated April 9, 1906, he specified perforated risers for his stairwell from the first floor to the mezzanine, whereas this specification was deleted in favor of solid stairs dated July 2, 1906.

11. Meyer, "The Humboldt Bank Building," 3-13; Charles Derleth Jr., "The Humboldt Savings Bank Building, San Francisco," *Engineering Record* 58(21) (Nov. 21, 1908), 584

12. Derleth, "The Humboldt Savings Bank Building," 581

13. "Royal Globe Insurance Company Building, San Francisco" (pamphlet), n.d., California Historical Society, San Francisco

14. Ibid.

15. Corydon T. Purdy, "Special Structural Details of the Old Colony Building, Chicago, Ill.," *Engineering News* (Dec. 21, 1893), 486

16. "Earthquake-Proof Royal Globe Building, San Francisco," *Insurance Engineering* 22(3) (Sept. 1911), 164-168

17. Building Ordinance No. 2704 (new series, April 16, 1914), *General Orders and Ordinances of the Board of Supervisors of the City and County of San Francisco* (1915)

18. "The Phelan Building Floor Construction," *Engineering Record* 57 (May 2, 1908), 585

19. Ibid., 366-367

20. Ibid., 585

21. Henry J. Brunnier's office is still intact in the Sharon Building. The library and the calculations and plans for Brunnier's buildings are preserved there, along with many photographs.

22. Henry J. Brunnier, interview by Frank Killinger, August 31, 1955, later published in *Connections: The EERI Oral History Series, Henry J. Brunnier and Charles de Maria* (Oakland, Calif.: Earthquake Engineering Research Institute, 2001)

23. Henry J. Brunnier, "Development of Aseismic Construction in the United States," in *Proceedings of the World Conference on Earthquake Engineering,* 26-2 (Berkeley: Earthquake Engineering Research Institute, June 1956)

24. Ed Zacher, a longtime member of Brunnier's office, told me this in the 1990s. Also see Pregnoff's comments on Brunnier's idea of tying buildings together in *Connections*, 17.

25. Brunnier, "Development of Aseismic Construction." The lateral force of 20 psf is calculated on sheets with the plans (Henry J. Brunnier, office files, the Sharon Building)

26. Charles Derleth Jr., "The Sather Campanile," *University of California Chronicles* 16 (July 1914), 306-310; Charles Derleth Jr., "The Campanile," *California Journal of Technology* 17 (Nov. 1913), 51-56; Charles Derleth Jr. "Sather Campanile, University of California, Berkeley," *The Bridgemen's Magazine* 14 (April 1914), 244-245; Erle L. Cope, "An Earthquake-Proof Tower: Semi-Flexible Construction Selected for Steel and Concrete Structure, 303 Feet High, at Berkeley, California," *Engineering Record* 69(11) (March 14, 1914), 312-313

27. R. S. Chew, "The Effect of Earthquake Shock on High Buildings," *Transactions of the American Society of Civil Engineers* 61 (Dec. 1908), 238-252

28. Ibid., 245

29. Derleth, "The Sather Campanile," 306-310; Frank P. Ulrich, "Vibration and Tilt Studies on Sather Tower, University of California, Berkeley, California," Dept. of Commerce, U.S. Coast and Geodetic Survey, San Francisco, Jan. 17, 1950

30. Derleth, "The Sather Campanile," 308-309

31. "Jury Selects Design for the New City Hall," *San Francisco Chronicle,* June 21, 1912, 19; "Plans for New City Hall Are Decided Upon," *San Francisco Call,* June 21, 1912, 12

32. George Wagner, "The Builder of San Francisco City Hall," interviewed by Bea Sebastian (1978), 7, 11

33. Christopher H. Snyder, "Some of the Engineering Features of the San Francisco City Hall," *Architect and Engineer* 46(2) (August 1916), 79

34. Ibid., 80

35. See note 17

36. "Chronicle Building Annex, 690 Market, Burnham and Root Architects," *Architect and Engineer* 10 (Oct. 1907), 86

37. First National Bank Building, now Crocker Bank, Willis Polk for D. H. Burnham and Company at 1 Montgomery Street. The upper part of this structure is now dismantled. See "The First National Bank Building, San Francisco," *Architect and Engineer* 14 (Sept. 1908), A.

38. For the Olympic Club see Henry A. Schulze, "Some Interesting Features of the New Olympic Club Building," *Architect and Engineer* 9(1) (May 1907), 34-41. In the end Schulze did not complete the building.

39. The engineer of the Commercial Building, 853 Market St., was Christopher Snyder. Snyder identifies himself as the "contracting engineer" in a letter to William H. Crocker now in the library of the California Academy of Sciences, San Francisco.

40. Photographs of the Alaska-Commercial Building and the Kohler-Chase Building are held in the Environmental Design Archives, University of California, Berkeley. For the Clunie Building (southwest corner, Montgomery and California Streets)—architect T. Paterson Ross, engineer A. W. Burgren, constructor C. A. Blume, steel manufacturers Whitehead & Kales Iron Works, San Francisco and Detroit—see "The Work of T. Paterson Ross and A. W. Burgren," *Architect and Engineer* 13(1) (May 1908), 36, A2. See also Michael R. Corbett, *Splendid Survivors: San Francisco's Downtown Architectural Heritage* (San Francisco: California Living Books, 1979), 47, 164, 172.

41. William H. Hall, "The Rebuilding of San Francisco—Reinforced Concrete Buildings," *Architect and Engineer* 9 (July 1907), 61

42. Edward Allen, *Fundamentals of Building Construction: Materials and Methods* (New York: Wiley, 1985), 399-400

43. T. E. Keough, "Failure of the Bixby Hotel," *Architect and Engineer* 7(2) (Dec. 1906), 67-70; Henry A. Schulze, "Says Bixby Hotel Construction Was Not True Reinforced Concrete," *Architect and Engineer* 7(2) (Dec. 1906), 71-73; Joseph Simons, "The Bixby Hotel Disaster as Viewed by a Brick Man," *Architect and Engineer* 7(2) (Dec. 1906), 51-52; Louis H. Gibson, "The Recent Failures of Reinforced Concrete, *Engineering Magazine* 33 (1907), 118

44. Ibid., 61-67; William H. Hall, "Reinforced Concrete Practice in San Francisco—Column Design," *Architect and Engineer* 9 (May 1907), 49-58

45. For load tests see Kenneth MacDonald, "The Strength of Reinforced Concrete," *Architect and Engineer* 22(1) (August 1910), 90-93; "Test on Ransome Concrete Construction," *Insurance Engineering* 2 (Nov. 1901), 570-571

46. Personal observation as the buildings were being destroyed, late 1980s

47. Walter J. Kenyon, "Concrete in San Francisco," *Construction News* 24 (August 17, 1907), 113

48. "Hotel and Depot at Albuquerque," *Architect and Engineer* (August 1905), 44-45

49. Corbett, *Splendid Survivors*, 89

50. Ibid.

51. Kenyon, "Concrete in San Francisco," 113

52. Charles F. Whittlesey, "Reinforced Concrete Construction—Why I Believe in It," *Architect and Engineer* 12 (March 1908), 45

53. Robert W. Gardner, "Reinforced Brick Work," *Architect and Engineer* 5(2) (June 1906), 43

54. Joseph A. Hofmann, U.S. Patent No. 893,924, July 21, 1908. The Hofmann invention was in fact used at least once and the plaque commemorating that use can be see on an apartment building on Bush Street near Polk. Reinforced brickwork also appears in the steel-frame apartment house at Post and Jones Streets by T. Paterson Ross and A. W. Burgren pictured in Nathaniel Ellery, *Permanency in Building Construction* (San Francisco: Brick Builder Bureau, 1913), 109, 111 (my figures 13.24a and 13.24b).

55. P. J. Walker, "Notes on Brickwork," *American Builders Review* 6(1) (Jan. 7, 1907), 32

56. See William Holmes and Robert Reitherman in Rutherford and Chekene Consulting Engineers, "Seismic Retrofitting Alternatives for San Francisco's Unreinforced Masonry Buildings: Estimates of Construction Cost and Seismic Damage, Prepared for the City and County of San Francisco Department of City Planning," May 1990.

57. Walker, "Notes on Brickwork," 32; *Building Ordinances*, Sec. 137 (Dec. 22, 1909)

58. "Will Rush Work on Business Block," *San Francisco Evening Globe*, Nov. 5, 1908, 5; "Lincoln Realty Building," *San Francisco Chronicle*, Oct. 14, 1908, 5; "Fire Chief Probes Market St. 'Eye-Sore'," *San Francisco Evening Globe*, May 24, 1908, 8; "White Building for Lincoln School Site," *San Francisco Chronicle*, Oct. 14, 1908

59. Lists of special construction schools as well as Class C schools appear in five years of the *San Francisco Municipal Reports*. A good list with descriptions appears in *San Francisco Municipal Reports for the Fiscal Year 1910-11*, 534-545.

60. "New Semi-Fireproof Type of Buildings for the 'Fire Limits' Area," *Coast Review* 73 (Jan. 1908), 131

61. "Three San Francisco Schools Closed as Earthquake Hazards," *Engineering News Record* 3 (Nov. 30, 1933), 665. The San Fernando earthquake in 1971 was the impetus for yet another investigation of San Francisco schools. The result was that several schools, Madison among them, were demolished in the 1970s. "62 S.F. Schools Below Safety Standards," *San Francisco Chronicle* (Feb. 11, 1971), 2; "An Order to Close Six S.F. Schools," *San Francisco Chronicle* (April 22, 1971); "7 'Hazardous' Schools May Be Closed," *San Francisco Chronicle* (July 16, 1971), 1

62. The extensive problems of early San Francisco firehouses in relation to seismic activity have been investigated by EQE International Inc., Charles Scawthorn, principal investigator.

63. "Temporary City Hall," Central Emergency Hospital Building B, South Side of Market Street (between 8th and 9th Streets) for James Otis Trustee; Wright, Rushforth and Cahill Architects Collection, University of California, Berkeley

64. Sally B. Woodbridge, *Bernard Maybeck, Visionary Architect* (New York: Abbeville Press Publishers, 1992), 129-136

65. The Tobriner house, formerly 3494 Jackson, was built in 1907 by Newsom and Newsom. It was published in two articles: "Practical Plans for the Home Builders—I," *Overland Monthly* 51 (April 1908), 314-317; and "Fine Residence for Jackson Street Begun," *San Francisco Chronicle* (April 23, 1907), 3.

    I contend that the Tobriner house, the 1911 house on Woolsey in Berkeley, and other houses, like the 1914 house at 1334 Bonita Ave. in Berkeley, are unusual because they have more diagonal bracing than other balloon-frame derivatives in the United States. The 1901 San Francisco codes (Section 20) include the unusual provision that "all outside walls and cross partitions shall be thoroughly and angle braced," which might indicate that this framing tradition derived from code requirements. According to Professor Dell Upton of the University of Virginia, an expert on early American construction, it is a rarity to see internal walls diagonally braced. If this practice was widespread in the San Francisco Bay Area, it might be related to an earthquake-resistant tradition in wooden structures. For typical balloon-frame construction, see Paul R. Sprague, "Chicago Balloon Frame; The Evolution During the 19th Century of George W. Snow's System for Erecting Light Frame Buildings from Dimension Lumber and Machine-Made Nails" in H. Ward Jandl, ed., *The Technology of Historic American Buildings: Studies of the Materials, Craft Processes, and the Mechanization of Building Construction* (Washington, D.C.: Foundation for Preservation Technology, 1983), 33-61.6

Chapter 14

1. For one of the most lyrical mythic expositions on the fill and PPIE as symbolic of San Francisco's denial and future fate, see Philip L. Fradkin, *The Great Earthquake and Firestorms of 1906: How San Francisco Nearly Destroyed Itself* (Berkeley: University of California Press, 2005), 341-344. See chapter 9 of this volume and note 9 below for the factual problems with the debris from 1906.

2. "Loma Prieta Earthquake, October 18, 1989, Part 2, National Geophysical Data Center (NGDC), *http://www.ngdc.noaa.gov/seg/hazard/slideset/13/13_thumbs.shtml,* publishes the following information, which is at best misleading and at worst completely wrong: "Some sand boils in the area spewed forth charred wood and other debris, remnants of the San Francisco that was destroyed by earthquake and fire in 1906." Not likely—see below.

3. See, for example, *San Francisco, the Financial, Commercial and Industrial Metropolis of the Pacific Coast* (San Francisco Chamber of Commerce, 1915).

4. Nancy Olmsted, Roger Olmsted, and Allen Pastron, *San Francisco Waterfront: Report on Historical Cultural Resources for the North Shore and Channel Outfalls Consolidation Projects* (San Francisco: San Francisco Wastewater Management Program, December 1977), 689; Frank Morton Todd, *The Story of the Exposition,* vol. 1 (New York: G. P. Putnam's Sons, 1921), 129-133, 282-286, 299-301

5. Raymond H. Clary, *The Making of Golden Gate Park, the Growing Years: 1906–1950* (San Francisco: Don't Call It Frisco Press, 1987), 37-41; Olmsted, *San Francisco Waterfront,* 686-689

6. Todd, *Story of the Exposition,* 300

7. Ibid., 301; A. H. Markwart, "The Panama-Pacific Exposition," *Engineering News* 70(19) (November 6, 1913), 898

8. Olmsted, *San Francisco Waterfront,* 690; James K. Mitchell, Tahir Masood, Robert E. Kayen, and Raymond B. Seed, "Soil Conditions and Earthquake Hazard Mitigation in the Marina District of San Francisco," Earthquake Engineering Research Center Report No. UCB/EERC 90/08, May 1990. For consciousness of the danger in the mid-1970s, see p. 21 documenting the 1975 USGS Miscellaneous Field Studies Map MF-709, which includes the Marina as one of the areas most likely damaged by future large earthquakes. The report states (p. 29), "Hydraulic fills of this type are now known to be vulnerable to soil liquefaction during earthquakes, but this was not understood in 1912." This statement corroborates Todd and Markwart. They were careful and did their best but did not understand the hazard they were creating. On the issue of fill, Mitchell et al. have this to say: "It is important to keep in mind when considering both existing conditions and remedial treatments that 'fill' is not synonymous with 'bad.' The term fill refers to earth that is brought to an area and placed. Whether the resulting material is good or bad relative to its suitability as a foundation material or in terms of its liquefaction resistance depends on what it is, how it was placed, and what was done to it after it was placed but before it was built upon. Thus the same sandy hydraulic fill material that was dumped in Marina Cove in 1912 and which liquefied in the Loma Prieta Earthquake could, if densified during or after placement, be used for earthquake resistant support for structures or for construction of barrier walls against lateral spreading of adjacent liquefied soil."

9. M. G. Bonilla, "The Marina District, San Francisco, California: Geology, History, and Earthquake Effects," *Bulletin of the Seismological Society of America* 81(5) (October 1991), 1969-1970; Bonilla, "Geologic and Historical Factors Affecting Earthquake Damage," in *The Loma Prieta, California, Earthquake of October 17, 1989—Marina District,* ed. Thomas D. O'Rourke, USGS Professional Paper 1551-F (Washington, D.C.: Government Printing Office, 1992), 7-34; Mitchell et al., "Soil Conditions and Earthquake Hazard Mitigation."

   Bonilla is the seismologist whose work is most frequently cited to prove the dumping of 1906 debris. In "Geologic and Historical Factors," he is very cautious about the evidence of 1906 debris and makes it clear that no conclusive evidence exists: "How much debris from the 1906 earthquake and fire was incorporated into fills in the Marina is unknown, and, as described in the subsection below...1906 debris would be difficult to distinguish from the Panama-Pacific Exposition debris...Two historical accounts that cover the 1906 earthquake made no mention of any dumping of 1906 debris at Harbor View...Two general reports on the 1906 earthquake stated that debris from the main part of San Francisco was dumped in the Mission Bay...and that some was hauled by barge to the vicinity of Mile Rock...Considering its age, however, the 1906–12 fill *could* include debris from the 1906 earthquake." Bonilla continues, "The source and method of emplacement of the 1895–1906 fill are largely unknown... The 1906–12 fill *probably* contains debris from the 1906 earthquake and fire, but exactly how much is problematic." [emphasis added]

10. Ibid., 1969; William Bronson, *The Earth Shook, the Sky Burned* (1959; reprint, New York: Crown, 1986), 170. Also see chapter 9 notes.

11. Markwart, "The Panama-Pacific Exposition," 902. According to Mitchell et al. (p. 29), a total of 500,000 lineal feet (almost 100 miles) of timber piles were driven to support the buildings of the PPIE.

12. Ibid., 900-901

13. Building Ordinance No. 1008, Sec. 89 (December 22, 1909), *General Orders and Ordinances of the Board of Supervisors of the City and County of San Francisco.*

14. Stephen K. Harris and John A. Egan, "Effects of Ground Conditions on the Damage to Four-Story Corner Apartment Buildings," in O'Rourke, *The Loma Prieta, California, Earthquake,* 181-194

15. Ground shaking was responsible for most of the building damage in the Marina: "It is very important to note, however, that much of the structural damage in the Marina District caused by the Loma Prieta earthquake is a direct result of ground shaking and the inability of the structures to withstand it. Liquefaction and the associated permanent

ground displacements were primarily responsible for building settlements, utility line and pavement breaks and some structural damage. However, liquefaction was neither the cause of amplified ground shaking…nor can ground treatment to prevent liquefaction in future earthquakes be expected to mitigate ground surface shaking." Mitchell et al., 20

16. Arthur C. Alvarez, in "An Earthquake-Proof Dwelling," *Architect and Engineer* 84(3) (March 1926), 105-108, illustrates that engineers recommended diagonal bracing for wood frames.

17. The Robert Dudman plans are for a small house on Greenwood Terrace owned by Beverley Bolt and the late Bruce Bolt.

18. Gladys Hansen and Emmet Condon, *Denial of Disaster* (San Francisco: Cameron and Co., 1989), 137-138

19. "Engineers Discuss Earthquake Experiences and Cautions," *Engineering News-Record* 95(7) (August 13, 1925), 271-272

20. "Lessons of the Santa Barbara Earthquake," *Architect and Engineer* 82(2) (August 1925), 104

21. "Earthquakes and Building Codes," *Engineering News-Record* 98(17) (April 28, 1927), 677

22. C. H. Snyder, "Earthquake and Building Codes," *Engineering News-Record* (June 16, 1927), 995

23. Ibid.

24. Ibid.

25. L. H. Nishkian, "Design of Tall Buildings for Resistance to Earthquake Stresses," *Architect and Engineer* 88(3) (March 1927), 81-83

26. H. J. Brunnier Associates office files, Sharon Building, San Francisco, California

27. For Rosseter Gardens, Green and Taylor Streets, San Francisco, see "A Quake-Proof Frame for an Apartment House," *Architect and Engineer* 29(1) (June 1928), 57-58.

28. Randolph Delehanty, *San Francisco: The Ultimate Guide* (San Francisco: Chronicle Books, 1995), 247; Madeleine Victoria Brocklebank MacAdam, *Fortune in My Own Hands* (Boston: Christopher Publishing House, 1940), 339-351

29. John R. Freeman, *Earthquake Damage and Earthquake Insurance* (New York: McGraw-Hill, 1932), 379

30. Arthur C. Alvarez, "The Santa Barbara Earthquake of June 19, 1925," *University of California Publications in Engineering* 2(6) (Nov. 17, 1925), 205-210

31. For buildings by Charles Peter Weeks and William P. Day, see Michael R. Corbett, *Splendid Survivors: San Francisco's Downtown Architectural Heritage* (San Francisco: California Living Books, 1979), 162, 204, 217, 224, 225, 229.

32. Delehanty, *San Francisco*, 247

33. Tachu Naito, *Earthquake-Resisting Construction* (New York: American Society of Civil Engineers, 1930)

34. Robert King Reitherman, "Frank Lloyd Wright's Imperial Hotel: A Seismic Re-evaluation," in *Proceedings of the Seventh World Conference on Earthquake Engineering* (Istanbul, Turkey: September 8–13, 1980), vol. 4

35. Wai-Fah Chen and Charles Scawthorn, eds., *Earthquake Engineering Handbook* (Boca Raton, Fla.: CRC Press, 2003), 1-22, 1-23; Henry Dewell and Bailey Willis, "Earthquake Damage to Buildings," *Bulletin of the Seismological Society of America* 15(4) (December 1925), 282

36. "Earthquake Research," Walter T. Steilberg Collection, MS, Environmental Design Archives, College of Environmental Design, University of California, Berkeley

37. Carl-Henry Geschwind, *California Earthquakes: Science, Risk & the Politics of Hazard Mitigation* (Baltimore: Johns Hopkins University Press, 2001),79-80

38. Chen and Scawthorn, *Earthquake Engineering Handbook*, 11-4, 11-5

39. Freeman, *Earthquake Damage*, 371-375

40. Ibid., 371

41. Jon P. Kiland and Thomas G. Atkinson, "The Evolution and History of SEAOC, A Celebration of 75 Years of History, 1929 to 2004," *Structural Engineers Association of California (SEAOC) 2004 Convention Proceedings* (Monterey, Calif.: August 25-28, 2004), 329-339; Donald R. Strand, "Code Development Between 1927 and 1980," *SEAOC 1984 Convention Proceedings* (Monterey, Calif.: October 18-20, 1984), 60

42. Chen and Scawthorn, *Earthquake Engineering Handbook*, 16-10

43. "Earthquake Conscious?" *Engineering News-Record* 113(1) (July 5, 1934), 21

Selected Bibliography

Books

Aitken, Frank W., and Edward Hilton. *A History of the Earthquake and Fire in San Francisco: An Account of the Disaster of April 18, 1906, and Its Immediate Results*. San Francisco: The Edward Hilton Co., 1906.

Alexander, James Beach, and James Lee Heig. *San Francisco: Building the Dream City*. San Francisco: Scottwall Associates, 2002.

Allen, Edward. *Fundamentals of Building Construction: Materials and Methods*. New York: Wiley, 1985.

Ambrose, James, and Dimitry Vergun. *Simplified Building Design for Wind and Earthquake Forces*. New York: John Wiley & Sons, 1980.

Atkinson, Edward. *The Prevention of Loss by Fire*. Boston: Damrell and Upham, 1900.

Babrauskas, Vytenis. *Ignition Handbook, Principles and Applications to Fire Safety Engineering, Fire Investigation, Risk Management and Forensic Science*. Issaquah, Wash.: Fire Science Publishers, 2003.

Bancroft, Hubert H. *History of California*. San Francisco: A. L. Bancroft and Co., 1884.

Barker, Malcolm E., comp. *Three Fearful Days, San Francisco Memoirs of the 1906 Earthquake and Fire*. San Francisco: Londonborn Publications, 1998.

Barth, Gunther. *Instant Cities: Urbanization and the Rise of San Francisco and Denver*. New York: Oxford University Press, 1975.

Bean, Walton. *Boss Ruef's San Francisco*. Berkeley: University of California Press, 1952.

Berger, Richard W., with Roland Mainstone. *Palace of the Sun, the Louvre of Louis XIV*. University Park, Penn.: Pennsylvania State University Press, 1993.

Birkmire, William H. *The Planning and Construction of High Office Buildings*. 3d ed. New York: John Wiley & Sons, 1903.

Birt, Rodger C. "The San Francisco Album and Its Historical Moment: Photography, Vigilantism, and Western Urbanization." In *San Francisco Album: Photographs of the Most Beautiful Views and Public Buildings* by G. R. Fardon. San Francisco: Chronicle Books, 1999.

Bogart, Sewall. *Lauriston: An Architectural Biography of Herbert Edward Law*. Portola Valley, Calif.: Alpine House Publications, 1976.

Bolt, Bruce A. *Earthquakes*. New York: W. H. Freeman and Co., 1993.

Bolton, Herbert E., ed. *Anza's California Expeditions*. Berkeley and Los Angeles: University of California Press, 1930.

———. *Historical Memoirs of New California*. Berkeley: University of California Press, 1926.

Boutelle, Sara H. *Julia Morgan, Architect*. New York: Abbeville Press Publishers, 1988.

Brechin, Gray. *Imperial San Francisco; Urban Power, Earthly Ruin*. Berkeley & Los Angeles: University of California Press, 1999.

Bronson, William. *Still Flying and Nailed to the Mast*. Garden City, N.Y.: Doubleday, 1963.

———. *The Earth Shook, the Sky Burned*. 1959. Reprint, New York: Crown, 1986.

Brooks, George. *The Spirit of 1906*. San Francisco: California Insurance Company, 1921.

Browning, Peter, ed. *San Francisco/Yerba Buena: From the Beginning to the Gold Rush, 1769–1849*. Lafayette, Calif.: Great West Books, 1998.

Burchell, R. A. *The San Francisco Irish, 1848–1880*. Manchester: University Press, 1979.

Burnham, Daniel H. and Edward H. Bennett. *Report on a Plan for San Francisco*. San Francisco: Sunset Press, 1905.

Chen, Wai-Fah, and Charles Scawthorn, eds. *Earthquake Engineering Handbook*. Boca Raton, Fla.: CRC, 2003.

Clary, Raymond H. *The Making of Golden Gate Park, the Growing Years: 1906–1950*. San Francisco: Don't Call It Frisco Press, 1987.

Coleman, Charles M. *P. G. and E. of California: The Centennial Story of Pacific Gas and Electric Company, 1852–1952*. New York: McGraw-Hill, 1952.

Collier, Michael. *A Land in Motion; California's San Andreas Fault*. Berkeley and Los Angeles: Golden Gate Parks Association and University of California Press, 1999.

Collins, Peter. *Concrete: The Vision of a New Architecture*. London: McGill-Queen's University Press, 2004.

Comer, John P. *New York City Building Control, 1800–1941*. New York: Columbia University Press, 1942.

Condit, Carl W. *American Building Art: The Nineteenth Century*. New York: Oxford University Press, 1960.

———. *American Building: Materials and Techniques from the First Colonial Settlements to the Present.* Chicago: University of Chicago Press, 1968).

———. "The Two Centuries of Technical Evolution Underlying the Skyscraper. In *Second Century of the Skyscraper,* ed. Lynn S. Beedle. New York: Van Nostrand Reinhold, 1988.

Corbett, Michael R. *Splendid Survivors: San Francisco's Downtown Architectural Heritage.* San Francisco: California Living Books, 1979.

Cornell, James. *The Great International Disaster Book.* New York: Charles Scribner's Sons, 1976.

Cowan, Henry J. *Science and Building: Structural and Environmental Design in the Nineteenth and Twentieth Centuries.* New York and London: John Wiley and Sons, 1978.

Croker, Edward F. *Fire Prevention.* New York: Dodd, Mead, 1912.

Crouch, D. P., D. J. Garr, and A. I. Mundigo. *Spanish City Planning in North America.* Cambridge, Mass.: MIT Press, 1982.

Dalessandro, James. *1906.* San Francisco: Chronicle Books, 2004.

Davison, Charles. *Great Earthquakes.* London: T. Murby & Co., 1936.

Delehanty, Randolph. *San Francisco: The Ultimate Guide.* San Francisco: Chronicle Books, 1995.

Derleth Jr., Charles. *The Destructive Extent of the California Earthquake. Its Effect Upon Structures and Structural Materials within the Earthquake Belt.* San Francisco: A. M. Robertson, 1907.

Dillon, Richard. *North Beach, the Italian Heart of San Francisco.* Novato, Calif.: Presidio Press, 1985.

Douty, Christopher M. *The Economics of Localized Disasters: The 1906 San Francisco Catastrophe.* New York: Arno Press, 1977.

Draper, Joan. "Paris by the Lake: Sources of Burnham's Plan of Chicago." In *Chicago Architecture 1872–1922,* ed. John Zukowsky. Munich: Prestel–Verlag, 1987.

Dynes, Russell R. *Organized Behavior in Disaster.* Lexington, Mass.: D. C. Heath and Co., 1970.

Eldredge, Zoeth S. *The Beginnings of San Francisco.* San Francisco: Z. S. Eldredge, 1912.

Ellery, Nathaniel. *Permanency in Building Construction.* San Francisco: Brick Builder Bureau, 1913.

Elliott, Cecil D. *Technics and Architecture: The Development of Materials and Systems for Buildings.* Cambridge, Mass.: MIT Press, 1994.

Fleming, Robins. *Wind Stress in Buildings.* New York: John Wiley & Sons, Inc., 1930.

Fradkin, Philip L. *Magnitude 8: Earthquakes and Life along the San Andreas Fault.* New York: Henry Holt, 1998.

———. *The Great Earthquake and Firestorms of 1906: How San Francisco Nearly Destroyed Itself.* Berkeley: University of California Press, 2005.

França, José–Augusto. *Une Ville des Lumières, La Lisbonne de Pombal.* Paris: SEV–PEN, 1965.

Freeman, Donald. *Historical Building Construction, Design, Materials and Technology.* New York and London: Norton, 1995.

Freeman, John Ripley. *Earthquake Damage and Earthquake Insurance.* New York: McGraw–Hill, 1932.

Gall, Henry R., and William George Jordan. *One Hundred Years of Fire Insurance.* Hartford: Aetna Insurance Company, 1919.

Geison, Gerald L., ed. *Professions and Professional Ideologies in America.* Chapel Hill, N.C.: University of North Carolina Press, 1983.

Geschwind, Carl–Henry. *California Earthquakes: Science, Risk & the Politics of Hazard Mitigation.* Baltimore: The Johns Hopkins University Press, 2001.

Greely, A. W. *Reminiscences of Adventure and Service.* New York: C. Scribner's Sons, 1927.

Haas, J. Eugene, Robert W. Kates, and Martyn J. Bowden. *Reconstruction Following Disaster.* Cambridge, Mass.: MIT Press, 1977.

Hansen, Gladys, and Emmet Condon. *Denial of Disaster.* San Francisco: Cameron and Co., 1989.

Harper, Roger H. *Victorian Building Regulations.* London: Mansell Publishing, 1985.

Himmelwright, A. L. A. *The San Francisco Earthquake and Fire.* New York: Roebling Construction Co., 1906.

Hittell, John S. *The Resources of California,* 3d ed. San Francisco: A. Roman and Co., 1867.

———. *A History of San Francisco and Incidentally of the State of California.* 1878 and 1888. Reprint Berkeley: Berkeley Hills Books, 2000.

Hittell, Theodore H. *History of California.* San Francisco: N. J. Stone & Co., 1898.

Hoffman, Frederick L. *Earthquake Hazards and Insurance.* Chicago: Spectator Co., 1928.

Issel, William, and Robert W. Cherny. *San Francisco 1865–1932: Power, Politics, and Urban Development.* Berkeley: University of California Press, 1986.

James, Marquis. *Biography of a Business.* Indianapolis & New York: Bobbs–Merrill, 1942.

Jordan, David S., ed. *The California Earthquake of 1906.* San Francisco: A. M. Robertson, 1907.

Kagan, Richard. *Urban Images of the Hispanic World, 1492–1793.* New Haven: Yale University Press, 2000.

Kahn, Judd. *Imperial San Francisco: Politics and Planning in an American City, 1897–1906.* Lincoln: University of Nebraska Press, 1979.

Kahn, Julius. "The San Francisco Disaster—Honest and Dishonest Insurance." In *Speeches of Julius Kahn of California in the House of Representatives.* Washington, D.C.: Government Printing Office, 1906.

Kenlon, John. *Fires and Firefighters.* New York: George H. Doran Co., 1913.

Killinger, Frank, interviewer. *Connections: The EERI Oral History Series, Henry J. Brunnier and Charles de Maria.* Oakland, Calif.: Earthquake Engineering Research Institute, 2001.

Kirker, Harold. *California's Architectural Frontier.* 3d ed. Salt Lake City: Gibbs M. Smith, 1986.

Knowles, C. C., and P. H. Pitt. *The History of Building Regulation in London 1189–1972.* London, Architectural Press, 1972.

Kurzman, Dan. *Disaster! The Great San Francisco Earthquake and Fire of 1906.* New York: William Morrow, 2001.

Landau, Sarah, and Carl Condit. *Rise of the New York Skyscraper 1865–1912.* New Haven: Yale University Press, 1996.

Langley, Henry G. *San Francisco Directory for the Year Commencing December, 1860.* San Francisco: Henry G. Langley, 1860.

———. *San Francisco Directory for the Year Commencing December, 1864.* San Francisco: Henry G. Langley, 1864.

———. *San Francisco Directory for the Year Commencing December, 1869.* San Francisco: Henry S. Langley, 1869.

———. *San Francisco Directory for the Year Commencing December, 1876.* San Francisco: Henry G. Langley, 1877.

Laurie, Bruce. "Fire Companies and Gangs in Southwark: the 1840s." In *The Peoples of Philadelphia: A History of Ethnic Groups and Lower-Class Life, 1790–1940,* ed. Allen F. Davis and Mark H. Haller. Philadelphia: Temple University Press, 1973.

Lavender, David. *California, Land of New Beginnings.* Lincoln and London: University of Nebraska Press, 1972.

LeCount & Strong. *San Francisco City Directory.* San Francisco: LeCount & Strong, 1854.

LeGrand, H. E. *Drifting Continents and Shifting Theories.* Cambridge: Cambridge University Press, 1994.

Leonard, John William. *Who's Who in Engineering; A Biographical Dictionary of Contemporaries.* 2d ed. New York: Who's Who, 1925.

Levy, Matthys, and Mario Salvadori. *Why the Earth Quakes: The Story of Earthquakes and Volcanoes.* New York and London: W. W. Norton, 1995.

Lewis, Oscar. *San Francisco: Mission to Metropolis.* San Diego: Howell–North Books, 1980.

Limpus, Lowell M. *History of the New York Fire Department.* New York: E. P. Dutton, 1940.

Lipsky, Florence. *San Francisco: La grille sur les collines/The Grid Meets the Hills.* Marseille: Editions Parenthèse, 1999.

Lloyd, B. E. *Lights and Shades in San Francisco.* 1876. Reprint Berkeley: Berkeley Hills Books, 1999.

Lockwood, Charles. *Suddenly San Francisco: The Early Years of an Instant City.* San Francisco: San Francisco Examiner Div. of Hearst Corp., 1978.

Lotchin, Roger W. *San Francisco 1846–1856: From Hamlet to City.* New York: Oxford University Press, 1974.

MacAdam, Madeleine Victoria Brocklebank, *Fortune in My Own Hands.* Boston: Christopher Publishing House, 1940.

MacKenzie, Robert K. *The San Francisco Earthquake and Conflagration, April, 1906.* Liverpool: Liverpool Daily Post and Mercury Printing Works, 1907.

MacPhail, Archibald. *Of Men and Fire: A Story of Insurance in the Far West.* San Francisco: Fire Underwriters Association of the Pacific, 1948.

Marryat, Frank. "The Trouble with Iron Houses." In *San Francisco Memoirs, 1835–1851: Eyewitness Accounts of the Birth of the City,* comp. Malcolm E. Barker. San Francisco: Londonborn Publications, 1994.

McGloin, John B. *San Francisco: The Story of a City.* San Rafael, Calif.: Presidio Press, 1978.

McKeon, Peter J. *Fire Prevention: A Treatise and Textbook on Making Life and Property Safe Against Fires.* New York: Chief Publishing Co., 1912.

Meli, Roberto. *Ingeriería Estructural de los Edificios Históricos.* Mexico, D.F.: Fundación ICA, 1998.

Moores, Eldridge M., ed. *Shaping the Earth: Readings from* Scientific American *Magazine.* New York: W.H. Freeman, 1990.

Moudon, Anne Vernez. *Built for Change, Neighborhood Architecture in San Francisco.* Cambridge, Mass.: MIT Press, 1986.

Muscatine, Doris. *Old San Francisco: The Biography of a City.* New York: Putnam, 1975.

Naitu, Tacho. *Earthquake-Resisting Construction.* New York: American Society of Civil Engineers, 1930.

Olsen, Donald J. *The City as a Work of Art.* New Haven: Yale University Press, 1986.

Parsons, William Barclay. *Engineers and Engineering in the Renaissance.* Cambridge, Mass.: MIT Press, 1968.

Pasley, C. W. *Observations on Limes, Calcareous Cements, Mortars, Stuccos, and Concrete.* 2d ed. London: J. Weale, 1847.

Peck, Ralph. "History of Building Foundations in Chicago, Structural and Civil Engineering Design." In *Studies in the History of Civil Engineering,* ed. William Addis. Vol. 12. Brookfield, Vt.: Ashgate Publishing, 1999.

Pelton, John Cotter. "San Francisco: Its Position in Architectural and Constructive Development." In *Modern San Francisco, 1907–1908* by Western Press Association. San Francisco: Western Press Association, 1908.

Powell, John Walker. *An Introduction to the Natural History of Disaster.* University of Maryland: Disaster Research Project, 1954.

Rabun, J. Stanley. *Structural Analysis of Historic Buildings: Restoration, Preservation, and Adaptive Reuse Application for Architects and Engineers.* New York: John Wiley & Sons, 2000.

Ransome, Ernest, and Alexis Saurbrey. *Reinforced Concrete Building.* New York: McGraw–Hill, 1912.

Ribeiro dos Santos, Maria Helena. *A Baixa Pombalina, Passado e Futuro.* Lisbon: Livros Horizonte, 2000.

Righter, Robert W. *The Battle over Hetch Hetchy: America's Most Controversial Dam and the Birth of Modern Environmentalism.* New York: Oxford University Press, 2005.

Ronneberg, Norman J. "The Bittersweet Life of John Wright." In *John Wright (1830–1915), Grandfather of West Coast Architecture.* Victoria, B. C.: Maltwood Art Museum and Gallery, 1990.

Rowlandson, Thomas. *A Treatise on Earthquake Dangers, Causes and Palliatives.* San Francisco: Dewey & Co. Publishers, 1869.

Saalman, Howard. *Haussmann: Paris Transformed.* New York: George Braziller, 1971.

———. *Filippo Brunelleschi. the Buildings.* University Park, Penn.: Pennsylvania University Press, 1993.

San Francisco Directory, May 1906

Sandweiss, Eric, "Claiming the Urban Landscape: The Improbable Rise of an Inevitable City." In *Eadweard Muybridge and the Photographic Panorama of San Francisco, 1850–1880* by David Harris. Montreal and Cambridge, Mass.: Centre Canadien d'Architecture and MIT Press, 1993.

Saul, Eric, and Don DeNevi. *The Great San Francisco Earthquake and Fire, 1906.* Millbrae, Calif.: Celestial Arts, 1981.

Scharlach, Bernice. *Big Alma, San Francisco's Alma Spreckels.* San Francisco: Shotwell Associates, 1999.

Schliemann, Heinrich. *Schliemann's First Visit to America.* Cambridge, Mass.: Harvard University Press, 1942.

Schussler, Herman. *The Water Supply of San Francisco, California, Before, During and After the Earthquake of April 18, 1906, and the Subsequent Conflagration.* New York: M. Brown Press, 1906.

Schuyler, Robert Livingston, ed. *Dictionary of American Biography.* Supplement II, "Charles Strobel." New York: C. Scribner's Sons, 1940.

Scott, Mel. *The San Francisco Bay Area: A Metropolis in Perspective.* Berkeley: University of California Press, 1959.

Scott, Stanley, interviewer. *Connections: The EERI Oral History Series, Michael V. Pregnoff and John E. Rinne.* Oakland, Calif.: Earthquake Engineering Research Institute; 1996.

Senkewicz, Robert M. *Vigilantes in Gold Rush San Francisco.* Stanford, Calif.: Stanford University Press, 1985.

Sherman, William T. *Memoirs of General W. T. Sherman.* New York: C. L. Webster & Co., 1894.

Shumate, Albert. *Rincon Hill and South Park, San Francisco's Early Fashionable Neighborhood.* Sausalito, Calif.: Windgate Press, 1988.

———. *A San Francisco Scandal: The California of George Gordon.* Spokane: Arthur H. Clark Co., 1994.

Soulé, Frank, John H. Gihon, and James Nisbet. *The Annals of San Francisco.* 1855; reprint, Berkeley: Berkeley Hills Books, 1998.

Sprague, Paul R. "Chicago Balloon Frame; The Evolution During the 19th Century of George W. Snow's System for Erecting Light Frame Buildings from Dimension Lumber and Machine–Made Nails." In *The Technology of Historic American Buildings: Studies of the Materials, Craft Processes, and the Mechanization of Building Construction,* ed. H. Ward Jandl. Washington, D.C.: Foundation for Preservation Technology, 1983.

Stein, Leon. *The Triangle Fire.* New York: A Carroll & Graf/ Quicksilver Book, 1962.

Steinbrugge, Karl. *Earthquakes, Volcanoes, and Tsunamis: An Anatomy of Hazards.* New York: Skandia America Group, 1982.

Stewart, George R. *Committee of Vigilance: Revolution in San Francisco, 1851.* New York: Ballantine Books, 1964.

Sutcliffe, Anthony. *Paris: An Architectural History.* New Haven: Yale University Press, 1993.

Thomas, Gordon, and Max Morgan Witts. *The San Francisco Earthquake.* New York: Stein and Day Publishers, 1971.

Timoshenko, Stephen P. *History of the Strength of Materials.* New York: Krieger, 1953.

Tobriner, Stephen. *The Genesis of Noto: An Eighteenth–Century Sicilian City.* Berkeley and London: University of California Press and A. Zwemmer, 1982.

———. "Safety and Reconstruction after the Sicilian Earthquake of 1693, the 18th–Century Context." In *Dreadful Visitations, Confronting Natural Catastrophe in the Age of Enlightenment,* ed. Alessa Johns. New York: Routledge, 1999.

———. "La casa baraccata: un sistema antisimico nella Calabria del xviii secolo." In *Per Costruire in Laterizio, antologia di saggii dalla rivista ufficiale,* ed. C. Latina. Rimini, 1999.

Todd, Frank Morton. *The Story of the Exposition.* Vol. 1. New York: G. P. Putnam's Sons, 1921.

Twain, Mark. *Roughing It.* New York: Harper & Row, 1899.

Vance, James E. Jr. *Geography and Urban Evolution in the San Francisco Bay Area.* Berkeley: Institute of Governmental Studies, University of California, 1964.

Vitruvius, Pollio. *Ten Books on Architecture.* Trans. Morris Hicky Morgan. Cambridge, Mass.: Harvard University Press, 1914.

von Dreble, David. *Triangle: The Fire that Changed America.* New York: Atlantic Monthly Press, 2003.

Wakabayashi, Minoru. *Design of Earthquake–Resistant Buildings.* New York: McGraw Hill, 1986.

West Publishing Co.. *Second Decennial Edition of the American Digest: A Complete Digest of all Reported Cases from 1906 to 1916.* Vol. 13. St. Paul: West Publishing, 1920.

Willard, Ruth Hendricks, and Carol Green Wilson. *Sacred Places of San Francisco.* San Francisco: Presidio Press, 1985.

Wisely, William H. *The American Civil Engineer 1852–1974: The History, Traditions and Developments of the American Society of Civil Engineers.* New York: American Society of Civil Engineers, 1974.

Woodbridge, Sally B. *Bernard Maybeck, Visionary Architect.* New York: Abbeville Press Publishers, 1992.

Wurm, Ted. *Hetch Hetchy and Its Dam Railroad.* Berkeley: Howell–North Books, 1973.

Yanev, Peter. *Peace of Mind in Earthquake Country: How to Save Your Home and Life.* San Francisco: Chronicle Books, 1991.

Yeats, Robert S. *Living with Earthquakes in California, A Survivor's Guide.* Corvallis: Oregon State University Press, 2001.

Young, John P. *San Francisco: A History of the Pacific Coast Metropolis.* San Francisco: S. J Clarke Publishing Company, 1912.

Zukowsky, John, ed. *Chicago Architecture 1872–1922: Birth of a Metropolis.* Munich: Prestel–Verlag, 1987.

Government Documents

Building Code of the City of New York. Sec. 134–140. April 12, 1906.

Building Law of the City and County of San Francisco. Sec. 5, 32A, 92–99, 102, 110 and 293, 310, 310A, 311 (1906).

Building Law of the City of Boston, Being Acts of 1907. October 1907, Ch. 550, Sec. 18.

California Code of Regulations, Sec. 1626, 2001.

City of New York Code of Ordinances, Chapter 5, Building Code, Sec. 472, 473, July 17, 1917.

Consolidation Act, or Charter of the City and County of San Francisco; with Other Acts Specially Relating to San Francisco, and of General Interest Therein; and the General Orders of the Board of Supervisors, Order 725, Sec. 9, August 21, 1866.

Foye, William H. U.S. Patent 92,033. Improved Submarine Foundation, June 29, 1869.

General Orders of the Board of Supervisors of the City and County of San Francisco. San Francisco: Board of Supervisors, including "Order No. 1,617: Defining the Fire Limits and Regulating the Construction of Buildings in the City and County of San Francisco," March 2, 1881; "Order No. 1,752: To Define the Fire Limits of the City and County of San Francisco, and Making Regulations Concerning the Erection and Use of Buildings in Said City and County," December 28, 1883; and Order No. 1,917: To Define the Fire Limits of the City and County of San Francisco, and Making Regulations Concerning the Erection and Use of Buildings in Said City and County," June 21, 1887; Building Ordinance No. 291, Sec. 18 and 130, May 8, 1901; Building Ordinance No. 645, Sec. 18, 96, 97, Feb. 5, 1903; Building Ordinance No. 1008, Sec. 75, 89, 137, and 233, Dec. 22, 1909; Building Ordinance No. 2704, new series, April 16, 1914; "State Tenement House Law," May 29, 1915.

Hofmann, Joseph A. U.S. Patent No. 893,924, July 21, 1908.

Kelly, John. U.S. Patent 74,547. Construction of Seawalls, February 18, 1868.

Proposed Building Ordinance of the City of Chicago, Sec. 252, 516–569, 1905, rev. June 27, 1910.

San Francisco Municipal Reports for Fiscal Years 1866–1867; 1871–1872; 1872–1873; 1874–1875; 1879–1880; 1898–1899; 1903–1904; 1904–1905; 1905–1906; 1907–08; 1908–09; 1909–10; 1910–11; 1911–12; 1912–13; 1913–14.

Touaillon, Jules. U.S. Patent 99,973. Improvement in Buildings, February 15, 1870.

U.S. Patent Office. *Annual Report of the Commissioner of Patents.* Vol 1375. Washington, D.C.: Government Printing Office, 1868.

U.S. Patent Office. *Official Gazette of the United States Patent Office.* Vol. 3. Washington, D.C.: Government Printing Office, 1887.

Journals, Magazines, and Newsletters

American Builders Review 4(1), 4(3), 5(1), 5(2), 6(1). *Engineering Supplement to the American Builders Review* June 30, 1906; July 7, 1906; Aug. 11, 1906; Sept. 15, 1906; Jan. 7, 1907.

Architect and Engineer Aug. 1905, May 1906, June 1906, Nov. 1906, Dec. 1906, Jan. 1907, May 1907, July 1907, Sept. 1907, Oct. 1907, March 1908, May 1908, June 1908, Sept. 1908, Aug. 1910, Aug. 1916, Aug. 1925, March 1926, March 1927, June 1928.

Architect and Engineer of California May 1906, August 7, 1906.

Architectural Record 20(1), 88(6).

Architectural Review 20.

Brickbuilder, The 15(5).

Bridgemen's Magazine 14.

Bulletin of the Imperial Earthquake Investigation Committee 1, Jan. 1907.

Bulletin of the Imperial Earthquake Investigation Committee in Foreign Languages 3(1), 1909.

Bulletin of the Seismological Society of America Dec. 1925, Dec. 1930, 1968, Dec. 1982, Feb. 1986, Oct. 1991, 1993, 1998.

Bulletin of the University of Wisconsin 1.

Burlington Magazine 138.

California Architect and Building News 10(11), 10(12).

California Geology 40.

California Historical Quarterly 16(3), 21(2), 23(2), 51.

California Journal of Technology 17.

Cement Age 1(6).

Chronicles of the University of California 1(1), 16.

Coast Review 69(3), 72(3), 73 (Jan. 1908).

Construction News, Feb. 2, 1907; Aug. 17, 1907.

Quaderni dell'istituto di Storia dell'Architettura, Nuova Serie, Fasc. 15–20.

Earthquake Spectra 1(1) and supplement to vol. 6.

Engineering Magazine 33 (1907).

Engineering News Dec. 21, 1893; May 17, 1906; Nov. 6, 1913; April 12, 1915.

Engineering News-Record Aug. 13, 1925; April 28, 1927; June 16, 1927; Nov. 30, 1933; July 5, 1934.

Engineering Record April 9, 1898; April 16, 1898; May 5, 1906; May 2, 1908; Nov. 21, 1908; March 14, 1914.

Insurance Engineering 2, 11(1), 12(1), 22(3), 24(3).

Iron Age, May 1907.

Journal of Electricity, Power and Gas 35.

Journal of the Society of Architectural Historians 18, 42(2).

Masonry Society Journal 5(1).

Mining and Scientific Press 17(19), 20, 24.

Monumentos 21.

Municipal Engineering 31.

Nature 32.

Newsletter, Center for Latin American Studies, winter 2000.

Observer, The 1(19).

Ohio Architect and Builder 9.

Overland Monthly 15(3), 51.

Pacific Reporter, July 5–August 16, 1909.

San Francisco Newsletter, May 7, 1870.

San Francisco Real Estate Circular, February 1873.

Scientific American, Feb. 1974.

Sunset June-July 1906, April 1907, Sept. 1907, April 1908, Nov. 1908.

University of California Publications in Engineering 2(6).

Newspapers

Alta California, Dec. 28, 1849; May 3, 1851; Jan. 8, 1863; Oct. 27, 1868; Feb. 2, 1869.

Argonaut, 1926: May 1, 15; June 5, 26; July 3, 10, 12; Aug. 7, 28; Sept. 11, 18; Oct. 9, 16, 23; Nov. 6, 13, 20, 27; Dec. 4, 18, 25. 1927: Jan. 1, April 23, July 23, Aug. 6, Aug. 13.

Daily Alta California, May 16, 1851; Nov. 27, 1852; Feb. 16, 1856; Feb. 18, 1856; June 12, 1857; Oct. 10, 1865; Aug. 31, 1866; Nov. 23, 1870.

Daily Morning Chronicle, October 24, 1868.

Morning Call, July 17, 1873; April 23, 1892; April 24, 1892.

Morning Daily Call, December 19, 1872.

San Francisco Bulletin, Oct. 10, 1865; March 22, 1869, April 29, 1906.

San Francisco Call, April 25, 1890; April 23, 1892; April 24, 1892; Sept. 23, 1895; June 21, 1897; April 1, 1898; May 4, 1906; May 11, 1906; May 20, 1906; May 29, 1906; June 18, 1906; Aug. 5, 1906; Dec. 1, 1906; May 30, 1907; Feb. 28, 1908; July 19, 1908; June 21, 1912; Aug. 10, 1912; Oct. 5, 1912; Oct. 12, 1912.

San Francisco Chronicle, Oct. 24, 1868; March 26, 1884; March 1, 1888; April 29, 1888; June 22, 1890; April 23, 1892; January 12, 1896; Nov. 9, 1905; April 23, 1906; April 27, 1906; May 5, 1906; May 6, 1906; May 7, 1906; May 11, 1906; May 12, 1906; May 16, 1906; May 19, 1906; May 22, 1906; May 26, 1906; May 28, 1906; May 29, 1906; June 6, 1906; June 12, 1906; June 19, 1906; February 16, 1907; March 3, 1907; April 23, 1907; May 26, 1907; Aug. 19, 1907; Oct. 14, 1908; June 21, 1912; Feb. 11, 1971; April 22, 1971; July 16, 1971; April 11, 2004.

San Francisco Evening Globe, May 24, 1908; Nov. 5, 1908.

San Francisco Examiner, April 27, 1906; July 25, 1915; May 2, 1906; May 5, 1906, Sept. 12, 1915.

Other: Dissertations, Government Reports, Proceedings, Technical Reports, Miscellany

San Francisco Gazette, Oct. 20, 1907.

American Society of Civil Engineers. *Proceedings,* October 1938, 64(8), part 2. 1887–1888.

American Society of Civil Engineers. *Transactions* 27 (1892). See also Chew, R. S.; Duryea, Edwin Jr.; Harris, Stephen K.; Rathburn, J. Charles; Whitworth, George F.

Basin Research Associates. *San Francisco Main Library Project: Archaeological Monitoring and Architectural Documentation, Site of the Former City Hall Completed in 1897.* Vol. 1. San Francisco: Basin Research Associates, December 1994.

Bonilla, M. G. "Geologic and Historical Factors Affecting Earthquake Damage." In *The Loma Prieta, California, Earthquake of October 17, 1989—Marina District,* ed. Thomas D. O'Rourke, U.S. Geological Survey Professional Paper 1551–F. Washington, D.C.: Government Printing Office, 1992.

Boschi, Enzo, et al. *Catalogo dei forti terremoti in Italia dal 461 a.C.al 1980.* Rome: Istituto Nazionale di Geofisica, 1995.

Bowden, Martyn J. "Reconstruction Following Catastrophe: The Laissez–Faire Rebuilding of Downtown San Francisco after the Earthquake and Fire of 1906." *Proceedings of the Association of American Geographers,* 1970.

Brown, A. A., et al. *See* Whitworth, George F.

Brunnier, Henry J. "Development of Aseismic Construction in the United States." In *Proceedings of the World Conference on Earthquake Engineering,* Berkeley: Earthquake Engineering Research Institute. June 1956.

Byerly, Perry. "History of Earthquakes in the San Francisco Bay Area." In *Geologic Guidebook of the San Francisco Bay Counties.* San Francisco: California Division of Mines, 1951.

California Office of Planning and Research. *Competing against Time, Report to Governor George Deukmejian from the Governor's Board of Inquiry on the 1989 Loma Prieta Earthquake,* 1990

California State Earthquake Investigation Commission. *See* Lawson, Andrew C. et al.

Cardwell, Kenneth. "Montgomery Block (Office Building)." Historic American Building Survey Cal–1228, August 1958.

Chew, R. S. "The Effect of Earthquake Shock on High Buildings." In *Transactions of the American Society of Civil Engineers* 61 (1908).

Clancey, Gregory. "Art Nation/Earthquake Nation: The Cultural Economy of Japanese Seismicity." Symposium: Architecture and Modern Japan, Columbia University. October 21, 2000.

Comerio, M., M. Green, and Stephen Tobriner. "Reconnaissance Report on the Umbria–Marche, Italy, Earthquakes of 1997." Earthquake Engineering Research Institute Special Report, December 1997.

Committee Appointed to Investigate the Affairs of the San Francisco City Hall Commission Report. Sacramento: T. A. Springer, 1874.

Duryea, Edwin Jr., et al. "The Effects of the San Francisco Earthquake of April 18th, 1906, on Engineering Construction." *Transactions of the American Society of Civil Engineers* 59 (1907).

Forell/Elsesser Engineers, Inc. *Final Report: Seismic Damage Evaluation and Upgrade Recommendation for U.S. Court of Appeals and Post Office, San Francisco, California.* San Francisco: Forell/Elsesser Engineers, Inc., June 1990.

Gilbert, Grove Karl, et al., eds. *The San Francisco Earthquake and Fire of April 18, 1906, and Their Effects on Structures and Structural Materials.* United States Geological Survey Bulletin 324. Washington, D.C.: Government Printing Office, 1907.

Greely, A. W. *San Francisco Relief Survey.* New York: The Russel Sage Foundation, 1913.

Hansen, Gladys, Frank Quinn, and F. E. McClure. "The 1906 Numbers Game." Abstract only. In *Seismological Research Letters* 58, Seismological Society of America, 82nd Annual Meeting. 1987.

Harris, Stephen K., and John A. Egan. "Effects of Ground Conditions on the Damage to Four–Story Corner Apartment Buildings." In *Transactions of the American Society of Civil Engineers* 61 (1908).

Hosford, A. R., et al. *Report of the Committee of Five to the "Thirty–five Companies" on the San Francisco Conflagration, April 18–21, 1906.* New York: Mail and Express Job Print, 1907.

Kiland, Jon P., and Thomas G. Atkinson. "The Evolution and History of SEAOC, A Celebration of 75 Years of History, 1929 to 2004." *Structural Engineers Association of California 2004 Convention Proceedings,* Monterey, Calif. August 25–28, 2004.

de Laveaga, E. I. Diary entry. Personal collection of Connie de Laveaga.

Lawson, Andrew C., et al. *The California Earthquake of April 18, 1906, Report of the State Earthquake Investigation Commission* vol. 1. Washington, D. C.: Carnegie Institution of Washington, 1908.

Levy, Richard M. "The Design of Tall Frames in America." Paper presented before the Mohawk–Hudson Chapter of the American Society of Civil Engineers. n.d.

———. "The Professionalization of American Architects and Civil Engineers, 1865–1917." Ph.D. Dissertation. University of California, Berkeley, 1980.

Manson, Marsden, et al. *Report, The Sub–Committee on Statistics to the Chairman and Committee on Reconstruction.* San Francisco, April 24, 1907.

Manson, Marsden, H. D. H. Connick, T. W. Ransome, W. C. Robinson. *Reports on an Auxiliary Water Supply System for Fire Protection for San Francisco, California.* Board of Public Works, 1908.

Milne, John. "On Construction in Earthquake Countries." *Minutes of Proceedings of the Institution of Civil Engineers* 83, 1886.

Mitchell, James K., Tahir Masood, Robert E. Kayen, Raymond B. Seed. "Soil Conditions and Earthquake Hazard Mitigation in the Marina District of San Francisco." Earthquake Engineering Research Center Report No. UCB/EERC 90/08, May 1990.

National Board of Fire Underwriters Committee on Fire Prevention. *Report on the City of San Francisco.* New York: National Board of Underwriters, 1910.

National Board of Underwriters Committee of Twenty. *Report on the City of San Francisco, California.* New York: National Board of Underwriters, October 1905.

National Conference of the Bay Area Regional Earthquake Preparedness Project. Proceedings, "Putting the Pieces Together," Hyatt Regency San Francisco Airport. October 15–17, 1990.

National Fire Protection Association. *The Baltimore Conflagration: Report of the Committee on Fire–Resistive Construction of the National Fire Protection Association.* Chicago: National Fire Association, 1904.

National Geophysical Data Center. "Loma Prieta Earthquake, October 18, 1989, Part 2. *http://www.ngdc.noaa.gov/seg/hazard/slideset/13/13_thumbs.shtml.* December 17, 2005.

National Park Service, Presidio of San Francisco, Golden Gate National Recreational Area. "Refugee Cottages." Presidio of San Francisco, *www.nps.gov/prsf/history/1906eq/cottages.htm.* December 17, 2005.

Neasham, Vernon Aubrey, ed. "The Montgomery Block, Registered Landmark #80." California Historical Landmark Series. Berkeley: State of California, Department of Natural Resources, 1936.

Nelson, Christopher H. "Classical California: The Architecture of Albert Pissis and Arthur Brown, Jr." Ph.D. Dissertation. University of California, Santa Barbara, 1986.

Olmsted, Nancy, Roger Olmsted, and Allen Pastron. *San Francisco Waterfront: Report on Historical Cultural Resources for the North Shore and Channel Outfalls Consolidation Projects.* San Francisco: San Francisco Wastewater Management Program, December 1977.

O'Rourke, Thomas D., ed. *The Loma Prieta, California, Earthquake of October 17, 1989—Marina District.* U.S. Geological Survey Professional Paper 1551–F. Washington, D.C.: Government Printing Office, 1992.

Pacific Submarine and Earthquake Proof Wall Company vs. P. H. Canavan, Joseph G. Eastland, and Charles E. McFane. Circuit Court of the United States of the Ninth Circuit in and for the District of California, "Testimony of William H. Foye." San Francisco, 1874–1877.

Quarles, Stephen L., Laurence G. Cool, and Frank C. Beall, *Performance of Deck Board Materials Under Simulated Wildfire Exposures,* Seventh International Conference on Wood Fiber–Plastic Composites (and Other Natural Fibers), May 19–20, 2003. Madison, Wisc.: Forest Products Society, 2003.

Rathbun, J. Charles. "Wind Forces on a Tall Building." In *Transactions of the American Society of Civil Engineers* 105 (1940).

Reed, S. Albert. *The San Francisco Conflagration of April, 1906: Special Report to the National Board of Fire Underwriters Committee of Twenty.* New York: The Committee, May 1906.

Reitherman, Robert King. "Frank Lloyd Wright's Imperial Hotel: A Seismic Re–evaluation." In *Proceedings of the Seventh World Conference on Earthquake Engineering,* vol. 4, Istanbul, Turkey. September 8–13.

Report of the Committee of Five to the Thirty-Five Companies. See Hosford, A. R., et al.

Robison, Elwin C. "St. Peter's Dome: The Michelangelo and Della Porta Designs." In *Proceedings of the IASS–MSU Symposium, Domes from Antiquity to the Present,* Istanbul, Mimar Sinan University. 1988.

Rutherford and Chekene Consulting Engineers. "Seismic Retrofitting Alternatives for San Francisco's Unreinforced Masonry Buildings: Estimates of Construction Cost and Seismic Damage, Prepared for the City and County of San Francisco Department of City Planning." May 1990.

Scawthorn, C., and T. D. O'Rourke. "Effects of Ground Failure on Water Supply and Fire Following Earthquake: The 1906 San Francisco Earthquake." Paper presented at Second Annual U.S.–Japan Workshop on Liquefaction, Large Ground Deformation and Their Effects on Lifelines. Grand Island and Ithaca, New York. September 26–29, 1989.

San Francisco Chamber of Commerce. *San Francisco, the Financial, Commercial and Industrial Metropolis of the Pacific Coast.* c.1915.

Sebastian, Bea. "George Wagner, the Builder of San Francisco City Hall. Interview. 1978.

Smith, Gretchen, and Robert Reitherman. *Damage to Unreinforced Masonry Buildings at Stanford University in the 1906 San Francisco Earthquake.* Redwood City, Calif.: Scientific Service, December 1984.

Special Committee of the Board of Trustees of the Chamber of Commerce of San Francisco. *Report on Insurance Settlements Incident to the San Francisco Fire.* San Francisco: Chamber of Commerce, November 13, 1906.

Stevenson, David. "Notice of Aseismic Arrangements, Adapted to Structures in Countries Subject to Earthquake Shocks." In *Transactions of the Royal Scottish Society of Arts* 7 (1868).

Stoltz, Nancy Elizabeth. "Disaster and Displacement: The Effects of the 1906 Earthquake and Fire on Land Use Patterns in San Francisco's South of Market District." Masters thesis. University of California, Berkeley, 1983.

Stover, Carl W., and Jerry L. Coffman. "Seismicity of the United States, 1568–1989." U.S. Geological Survey Professional Paper 1527 (Revised). Washington, D.C.: Government Printing

Office, 1993) and *http://neic.usgs.gov/neis/eq_depot/usa/1906_04_18.html.*

Strand, Donald R. "Code Development Between 1927 and 1980." In *Structural Engineers Association of California 1984 Convention Proceedings,* Monterey, Calif. October 18–20, 1984.

Thompson, Ervin. *Defender of the Gate: the Presidio of San Francisco.* Denver: U.S. Department of the Interior, National Park Service, Denver Service Center, 1997.

Tobriner, Stephen. "The Failure of San Francisco's City Hall in the Earthquake of 1906." In *Proceedings of the IASS–MSU Symposium, Domes from Antiquity to the Present,* Istanbul, Mimar Sinan University. 1988.

———. "The History of Building Codes to the 1920's." In *Proceedings of the Structural Engineers Association of California 1984 Convention,* Monterey, Calif. October 18-20, 1984.

Tocher, Don. "Seismic History of the San Francisco Region." In *San Francisco Earthquakes of March 1957.* California Division of Mines Special Report 57, 1959.

Todd, Frank Morton. *Eradicating Plague from San Francisco.* San Francisco: The Citizens' Health Committee, March 31, 1909.

Toppozada, Tousson R., et al. *Annual Technical Report Fiscal Year 1980–1981.* OFR 81–11 SAC. Sacramento: California Division of Mines and Geology, 1981.

Toppozada, Tousson R., C. R. Reale, and D. L. Park. "Preparation of Isoseismal Maps and Summaries of Reported Effects for Pre–1900 California Earthquakes." In *Annual Technical Report Fiscal Year 1979–1980.* OFR 80–15 SAC, 38, and 81–11 SAC, 1980. Sacramento: California Division of Mines and Geology, 1980.

U.S. Department of Interior, National Park Service, Western Region, Division of National Register Programs, and U.S. Department of Defense, Department of the Army. "Presidio of San Francisco, National Historic Landmark District," Historic American Buildings Survey Report, 1985.

U.S. War Department. *Earthquake in California, April 18, 1906, Special Report of Maj. Gen. Adolphus W. Greely.* Annual Report of the Secretary of War. Washington, D.C.: Government Printing Office, 1906.

Ulrich, Frank P. "Vibration and Tilt Studies on Sather Tower, University of California, Berkeley, California." San Francisco: Dept. of Commerce, U.S. Coast and Geodetic Survey, Jan. 17, 1950.

Underwriters Adjusting Bureau. Sub–Committee Report of the Underwriters Adjusting Bureau for the Rialto Building. Corporate Archives, CIGNA Service Company (1600 Arch Street, Philadelphia, Penn.), September 22, 1906.

Wallace, Robert, ed. *The San Andreas Fault System.* U.S. Geological Survey Professional Paper 1515. Washington, D.C.: Government Printing Office, 1990.

Western Neighborhoods Project. "1906 Earthquake Refugee Shacks." *www.outsidelands.org/shacks.html.* December 17. 2005.

Whitney, Albert W. *On Insurance Settlements Incident to the 1906 San Francisco Fire.* 1906. Reprint, Pasadena: California Institute of Technology, 1972.

Whitworth, George F., ed., *Subsidence and the Foundation Problem in San Francisco, A Report of the Subsoil Committee of the San Francisco Section* (San Francisco: American Society of Civil Engineers, September 1932)

Wire Glass Manufacturers. "Earthquake and Fire. Concerning the Fire Resistance of Building Materials." n.d.

Youd, T. L., and S. N. Hoose. *Historic Ground Failures in Northern California Triggered by Earthquakes.* U.S. Geological Survey Professional Paper 993. Washington, D.C.: Government Printing Office, 1978.

Archives

Henry J. Brunnier office file archive. Sharon Building, San Francisco.

California Academy of Sciences Library, San Francisco. Christopher Snyder, letter to William H. Crocker.

California Historical Society Library, San Francisco. Fred J. Bowlen. "A Summary of the Introductory Paragraphs," "Outline of the History of the San Francisco Fire," "The San Francisco Earthquake and Fire 1906." 1934.

———. "Royal Globe Insurance Company Building, San Francisco" (pamphlet). n.d.

———. "San Francisco Fire Department Historical Review 1849–1967." 1967.

Society of California Pioneers, Phelan Sullivan Library, San Francisco. O. P. Stidger, "The Montgomery Block: The Realized Dream of Henry W. Halleck." March 1947.

———. "Specifications of the Different Kinds of Material, Workmanship and Dimensions, April 20, 1853." General Information, Montgomery Block, Deeds Agreements, Folder 2.

———. "Iron Bill, Iron as per Contract Made and Signed May 12, 1853." General Information, Montgomery Block, Deeds Agreements, Folder 4.

Stanford University Library. Edwin Duryea, Jr., "Suggestion from a Preliminary Report of the Sub–Committee on Water Supply and Fire Protection, Committee of Forty on Reconstruction of San Francisco, on Present Condition of Water System Serving San Francisco, and Notes on the Particular Features in which this System Fails to Give an Efficient Fire Protection to this City." May 18, 1906.

U.S. National Archives. A. B. Mullet to J. F. Morse, October 26, 1868. Entry 7, Box 19, Folder 2, RG 121, 132.

———. A. B. Mullet to J. F. Morse, November 2, 1868. Entry 7, Box 19, Folder 2, RG 121, 279.

———. A. B. Mullet to W. P. C. Stebbins, September 20, 1869. Entry 7, Box 25, Folder 1, RG 121, 389 ½.

———. *Pacific Submarine Company vs. United States,* 16. Case file 12345, General Jurisdiction Case Files, Record Group 124, Records of the United States Court of Claims.

University of California, Berkeley, Bancroft Library. Bancroft Scraps: December 20, 1863; March 6, 1864; May 21, 1864; December 9, 1865; October 9, 1865; October 22, 1868; October 24, 1868; November 25, 1868; January 6, 1869; February 19, 1869.

———. Font, Padre Fray Pedro. "Extracts from the Journal of Padre Fray Pedro Font." BANC MSS C-E 101.

———. George Davidson Papers. Horace D. Dunn, "Statement Relative to the Effects of the Earthquake in San Francisco, Ca., July 19, 1906," "Fred G. Plummer to G. K. Gilbert, October 10, 1906." BANC MSS C-B 490.

———. Charles Derleth Papers. BANC MSS C-B 717.

———. Andrew C. Lawson Papers. Fred G. Plummer to G. K. Gilbert. October 1906. BANC MSS C-B 602.

———. George Davis Louderback Papers, 1900–1956. Harry O. Wood to George D. Louderback. February 12, 1947, and March 26, 1947. BANC MSS 83/65.

———. Materials Relating to the San Francisco Fire Department. "Reports of the Fire Officers of the San Francisco Fire Department on the Fire of 1906: Copies of the Reports in the Possession of Battalion Chief Fred J. Bowlen." BANC MSS C-R 68.

———. George Cooper Pardee Papers. "Proclamation of the Governor, June 2, 1906." BANC MSS C-B 400.

———. Clemens Max Richter. "Autobiography and Reminiscences." October 1922. BANC MSS 71/57c.

———. Henry Morse Stephens, Correspondence and Papers, 1890–1910. Cartons 1, 4, 5, 7. BANC MSS C-B 926.

———. James R. Tapscott, letter to Mrs. I. J. Tapscott, April, 22, 1906. BANC MSS 73/122 c:15.

———. United States, Custom House, San Francisco Records, 1847–1912, BANC MSS C–A 169 Pt.III.

———. "United States Custom House Damage Report." William Craine and George Cofran. October 17, 1865. MSS 2005/26 c.

———. Mariano Guadalupe Vallejo. "Historical and Personal Memoirs Relating to Alta California." Trans. Earl R. Hewitt. BANC MSS C-B 1-36.

University of California, Berkeley, College of Environmental Design Archives. Polk Scrapbook.

———. Bernard J. S. Cahill Papers. "Temporary City Hall," Central Emergency Hospital Building B, South Side of Market Street (between 8th and 9th Streets) for James Otis Trustee.

———. Drawings for the Rialto Building on loan from the San Francisco Public Library.

———. Environmental Design Archives. Walter T. Steilberg Collection.

University of California, Berkeley, University Archives. The Committee on Instruction to the Regents of the University of California, Committee file 1868–1879. CU–1, box 2, folder 11.

Credits

All photographs appear courtesy of The Bancroft Library, University of California, Berkeley, except as stated below:

1.1 Moulin Studios, San Francisco

1.2 From Division of Mines and Geology, Open File Report 2000-009, November 17, 2000

1.3 Adapted from USGS website http:/pubs.usgs.gov/gip/earthq3/where.html, June 24, 1997

2.1 From *The Annals of San Francisco* by Frank Soulé, John H. Gihon, and James Nisbet (1855; reprint, Berkeley: Berkeley Hills Books, 1998)

3.1 From *Buildings at Risk: Seismic Design Basics for Practicing Architects* (Washington, D.C.: AIA/ASCA Council on Architectural Research, NHRP, 1994)

3.2 From *Earthquakes* by Bruce A. Bolt (New York: W. H. Freeman and Co., 1993)

3.3 From *Earthquakes* by Bruce A. Bolt (New York: W. H. Freeman and Co., 1993)

3.4 From *Earthquakes* by Bruce A. Bolt (New York: W. H. Freeman and Co., 1993)

3.6 Drawn by Kimberly Butt, 2005

3.7 California Historical Society, San Francisco

3.8 *Dearborn Foundry Company* brochure, Chicago, 1887

3.9 Drawn by Kimberly Butt, incorporating map from *The California Earthquake of April 18, 1906, Report of the State Earthquake Investigation Commission* by Andrew Lawson et al. (Washington, D.C.: Carnegie Institution of Washington, 1908)

3.15 California Historical Society, San Francisco

3.19 Drawn by Anthony Vizzari, 2005

4.1 From *Recommended Minimum Requirements for Small Dwelling Construction*, Report of Building Code Committee (Washington, D.C.: Bureau of Standards, 1923)

4.2 From *Recommended Minimum Requirements for Small Dwelling Construction*, Report of Building Code Committee (Washington, D.C.: Bureau of Standards, 1923)

4.3 Redrawn and modified by Anthony Vizzari from *Peace of Mind in Earthquake Country* by Peter Yanev (San Francisco: Chronicle Books, 1991)

4.5 From "Seismic Retrofitting Alternatives for San Francisco's Unreinforced Masonry Buildings: Estimates of Construction Cost and Seismic Damage," by Rutherford and Chekene Consulting Engineers (City and County of San Francisco Department of City Planning, May 1990)

4.6a From *The Stonemason and Bricklayer (*London and New York: Ward, Lock and Co., 1891)

4.6b From *Bricklaying System* by Frank Gilbreth (New York: M. C. Clark and Co., 1909)

4.7 From *Memoir of the Life of Sir Marc Isambard Brunel* by Richard Beamish (London: Longman, Green, Longman and Roberts, 1862)

4.10 *The California Mail Bag*, August 1871

4.16 Photograph by the author

4.17 From *History and Report of the Construction of the New City Hall from April 4, 1870 to November 1889* by T. Spotts, City of San Francisco, Board of New City Hall Commissioners (San Francisco: W. M. Hinton and Company, 1889)

4.18 California Historical Society, San Francisco

4.19 California Historical Society, San Francisco

4.21 Photograph by Anthony Vizzari

4.22 Drawing Mike Z. Tobriner

4.23 Photograph by the author

4.24 Photograph by the author

5.2 From "The Chronicle's New Home," *San Francisco Chronicle*, June 1890

5.4 From "The Spreckel's Building, San Francisco," *Engineering Record* (April 9, 1898), with new overlay by Anthony Vizzari

5.5a Original drawing with overlay by Forell/Elsesser Engineers

5.9a Courtesy of James C. Flood

5.10 Photograph courtesy of The Bancroft Library altered to include section from *San Francisco Chronicle* of January 7, 1895

5.11 Redrawn by Ki Yeong Kim from *Simplified Building Design for Wind and Earthquake Forces* by James Ambrose and Dimitry Vergun (New York: John Wiley & Sons, 1980)

5.12 Redrawn by Ki Yeong Kim from *Why the Earth* Shakes by Mattys Levy and Mario Salvadori (New York: Norton, 1995) and *Simplified Building Design for Wind and*

Earthquake Forces by James Ambrose and Dimitry Vergun (New York: John Wiley & Sons, 1980)

6.1 From *Earthquake Damage and Earthquake Insurance* by John R. Freeman (New York: McGraw-Hill, 1932)

6.2 From *Messina, Le città nella storia d'Italia* by Amalia Ioli Gigante (Roma e Bari: Gius. Laterza e Figli, 1980)

6.3 From *Earthquake Damage and Earthquake Insurance* by John R. Freeman (New York: McGraw-Hill, 1932)

6.4 From *Victorian San Francisco: The 1895 Illustrated Directory* by Wayne Bonnett (Sausalito, Calif.: Windgate Press, 1996)

6.6 From *Victorian San Francisco: The 1895 Illustrated Directory* by Wayne Bonnett (Sausalito, Calif.: Windgate Press, 1996)

6.8 Drawn by Anthony Vizzari from map courtesy of San Francisco Public Library

6.14 San Francisco Public Library

6.21 From *The San Francisco Earthquake and Fire* by A.L. A. Himmelwright (New York: Roebling Construction Co., 1906)

6.22 Photograph courtesy of The Bancroft Library, altered by Anthony Vizzari, 2005

6.24 San Francisco Public Library and *Victorian San Francisco: The 1895 Illustrated Directory* by Wayne Bonnett (Sausalito, Calif.: Windgate Press, 1996)

6.25a Earthquake Engineering Research Center Library, University of California, Berkeley

6.25b Earthquake Engineering Research Center Library, University of California, Berkeley

6.29 Photograph by the author

7.1 Map courtesy of The Bancroft Library modified by Anthony Vizzari

7.2 Drawn by Kimberly Butt

7.2 Drawn by Anthony Vizzari; "The Spreckel's Building, San Francisco," *Engineering Record* (April 9, 1898)

8.1 Base map after *American Society of Civil Engineers* (1907)

8.9 F. A. Koetitz, "The Tower of the Union Ferry Depot, San Francisco," *American Builders Review* 6(1), January 1907

8.11 From *The San Francisco Earthquake and Fire of April 18, 1906, and Their Effects on Structures and Structural Material,"* by Grove Karl Gilbert, et al., USGS Bulletin 324 (Washington D.C.: Government Printing Office, 1907)

8.12 Photograph by the author

8.13 Photograph by the author

8.17 Henry J. Degenkolb Collection

9.3 From *Report on a Plan for San Francisco* by Daniel H. Burnham (San Francisco: Sunset Press, 1905)

9.4 From *Permanency in Building Construction* by Nathaniel Ellery (San Francisco: Brick Builders Bureau, 1913)

9.6 From *Lauriston: An Architectural Biography of Herbert Edward Law* by Sewall Bogart (Alpine House Publications, Portola Valley, 1976)

9.7 From *The San Francisco Earthquake and Fire* by A. L. A. Himmelwright (New York: Roebling Construction Co., 1906)

9.8 Environmental Design Archives, University of California, Berkeley

9.9 From *The San Francisco Earthquake and Fire* by A. L. A. Himmelwright (New York: Roebling Construction Co., 1906)

10.1 From Reports on an Auxiliary Water Supply System for Fire Protection for San Francisco, California by Marsden Manson, H. D. H. Connick, T. W. Ransom, and W. C. Robinson (San Francisco Board of Public Works, 1908)

12.1 Photograph by the author

12.2 From *Reports on the Water Supply of San Francisco, California, 1900–1908 Inclusive* (San Francisco Board of Supervisors, 1908)

12.3 From "The Effects of the San Francisco Earthquake of April 18, 1906, on Engineering Constructions," *Transactions of the American Society of Civil Engineers* 59, December 1907

12.4 From "Development of San Francisco's Water Supply in Case of Emergencies" by N. A. Eckart, *Bulletin of Seismological Society of America* 27(3), July, 1907

12.5 From "San Francisco High-Pressure Gravity Supply and Auxiliary Pumps—Earthquake Proof" by W. H. Ticknor, *Insurance Engineering* 24(3), September 1912

12.6 From "San Francisco High-Pressure Gravity Supply and Auxiliary Pumps—Earthquake Proof" by W. H. Ticknor, *Insurance Engineering* 24(3), September 1912

12.7 From *San Francisco Municipal Reports* (1908–1909)

12.8 From *The Water Supply of San Francisco, California, Before, During, and After the Earthquake of April 18, 1906 Earthquake and the Subsequent Conflagration* by Hermann Schussler (New York: Martin B. Brown Press, 1906)

12.9 From "San Francisco High-Pressure Gravity Supply and Auxiliary Pumps—Earthquake Proof" by W. H. Ticknor, *Insurance Engineering* 24(3), September 1912

12.10 *San Francisco Municipal Reports* (1908–1909)

12.11 From Reports on an Auxiliary Water Supply System for Fire Protection for San Francisco, California by Marsden Manson, H. D. H. Connick, T. W. Ransom, and W. C. Robinson (San Francisco Board of Public Works, 1908)

12.12 *San Francisco Municipal Reports* (1908–1909)

12.13 *Twenty-Ninth Convention of the Pacific Coast Fire Chiefs,* August 9-12, 1922, and *Golden Anniversary Congress of the International Association of Fire Engineers,* August 15-18, 1922 (San Francisco)

12.14 Photograph by the author

12.15 Photograph by the author

12.16 *San Francisco Municipal Reports* (1908–1909)

12.17 *San Francisco Municipal Reports* (1908–1909)

12.18 *Twenty-Ninth Convention of the Pacific Coast Fire Chiefs,* August 9-12, 1922, and *Golden Anniversary Congress of the International Association of Fire Engineers,* August 15-18, 1922 (San Francisco)

13.2a Facade design, Environmental Design Archives, University of California, Berkeley; photograph of building, The Bancroft Library

13.3 The Bancroft Library, with construction detail from "The Humboldt Savings Bank Building, San Francisco," by Charles Derleth Jr., *Engineering Record* 58(21), November 21, 1908

13.4 *American Builders Review* 6(5), May 1907

13.5 Environmental Design Archives, University of California, Berkeley

13.7 From "Earthquake-Proof Royal Globe Building, San Francisco," *Insurance Engineering* 22(3), September 1911

13.8 From "Earthquake-Proof Royal Globe Building, San Francisco," *Insurance Engineering* 22(3), September 1911

13.9 Photograph courtesy of California Historical Society, San Francisco, altered by Anthony Vizzari

13.10 Archives of H. J. Brunnier Associates

13.11a Photograph by Ki Yeong Kim

13.11b From *California Journal of Technology* 17(2), November 1913

13.12 *Transactions of the American Society of Civil Engineers* 61 (December 1908)

13.15 California Academy of Sciences, San Francisco

13.16a Environmental Design Archives, University of California, Berkeley

13.16b Environmental Design Archives, University of California, Berkeley

13.17a Environmental Design Archives, University of California, Berkeley

13.17b Environmental Design Archives, University of California, Berkeley

13.18 From *Architect and Engineer* 13(1), May 1908

13.19 From *Architect and Engineer* 9 (May 1907)

13.21 *Sunset* 20(6), April 1908

13.23 Photograph by the author

13.24a From *Permanency in Building Construction* by Nathaniel Ellery, vol. 2 (San Francisco: Brick Builders Bureau, 1913)

13.24b From *Permanency in Building Construction* by Nathaniel Ellery, vol. 2 (San Francisco: Brick Builders Bureau, 1913)

13.25 Photograph courtesy of George and Alexis Selland

13.26 Oliver Everett, Specifications for the Segalas and Plante Building, 1911, collection of George and Alexis Selland

13.27 Photograph by the author

13.28 Photograph by the author

13.29 Photograph by the author

13.30 Collection of the author

13.31 Collection of the author

13.32 *San Francisco Municipal Reports for the Fiscal Year 1910–11, ending June 30, 1911* (San Francisco Board of Supervisors, 1912)

13.33 Photograph by the author

13.34 Environmental Design Archives, University of California, Berkeley

13.35 Environmental Design Archives, University of California, Berkeley

13.36 Photograph by the author

13.37 *Overland Monthly* 51 (1908)

14.1 Earthquake Engineering Research Library, University of California, Berkeley

14.2a From "The Panama-Pacific Exposition" by A. H. Markwart, *Engineering News* 70(19), November 6, 1913

14.2b From "The Panama-Pacific Exposition" by A. H. Markwart, *Engineering News* 70(19), November 6, 1913

14.2c Environmental Design Archives, University of California, Berkeley

14.3 From "The Panama-Pacific Exposition" by A. H. Markwart, *Engineering News* 70(19), November 6, 1913

14.4 Collection of Bruce A. and Beverley Bolt

14.5 From *Earthquake Damage and Earthquake Insurance* by John R. Freeman (New York: McGraw-Hill, 1932)

14.6 Archives of H. J. Brunnier Associates.

14.7 From "A Quake-Proof Frame for an Apartment House," *Architect and Engineer* 29(1), June 1928

14.8 From "A Building Designed and Built to Withstand Earth Shocks" by Will P. Day, *Architect and Engineer* 85(2), May 1926

14.9 From *Earthquake Damage and Earthquake Insurance* by John R. Freeman (New York: McGraw-Hill, 1932)

Index

Stephen Tobriner is a professor of architectural history at the University of California, Berkeley. He has written extensively on architecture and the history of reconstruction after earthquakes, and has investigated damage in earthquakes around the world with teams sponsored by the United Nations, the National Science Foundation, the Earthquake Engineering Research Center, and the Earthquake Engineering Research Institute.

Heyday
HEYDAY INSTITUTE

Since its founding in 1974, Heyday Books has occupied a unique niche in the publishing world, specializing in books that foster an understanding of California history, literature, art, environment, social issues, and culture. We are a 501(c)(3) nonprofit organization committed to providing a platform for writers, poets, artists, scholars, and storytellers who help keep California's diverse legacy alive.

We are grateful for the generous funding we've received for our publications and programs during the past year from various foundations and more than three hundred individuals. Major recent supporters include:

Anonymous; Anthony Andreas, Jr.; Arroyo Fund; Bay Tree Fund; California Association of Resource Conservation Districts; California Oak Foundation; Candelaria Fund; CANfit; Columbia Foundation; Colusa Indian Community Council; Wallace Alexander Gerbode Foundation; Richard & Rhoda Goldman Fund; Evelyn & Walter Haas, Jr. Fund; Walter & Elise Haas Fund; Hopland Band of Pomo Indians; James Irvine Foundation; Guy Lampard & Suzanne Badenhoop; Jeff Lustig; George Frederick Jewett Foundation; LEF Foundation; David Mas Masumoto; James McClatchy; Michael McCone; Gordon & Betty Moore Foundation; Morongo Band of Mission Indians; National Endowment for the Arts; National Park Service; Ed Penhoet; Rim of the World Interpretive Association; Riverside/San Bernardino County Indian Health; River Rock Casino; Alan Rosenus; John-Austin Saviano/Moore Foundation; Sandy Cold Shapero; Ernest & June Siva; L.J. Skaggs and Mary C. Skaggs Foundation; Swinerton Family Fund; Susan Swig Watkins; and the Harold & Alma White Memorial Fund.

For more information about Heyday Institute, our publications and programs, please visit our website at www.heydaybooks.com.